The Boeing 737 Technical Guide

Chris Brady

The Boeing 737 Technical Guide

First published July 2006

Version 69, October 2014

Published by

Tech Pilot Services Ltd

ISBN 978-1-291-75644-9

Please address all enquiries and order additional copies of this book through the website at:

FOREWORD

This book is a plain English, illustrated, technical guide, intended to fill in the gaps left by existing publications. It contains facts, tips, photographs and points of interest, rather than simply being a reproduction of the manuals. Its broad scope will hopefully make it as interesting to students doing their type rating as it will be to Training Captains fielding unusual and searching questions from colleagues.

It is in no way intended to replace any of your FCOMs, company Ops Manuals or indeed any other official source of material. Nor is it a stand-alone study guide for the type rating - although it will offer considerable help. But rather it is intended to supplement all of the above, whilst hopefully being an interesting read.

Fly safely and enjoy the 737.

Chris Brady

CONTENTS

HISTORY AND DEVELOPMENT

The Idea

In late 1958 Boeing announced a design study for "A twin-engined feeder airliner to complete the family of Boeing passenger jets". In Feb 1965 the first order was placed and the project went ahead. The 737 has since become the best-selling commercial aircraft in aviation history. Here are some statistics which illustrate its success:

- The Boeing 737 is the best-selling commercial jetliner in history, with, as of August 2013, orders for more than 11,150 aircraft from 265 customers. Over 7,700 737s have been delivered.
- On Feb. 13, 2006, Boeing delivered the 5,000th 737 to Southwest Airlines. Guinness World Records acknowledged the 737 as "the most-produced large commercial jet" in aviation history.
- With approximately 5,500 airplanes in service, the Boeing 737 represents a quarter of the total worldwide fleet of large commercial jets flying today.
- 31% of all commercial flights are operated by the 737s (24,000 every day).
- More than 338 airlines in 112 countries fly the 737.
- On average, over 2,000 737's are in the air at any given time.
- On average, one 737 takes off or lands every 2 seconds.
- Boeing 737's have carried more than 16.8 billion passengers; that is equivalent to every single man, woman and child flying at least twice. (2012 world population was 7 billion).
- Boeing 737's have flown more than 115 billion miles; equivalent to approximately 624 round trips from the earth to the sun.
- Boeing 737's have flown more than 180 million flights.
- Boeing 737's have flown more than 257.6 million flight hours; the equivalent to one airplane flying more than 29,416 years nonstop.

Design

Boeing wanted a true short-haul jet to compete with the Caravelle, BAC One-Eleven & DC-9 but was way behind them. The DC-9 was about to fly, the One-Eleven was well into its flight test program and the Caravelle had been in service for 5 years. They had some catching up to do. Designers Joseph Sutter and Jack Steiner began work on the 737 in November 1964.

The original 1964 specification was for a capacity of about 60-85 passengers, an economical operating range of between 100 and 1000 miles and to be able to break even at a 35% load factor. As a result of final design talks with launch customer Lufthansa the capacity was increased to 100, but the range and load factor figures still stand.

Where to put the Engines

Joe Sutter (also considered to be the father of the 747) knew that Boeing was competing in broadly the same market as the Caravelle, BAC One-Eleven & DC-9 and needed something different. Sitting at his desk one day, Sutter took some scissors and cut up a drawing of the initial design for the 737 which also had a T-tail with aft mounted engines. He began moving the engines around to find a better layout. But putting the engines on struts under the wing like the 707 would block boarding access to the main cabin door on the shorter fuselage of the 737.

"I slid the cut-out tight under the wing and felt a sudden flash of excitement. Instead of mounting the engines away from the wing on struts, why not mount them hard against the underside of the wing itself?"

The wing mounted engines gave the advantages of reduced interference drag, a better C of G position, quieter aft cabin, more usable cabin space at the rear, fore & aft side doors, easier access to engines for maintenance and required less pipe work for fuel & bleeds. The weight of the engines also provides bending relief from the lift of the wings. Apparently this benefit was over-estimated and a set of wings failed in static tests at 95% of max load so the wing had to be redesigned.

1

The disadvantage of wing-mounted engines was that the size of the fin had to be increased for engine-out operation over centreline thrust aircraft. Also, due to the reduced ground clearance, the engines had to be almost an integral part of the wing, which in turn necessitated a short chord. The engines extended both forward and aft of the wing to reduce aerodynamic interference (further improved by the longer tailpipe of the target thrust reversers in 1969) and the straight top line of the nacelle formed a "stream tube" (aka streamline flow) over the wing to further reduce drag.

Initial worries about the low mounted engines ingesting debris proved unfounded, this was demonstrated by the Boeing 720B whose inboard engines are lower than the 737's and had been in service for four years without significant problems.

The final wing was a work of art as the specification required both good short field performance and economy at altitude.

Overall, the wing-mounted layout had a weight saving of 700kgs (1500lbs) over the equivalent "T-tail" design and had performance advantages. A further advantage of the wing-mounted engine design was its commonality with earlier Boeings such as the 707.

The underslung engines of a 737-200. Notice the sandbags against the fan blades of this aircraft in open storage

Heritage

Jack Steiner had helped design the B29 & 707 and was Chief Designer of the 727. His major contribution to the 737 project was to use as much as possible of the 727 in the 737, in particular the fuselage cross section. This gave not only cost savings in tooling commonality, but also the payload advantage of 6 abreast seating, one more than the DC-9 or BAC-111 and allowed it to carry standard sized cargo containers on the main deck. It gave the interior a spacious look and allowed Boeing to use standard cabin fittings from the 727 & 707 such as toilets and galleys. This together with its large hold capacity, gave it scope for using the aircraft as a freighter, a role to which many of the older 737s are now being converted, whilst the competitors are being scrapped.

In fact the 737 had a 60% parts commonality with the 727 which included the doors, leading edge devices, nacelles, cockpit layout, avionics, components and other fittings. The 727 in turn had a similar commonality with the 707, so parts of the 737 can be traced back to the early 1950's, eg the fuselage cross-section above the floor.

Using off-the-peg components was quick and cheap for both design and production and also helped pilots and engineers convert to the new type, but they also dated the aircraft, a feature which was never really addressed until the NGs 30 years later.

On 19 Feb 1965, Boeing announced the 737-100 after an initial order of 10 from Lufthansa, who became the first foreign airline to launch a US airliner. Assembly started in the summer of 1966.

Originals

First Flight

The maiden flight of the 737 was on 9 April 1967, just two years after the project launch. Boeing's assistant director of flight operations, Brien Wygle was in command and Lew Wallick the chief test pilot was co-pilot. After the uneventful two and a half hour flight from Boeing Field to Paine Field, Wygle said "I hate to quit, the airplane is a delight to fly."

Flight testing continued at a blistering pace with the prototype clocking up 47hr 37min in the first month. Soon six aircraft, including the first -200, were on the flight test & certification program. Between them they flew 1300hrs of flight tests. Many changes were made to the aircraft in this time, e.g. trying inflatable main landing gear door seals, although these were soon changed to the present rubber strips.

However the earliest 737s had some problems, including clamshell door thrust reversers (from the 727) which didn't work properly, and a shimmy in the landing gear, "but it was a good airplane from the start", recalls Brien. FAA type certification A16WE was gained on 15 Dec 1967.

The prototype 737, N515NA, formerly N73700, on display at Boeing Field in 2006
Although the oldest 737 it only accumulated 3297 flight hours. Photo: Bob Bogash

The prototype went to NASA in 1974 where it became known as the Transport Systems Research Vehicle (TSRV). It was in regular experimental use until 1997 and is now stored at Boeing Field where it is on permanent display. NASA 515 was involved in numerous pioneering flight investigations including control systems, 3D and 4D navigation, in-flight energy management, computerized flight management systems, electronic displays, Microwave Landing System (MLS) development (overrun in late development by the advent of GPS), slippery runway studies, and clear air turbulence and wind shear detection and warning. The airplane has a second flight deck, fully functional, installed in the main cabin, which was used for much of the flying. Control systems used included the original Boeing control column, Brolly handles (like bicycle handlebars), and the current side-stick controllers. Numerous glass cockpit CRT display configurations were tried, that attacked many questions regarding display arrangements, colours and symbology. Much of this work ended up on aircraft from Boeing, Douglas, Airbus and the Space Shuttle. The airplane has a unique fourth hydraulic system with a reservoir, pump and filter system. Other studies were performed of drag-reducing external coatings, cockpit displayed traffic information, takeoff performance monitoring, and precision flare guidance during landing touchdown. See page 60 to see the master MCP of this two cockpit aircraft!

The 2 / 3 Crew Issue

The 737 was the first 2 crew aircraft produced by Boeing, all others had a flight engineer station which was necessary as early airliners had been more complex and less reliable. The 3 crew issue had been around since the late 1950's when the Lockheed Electra sparked disputes between airlines and the pilots' and engineers' unions which lead to a high profile four month strike that in the end only deferred a new policy decision. When the 737 was announced ALPA and the FAA were on the case from the outset as the 3 crew issue had still not been resolved.

To finally determine if the aircraft was capable of safe 2 crew operation, a 737 was flown with an FAA pilot and a Boeing pilot over the busiest week of the year (Thanksgiving) in the busy Boston - Washington corridor. They flew 40 sectors in 6 days, including approaches to minimums, go-arounds, diversions, simulated instrument failures and crew incapacitation. In December 1967 the FAA issued a statement declaring that "the aircraft can be safely flown with a minimum of two pilots."

Even after the FAA statement, American, United and Western continued to operate with 3 pilots until 1982. Fortunately the rest of the world was not so limited and this helped sales to recover.

737-100. Notice the short leading edge flap   *Photo: Lufthansa*

Teething Troubles

The first 737 went into service with Lufthansa on 10 February 1968. Generally operators were very impressed with the reliability of the aircraft from the start, although inevitably there were some technical issues found during line work. The APU had a tendency to shut down under load; this was solved by developing a new acceleration control thermostat. Engine starter valves were found to clog with sand from treated runways; this was fixed by using a finer mesh on the filter screen. Perhaps most troublesome were the integral airstairs which have a complicated way of folding and simultaneously collapsing the handrail as they retract or extend. The airstairs can still be very frustrating and in 2005 at least two UK based 737 operators decided to remove all airstairs from their fleets to avoid despatch delays, further maintenance costs and giving a 177kg (390lbs) weight saving every sector. The first 737s had twice the problems because many aircraft were also fitted with rear airstairs which were, if anything, more complicated. Other minor issues were nosewheel corrosion, ram air inlet problems and hydraulic line failure, all of which were sorted out within a couple of years.

737 Originals Key Dates:

11 May 1964	*Formal design begins*
9 Nov 1964	*737 Program go-ahead*
19 Feb 1965	*First order from Lufthansa*
5 Apr 1965	*First order for 737-200 from United*
9 Apr 1967	*First flight of 737-100*
8 Aug 1967	*First flight of 737-200, the 5th 737 to fly*
15 Dec 1967	*FAA Type certification of 737-100 and -200*
10 Feb 1968	*First revenue flight of 737-100 with Lufthansa*
18 Sep 1968	*First flight of 737-200QC*
20 Feb 1969	*Gravel runway certification*
4 Mar 1969	*First 737-200 delivered with flaps & thrust reverser modifications*
15 Apr 1971	*First flight of 737-200Adv*
15 Oct 1973	*Quiet nacelle modification goes into service with Quebecair*
2 Aug 1988	*Last delivery of a 737-200Adv*

737-100

FF 9 April 1967, 30 built, none remaining in service.

The -100 was 94ft (28.65m) long, carried 115 passengers and had an MTOW of just 42,411kgs (93,500lbs), less than half that of the current -900 series. The original choice of powerplant was the Pratt & Whitney JT8D-1 at 14,000Lbs thrust, but by the time negotiations with Lufthansa had been completed the JT8D-7 was used. The -7 was flat rated to develop the same thrust at higher ambient temperatures than the -1 and became the standard powerplant for the -100.

Just 30 series 100s were built, with 22 going to Lufthansa, 5 for Malaysia Airlines and 2 for Avianca. The last airworthy 737-100, L/N 3 which first flew 12 June 1967 was finally retired from Aero Continente in Peru as OB-1745 in 2005.

737-200 Basic

FF 8 August 1967, 1114 built, 116 remain in service. (All -200 variants)

An original 737-200, L/N 54. Notice the short nacelles and clamshell thrust reversers.   Photo: Steve Williams 1969

5

It was immediately realised that most airlines wanted a slightly higher passenger load, to which Boeing responded with the 737-200. Two sections were added to the fuselage; a 36in section forward of the wing (STA 500A&B) and a 40in section aft of the wing (STA 727A&B), giving a maximum capacity of 130 passengers with a 28in seat pitch. All other dimensions remained the same. The JT8D was increased to 14,500lbs with the -9. Six weeks later on the 5 April 1965 the -200 series was launched with an order for 40 from United Air Lines. Development and production of the two series ran simultaneously.

Flight testing had shown a 5% increase in drag over predicted figures, this equated to a 30kt reduction in cruise TAS. After almost a year of wind tunnel and flight testing several aerodynamic modifications were made. Flaps and thrust reversers were improved from aircraft number 135 (March 1969) and free mod kits were made available for existing aircraft. The thrust reversers were totally redesigned by Boeing and Rohr since the aircraft had inherited the same internal pneumatically powered clamshell thrust reversers as the 727 which were ineffective and apparently tended to lift the aircraft off the runway when deployed! The redesign to external hydraulically powered target reversers cost Boeing $24 million but dramatically improved its short field performance, which boosted sales to carriers proposing to use the aircraft as a regional jet from short runways. Drag reduction measures included extending the engine nacelles by 1.14m (3ft 9in) and widening the strut fairings; enhanced flap, slat and panel seals; eliminating 13 inboard wing upper surface vortex generators and shrinking the rest.

The original series 200 had narrow engine pylons and smaller inboard leading edge Krueger flaps.

The later series 200 had broad engine pylons and the inboard leading edge Krueger flaps extended to the fuselage.

The MTOW of 49,440kgs (109,000lbs) and MLW of 44,450kgs (98,000lbs) were often limiting so Boeing made structural changes to increase these weights and called the redesigned aircraft the 737-200 Advanced.

737-200 Advanced

FF 15 Apr 1971

As well as incorporating all of the later -200 modifications, the -200 Adv included major wing improvements such as new leading edge flap sequencing, increase in droop of outboard slats, extension of the inboard Krueger Flap, to produce a significant increase in lift and a reduction of take-off & approach speeds for better short field performance or an MTOW increase of 2268kg (5,000lbs). Autobrake, improved anti-skid, automatic speedbrake for RTO, automatic performance reserve and even nose-brakes became available. Again, kits were available for existing operators of the -200. With the JT8D-15 at 15,500lbs the MTOW was now up to 52,390kgs (115,500lbs) and MLW 48,534kgs (107,000lbs).

A 737-200Adv shortly after take-off

These performance improvements increased the service ceiling by 2,000ft to 37,000ft. The maximum cabin differential pressure was increased from 7.5 to 7.8psid to accommodate this.

737-200 Advanced – Summary of Improvements	
Stopping Improvements:	**High Lift Improvements:**
Automatic brakes	Extended Krueger flap
Improved antiskid	Wider nacelle strut
Revised main gear metering pin	Repositioned slats
	Smooth fixed leading edge
	Sealed trailing edge

For many years United remained the only major US carrier to order large numbers of 737s, because although the aircraft was designed to be flown by 2 crew, the US flight-crew union ruled that aircraft in that class had to be flown by three crew. United were forced to fly their 737s with three crew until 1981. Air France had also been trying to order the aircraft for several years but were unwilling to because of staff opposition until after 1981.

In 1973 when noise was becoming a factor, the nacelle was acoustically lined by Boeing and P&W swapped one fan stage for two compressor stages in the JT8D-17 while increasing thrust to 16,000Lbs. The JT8D got up to 17,400Lbs thrust on the -17R.

Most pilots who have flown different generations of the 737 say that the -200Adv is by far the best for handling.

737-200 Convertible (C), Quick Change (QC) & Combi

FF 18 Sept 1968, 96 (C) built

The Convertible passenger / freight version had a 3.4m x 2.18m (138in x 86in) side cargo door (SCD) on the forward port side for pallet loading. They also had strengthened floors and additional seat tracks. As a freighter it could accommodate seven LD7 (88in x 125in) palettes on the main deck plus any loose cargo in the two holds. Conversion time was approximately 3hrs but this was later reduced to about 1hr with the QC which had 12 passenger seats ready mounted on each pallet. This realistically allowed the aircraft to be used for both roles allowing it to earn money around the clock by carrying passengers by day and freight overnight. Some airlines even operated them as Combi's with pallets at the front and pax at the rear.

737-200 Executive

Originally designated the Corporate 77-32 or the Corporate 200, this was an executive jet version of the -200 and -200 Adv, similar in concept to the current BBJ. These were either fitted with one of the many Boeing interiors or were delivered green for customer installation of special interiors. A 3,065ltr (810USGal) auxiliary fuel tank was also fitted to give a maximum range of up to 4,000nm with a 1,134kg (2,500lbs) payload.

An anonymous VIP 737, notice the window arrangement

T-43A, CT-43A, NT-43A

FF 10 Apr 1973, 19 built

The T-43A is a US Air Force version of the Boeing 737-200. The exterior differences between the military and commercial aircraft were that they only had 9 windows each side of the fuselage and door 1R and 2L were not fitted. There were also many small blade-type antennas for the UHF communications and 5 overhead sextant ports. The aircraft were fitted with an 800 US Gal auxiliary aft centre tank as standard.

The first T-43A was delivered to the Air Education and Training Command at Mather AFB, Ca, in September 1973. The fleet relocated to Randolph AFB, Texas, in May 1993 when Mather closed. The majority of the T-43As are used in the USAF's undergraduate navigator training program to train navigators for strategic and tactical aircraft. The rest are configured for passengers as the **CT-43A** and are assigned to the Air National Guard at Buckley, where they are used for the USAF Academy's airmanship program and to provide travel service for academy sports teams. In addition, U.S. Southern Command has a CT-43 for commander transport.

T-43A in open storage at AMARC, Davis-Monthan AFB

Inside each T-43A training compartment are two minimum proficiency, two maximum proficiency and 12 student stations. Two stations form a console, and instructors can move their seats to the consoles and sit beside students for individual instruction. The cabin floor was strengthened to take the weight of these consoles. The large cabin allows easy access to seating and storage yet reduces the distance between student stations and instructor positions.

The student training compartment is equipped with avionics identical to that of Air Force operational aircraft. This includes Doppler and mapping radar; LORAN, VOR, TACAN, INS, radar altimeter; UHF & VHF Comms. Five periscopic sextants are spaced along the length of the training compartment for celestial navigation training.

Gradually most of the T-43s went into storage at the Aircraft Maintenance And Reutilisation Center (AMARC) at Davis Monthan AFB near Tucson, Arizona. However one (tail number 73-1155) was recovered in March 2000 and flown to an aircraft maintenance and modification facility at Goodyear, Arizona for conversion to a radar test bed and became the **NT-43A**.

T-43A Navigator Training Console

737 Testbeds

The NT-43A had two oversized radomes on the nose and tail. Its first flight in this new configuration was on 21 March 2001. The radomes were built by the Lockheed Martin Advanced Prototype Center, part of the Advanced Development Programs' (ADP) organization for Denmar which is a company specializing in stealth technology. The "Den" stands for President Denys Overholser, the former Skunk Works engineer credited with devising the shape of the first stealth aircraft. The design, fabrication and machining of the structure's components were all performed at Palmdale. The radome structure is about 6.2 feet in diameter and 16.5 feet in length and made of a 90-percent carbon epoxy/honeycomb sandwich material, with machined aluminium parts.

A rare sighting of the sole NT-43A
Photo: Brian Lockett, Goleta Air & Space Museum

The NT-43A can be seen flying in formation with various stealth aircraft, usually in the radar free environment of Death Valley. Its task is to make radar images of these aircraft to evaluate their stealth characteristics. The images can be used to reveal the rate of degradation of the radar deflecting and absorbing components as the aircraft age, and to determine the effectiveness of maintenance and repair methods.

IAI Elta Radar Testbed

The IAI Elta division 737-200A testbed has been used to develop systems since 1979. These have included maritime patrol signal intelligence, image intelligence using synthetic aperture radar, AEW and most recently Flight Guard, a commercial aircraft anti-missile protection system.

IAI Elta 737-200 Radar Testbed   *Photo: Elta*

Flight Guard is based around six miniaturised pulse-Doppler sensors, located to give all round coverage. On the Boeing 737 demonstrator two are fitted below the nose radome, with two more further aft on the forward fuselage and two on the tailcone. These are used to automatically trigger the release of IR decoy flares in the event of any attack. The system gives greater than 99% probability of missile detection, and has a very low false alarm rate.

Flare dispensers would usually be fitted in the wing-fuselage fairing as this does not involve penetrating the aircraft's pressure hull, and gives a minimum drag configuration. The system is armed/disarmed at the Cockpit Control / Display Unit.

Elta pulse-Doppler sensors   *Photo: Elta*

Boeing Avionics Flying Laboratory

First flown as the AFL on 26 March 1999, this highly modified 737-200 originally flew in 1968. It was modified by Boeing Aerospace Support to accommodate special avionics and instrumentation for development of the F-35 JSF.

The reason for the AFL according to Dan Cossano, manager of Boeing JSF Mission Systems: "We will save development time and costs because the AFL allows us to test more efficiently than with a fighter platform".

The modified nose of the Boeing AFL   Photo: Jeffrey Phinney

The AFL heat exchanger   Photo: Jeffrey L Phinney

The design and modification program took less than a year. Boeing Military Programs fitted a 48-inch nose and radome assembly to the forward pressure bulkhead of the aircraft. The radome housed the JSF synthetic aperture radar and forward-looking infrared sensors used for targeting. The aircraft also was fitted with a JSF cockpit in the cabin, several antennas, a heat exchanger on the port side of the fuselage and provisions for a supplemental power system. The AFL was broken up in 2004.

Co-operative Avionics Test Bed "CATBird"

BAE Flight Systems spent 3 years building a replacement AFL for Lockheed Martin on a 737-300 airframe at Mojave. The new aircraft, called Co-operative Avionics Test Bed (CATB) or "CATBird", first flew on 23rd Jan 2007. Structural modifications include a nose extension and canards to simulate JSF aerodynamics, a 42-foot long spine on the top and a 10-foot canoe on the bottom to house the line-replaceable units. It was based at Lockheed Martin's Fort Worth plant for test operations. According to BAe "The CATB will develop and verify the F-35's capability to collect data from multiple sensors and fuse it into a coherent situational awareness display in a dynamic airborne environment."

The CATBird at Mojave, Dec 2006   *Photo: Alan Radecki*

MP Surveiller

FF 21 Apr 1982, 3 built

Maritime Patrol aircraft for the Indonesian Air Force. Fitted with Motorola AN/APS-135(V) Side Looking Airborne Modular Multi-Mission Radar (SLAMMR), the antenna of which was mounted in two 16ft housings on the upper rear fuselage. This system could spot small ships at ranges of 100nm. At least one Surveiller was still in service with the Indonesian Air Force in 2007.

A Maritime Patrol Surveiller on task   Photo: Indonesian AF

Last of the Originals

Such was the popularity of the -200 that its production continued for over four years after the introduction of the first -300. The last -200Adv (l/n 1585) was delivered on 2 August 1988 to Xiamen Airlines. 1,144 originals were built, many of which are still flying today although noise restrictions have made necessary the installation of stage 3 hush kits available from Nordam or AvAero. P&W was even considering a re-engining program with the PW6000, but with the post Sept 11th economic downturn sending thousands of aircraft into storage, the idea was dropped.

737-200 with hushkits

The 737-200 is still being developed by third party companies. In June 2005 Quiet Wing gained FAA certification for a flap modification package to increase take-off performance by 3,200kg (7,000lbs), reduce fuel consumption by 3% and reduce stalling (and hence take-off & approach) speeds by 5kts. It works by drooping the trailing edge flaps by 4 degrees and the ailerons by 1 degree to increase the camber of the wing. Whilst this slightly increases drag, it does give much more lift thereby increasing the aerodynamic efficiency of the wing. One unusual benefit is that operators may want to replace their JT8D-15 engines with older but lower thrust, lower fuel consumption -9As. Quiet Wing are now developing the same package for Classics.

Classics

The 737-200 was succeeded in 1984 by the 737-300. This was a much quieter, larger and more economical aircraft and contained a host of new features and improvements. The new model featured many aerodynamic, structural, cockpit and cabin features developed for the new generation 757/767.

Commonality

One of the objectives was to have a high degree of commonality with the 737-200; the achieved figure was 67% by part count. This gives saving for airlines in maintenance, spares, tools for existing 737-200 operators. Also the aircraft was designed to have similar flying qualities, cockpit arrangements and procedures to minimise training differences and permit a common type rating.

Engines

The sole powerplant was the CFM-56, the core of which is produced by GE and is virtually identical to the F101 as used in the Rockwell B-1. SNECMA produced the fan, IP compressor, LP turbine, thrust reversers and all external accessories. The main problem was the size of the engine for ground clearance; this was overcome by mounting the accessories on the lower sides to flatten the nacelle bottom and intake lip to give the "hamster pouch" look. The engines were moved forward and raised level with the upper surface of the wing and tilted 5 degrees up. This not only helped the ground clearance, but also directed the exhaust downwards, reducing the pylon heating and giving some vectored thrust to assist take-off performance. The CFM56-3 proved to be almost 20% more efficient than the JT8D.

Airframe

Two sections were added to the basic -200 fuselage; a 44in section forward of the wing and a 60in section aft of the wing. Composite materials were used on all flight controls to reduce weight. Aluminium alloys were used in areas such as wing spars, keel beams and main landing gear beams to improve their strength by up to 12 percent, thereby increasing service life.

The wings were extensively redesigned to enhance low speed performance and cruise efficiency. The chord of the leading edge outboard of the engines was extended by 4.4%; this reduced the wing upper surface camber forward of the front spar to increase Mcrit, thereby giving better transonic airflow characteristics and improved buffet margins. The span was increased by a wingtip extension of 27.9cm (11in). These two changes had the greatest impact on high speed performance and as a result, the turbulent air penetration speed was increased to 280kts/0.73M on Classics from 280kts/0.70M on the originals.

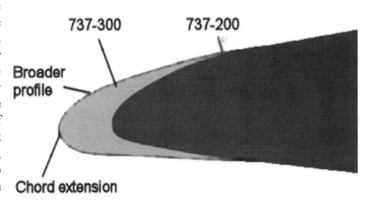

Comparison of 737-200 and 737-300 leading edge slat profiles

High lift characteristics were also enhanced by re-sequencing the slats and flaps. The leading edge slat radius was also increased which gave a 2.5kt reduction in Vref over a -200 at the same weight. Other changes to the wing structure included strengthened materials and corrosion protection. Rumour has it that the fin was also "given" to Shorts for use on their SD-360 as a reward for their good contract work for Boeing. Whether this is true or not, the SD-360 fin certainly looks identical, albeit scaled down.

Flightdeck

Flight Management System with fully integrated Digital Flight Control System, Autothrottle, Flight Management Computer, Dual laser gyro Inertial Reference System. EFIS CRT displays.

- **Performance**: Higher ceiling 37,000ft; Higher MTOW; 20% lower fuel burn than -200.
- **Engines:** High bypass (5:1), 18,500-23,500lb, CFM56-3.
- **Fuselage:** 44in fwd body extension, 60in aft body extension. Strengthened body skins & stringer.
- **Wings:** Span increased by 11in wingtip extensions; Modified slat aerofoil, flipper flaps & flap track fairings. Increased fuel capacity to 16,200kg (35,700lbs).
- **Tail:** Dorsal fairing added for stability during asymmetric conditions; Stabiliser extended 30in due to increased fuselage length.
- **Flight Controls:** New slats to reduce approach speeds to within a few knots of the much lighter -200. Additional ground spoiler. Stab tip extension.
- **Nose Gear:** Lengthened 6in and repositioned to help provide the same engine inlet ground clearance as the -200.
- **Main Gear:** Strengthened wheels, 40 or 42 inch tyres & better brakes for the increased MTOW.
- **Flightdeck:** EFIS CRT displays replacing many conventional instruments.

Various changes and customer options became available on the Classics over their production run, these include:

- Taxi light extinguishing automatically with gear retraction.
- Addition of a strobes AUTO position.
- Light test not illuminating engine, APU & wheel-well fire warnings.
- OVHT/FIRE system test will only show a fault, ie no method of determining whether open or closed circuit.
- Cabin pressure control system (CPCS) & indicators replaced by digital system (DCPCS) which has no standby mode but has a second (alternate) automatic mode.
- A Sundstrand APU with no EGT limits and shorter restart intervals.
- Extra automatic DC fuel pump for APU starts.
- Various Comms options such as VHF3, HF, SELCAL, ACARS & TCAS.
- Option of VSCF instead of CSD.
- Dripsticks replaced by floatsticks.
- Option of a fourth fuel tank.
- Refuel preset facility on refuelling panel.
- Alternate nose wheel steering (from hydraulic system B).
- Side window heat now also heats window 3.
- Third automatically tuned DME radio for FMC position.
- Round dial engine instruments replaced by EIS panels.
- APU battery.
- Aspirated TAT probe

Test Pilot Jim McRoberts had the honour of making the first flights of all three of the Classics series. The flight test program lasted nine months and the three aircraft flew a total of 1,294 hours. The program was largely uneventful. There were some problems with engine/wing flutter, but these were overcome with the addition of a 50kg (100lb) mass balance at each wingtip. The 737-400 flight test program was a further 400 hours and the 737-500 just 375 hours.

Future

There are now three separate cargo-conversion programs (Pemco, IAI Bedek & Goodrich) for the -300 / -400, which will give capacity for 8 / 9 pallets and an 18,800/20,900kg (41,500/46,000lbs) payload respectively.

CFMI are now offering a core upgrade kit which they claim will save up 1% specific fuel consumption. The kit also increases EGT margins by 15C thereby giving up to 1400 additional cycles.

Aviation Partners Boeing APB are offering retrofit kits to add winglets to any 737-3/4/500. Known as Special Performance (SP), the winglets give a 4.5% drag reduction over the standard model. The first 737-300SP was delivered 9 Jul 2003 and the retrofit kits became available for the -400SP in 2005 and -500SP in 2007.

737 Classics Key Dates:

15 Mar 1981	*737-300 announced*
17 Jan 1984	*Prototype 737-300 rolled out*
24 Feb 1984	*First flight of 737-300*
24 Jul 1986	*EADI & EHSI certified by the FAA*
19 Feb 1988	*First flight of 737-400*
25 Feb 1991	*2000th 737 delivered*
15 Jun 1997	*A CFM56-3C1 engine mounted on a Boeing 737-500 with Braathens S.A.F.E. reached 19,855 cycles without a single shop visit, setting a new world record for time on wing. The previous record of 19,841 cycles was held by a CFM56-3 engine in service with Southwest Airlines. The engine was removed after nearly six years of service due to life-limited parts in the core.*
9 Dec 1999	*Final 737 "Classic" - 737-400, OK-FGS, L/N 3132, is rolled out.*
14 May 2002	*First 737-300 cargo conversion by Pemco Aviation Group at its Dothan, Alabama facility*
4 Apr 2005	*Israel Aircraft Industries (IAI), Bedek Aviation Group delivers first 737-300SF conversion to Kitty Hawk, Inc*
14 Mar 2012	*Classics gain FAA (Stage 4) & EASA (Chapter 4) noise recertification.*

737-300

FF 24 Feb 1984, 1113 built, 514 remain in service.
Overall length 33.4m (105ft 7in). Maximum capacity 149 passengers. MTOW 63,275kg (139,500lbs)

The author departing in a 737-300 on a post maintenance airtest.   Photo: Keith Burton

737-400

FF 19 Feb 1988, 489 built, 266 remain in service.

A 737-400 at rotation, just visible is the tailskid for tailstrike protection on take-off

The Boeing 737-400 was given a further stretch of 2.8m (9ft 6in) over the -300, giving a length of 36.45m (119ft 7in). A ventral tailskid was installed to prevent tailstrike damage on take-off (not landing!); the FCTM shows that the -400 has the greatest risk of tailstrike of all the series of 737. Passenger capacity was now up to 174, this required two extra overwing exits and a boost to the air-conditioning system to give the necessary increase in air flow rate. Standard MTOW now up to 65,000kg (143,500lbs) with a 68,000kg (150,000lbs) HGW option. The wings of all -400s were strengthened and flap limiting speeds increased over the -300.

Differences from the -300 include:
- Fuselage plugs of 72in (183cm) fwd and 48in (122cm) aft.
- Standard engine CFM56-3B2 rated at 22,000lbs with optional 3C1 rated at 23,500lbs.
- Additional pair of overwing exits added.
- Tailskid added to section 48.
- Improved Environmental Control System and extra riser ducts added.
- Re-gauged skins and stringers to strengthen overwing body, keel beam and wings.
- Strengthened slats and Krueger flaps giving higher operating and limiting speeds.
- 44.5in (1.13m) tyres, heavy-duty wheels and brakes.
- Strengthened main landing gear structure.

737-500

FF 30 Jun 1989, 388 built, 199 remain in service.

The 737-500 (originally known as the 737 Lite or the 737-1000) combined the original length fuselage of the -200, with the various improvements of the -300 and -400. Overall length 31.0m (101ft 9in). MTOW up to 60,555kg (133,500lb), although many operators in Europe declared it as 49,999kg to reduce navigation charges & landing fees. 3B1 engines rated at 20,000 or 18,500lbs.

A British Midland 737-500 at Heathrow c1995

Cargo Versions

Perhaps inevitably as the world's fleet of Classics get older, so more are being converted into freighters. By 2006 there were 70 cargo Classics (and 40 Originals) in service.

737-300SF conversion by IAI-Bedek

In Oct 2003 Pemco of Florida completed conversion of the first **737-300QC**. It has a main cargo door and the seats can be removed in 45mins to allow 8 standard pallets with approx 17,000kgs (37,500lbs) payload. Pemco also converts to pure freighters as the **737-300F**, these have a 9 pallet / 18,000kg (39,700lbs) capacity.

IAI-Bedek Aviation Group have converted several **737-300SF**s. Pemco and IAI both hold STCs for 737-300/400 cargo conversions, but are not supported by Boeing. Boeing tried to offer their own 737-300/400 SF freighter conversion with InterContinental Aircraft Services of Taiwan, but dropped the service after five years without a launch customer. AEI of Miami also offers a 9 pallet 737-300 cargo conversion. The **737-400SF** freighter conversion by ICAS has a 9 pallet capacity although other conversions offer 10. There are unlikely to be any 737-500 cargo conversions as the short forward fuselage gives insufficient access to any cargo door past the engines.

Military Versions

This Command & Control 737-300 was developed and converted by Xian Aircraft Corporation in 2005, without Boeing or US approval. The aircraft are operated by China United Airlines, who are the VIP transport division of the Peoples Liberation Army Air Force have nine 737s although only two have been converted.

The aircraft have three large farings housing a large SATCOM antenna above the fuselage and two for data-link / communication below.

The Chinese Command & Control 737-300

Next Generation

The Boeing 737-X programme was launched on 29 June 1993, with a 63 aircraft order from Southwest Airlines for the 737-300X. This became the 737-700, 22cm (9in) longer than the original 737-300, seating up to 149. The main differences of the 737 Next Generation (NG) are as follows:

- **Performance**: Faster cruise M0.78, Higher ceiling 41,000ft, Lower take-off & approach speeds, Higher MTOW, lower fuel burn.
- **Engines:** FADEC controlled CFM56-7, 2.5deg nozzle tilt, redesigned struts, improved nacelles with increased airflow and improved noise treatment, 7% more fuel efficient than CFM56-3.
- **Fuselage:** Strengthened for increased tail loads and design weights, new wing-body strake.
- **Wings:** New airfoil section, 25% increase in area, 107" semi-span increase, 17" chord increase, raked wing-tip, larger inspar wing box with machined ribs.
- **Fuel Tanks:** Main tanks smaller at 3900kg / 8600lbs each but centre tank much larger giving total fuel capacity of 20,800kg / 45,800lbs. (c.f. 16,200kg / 35,700lbs on Classics).
- **Tail:** 4ft 8in taller, 60 sq ft root insert, modified rudder, segmented rudder seals, digital yaw damper.
- **Flight Controls:** Increased elevator PCU capability, aileron and tab span increase, new double slotted continuous span flaps, new leading edge Krueger flaps, additional slat, additional spoiler.
- **Nose Gear:** Stroke increased 3.5" to relieve higher dynamic loads and wheelwell extended 3" forward.
- **Main Gear:** Longer to reduce tailstrike risk, one piece titanium gear beam, 43.5" tyres, digital antiskid.
- **Flightdeck:** 6 programmable LCDs, replacing EFIS CRT displays and most conventional instruments.
- **Systems:** Most systems developed particularly: electrics, powerplant & navigation.

The NGs have 33% fewer parts than the Classics which reduces production time. One of the main production differences with the NG is the single moving assembly line. This has the capacity to produce 21 aircraft a month with a flow time of just 13 days.

In April 2009 Boeing announced a series of improvements to the 737NG, which included a 12% decrease in fuel burn from the new CFM56-7BE engine the first of which was delivered in July 2011; and also from various airframe drag-reduction improvements such as: refined wing control surfaces, redesigned wheel-well fairing, a re-shaped anti-collision light and an ECS inlet/exhaust modulation. The cabin was also improved with a new "Sky" interior.

737 NG Key Dates:

17 Nov 1993	*Boeing directors authorize the Next-Generation 737-6/7/800 program. Southwest Airlines launches the -700 program, with an order for 63 aircraft.*
5 Sep 1994	*The 737-800 is launched at the Farnborough Air Show.*
15 Mar 1995	*The 737-600 is launched with an order for 35 from SAS.*
2 Jul 1996	*Boeing launches the Boeing Business Jet (BBJ) based on the 737-700.*
9 Feb 1997	*The first Boeing 737-700 makes its maiden flight from Renton Municipal Airport to Boeing Field, Seattle with Boeing Captains Mike Hewett and Ken Higgins.*
15 Mar 1997	*Captains Mike Carriker and Paul Desrochers fly the second 737-700 flight-test airplane to 41,000 feet during certification testing, higher than any other 737.*
22 Apr 1997	*YA001, the first 737-700, makes its 100th flight weighing 172,900 pounds - the highest Boeing 737 takeoff weight ever - and with an engine thrust of 27,000 pounds. During the flight the airplane conducts pre-certification flight testing to capture data for the 737-700 Increased Gross Weight (IGW) airplane. Commenting on the flight, Capt. Mike Hewett said "the airplane's wings performed exceptionally well and the stability control data points looked very good for the flight-test conditions."*
31 Jul 1997	*The 737-800 makes its first flight, with Captains Mike Hewett and Jim McRoberts.*
3 Sep 1997	*Boeing launches the 737-700C with an order for two from the U.S. Navy. The Navy calls the model, a cargo version of the 737-700, the C-40.*
7 Nov 1997	*The 737-700, earns FAA type certification. The certification formally recognizes that*

	the newest 737 airplane has passed all the stringent testing requirements mandated by the FAA and is ready to enter passenger service.
10 Nov 1997	*Alaska Airlines orders 10 737-900s and 10 options, launching the series. The aircraft is the longest 737 built, with a length of 138 feet 2 inches.*
8 Dec 1997	*Exactly a year to the date after the world premier of the first Next-Generation 737-700, the first 737-600 rolls out of the Renton factory. The aircraft will be the first of three 737-600s that will participate in the 737-600 flight testing and certification program.*
17 Dec 1997	*Boeing delivers the first 737 NG, a 737-700, to launch customer Southwest Airlines.*
22 Jan 1998	*The Boeing 737-600 makes its first flight.*
19 Feb 1998	*Europe's Joint Aviation Authorities (JAA), which comprise the aviation regulatory authorities of 27 countries, recommends type validation of the 737-700. The individual countries will award actual type certificates.*
13 Mar 1998	*The 737-800 earns FAA type certification. JAA follow suit on 9 Apr 1998.*
14 Aug 1998	*The 737-600 earns FAA type certification. JAA follow suit on 10 Sep 1998.*
1 Sep 1999	*737 NGs are certified for 180 minute ETOPS Operation.*
11 Oct 1999	*Boeing launches the BBJ-2, a modified version of the Next-Generation 737-800.*
14 Feb 2000	*Aloha Airlines begins first 180-minute ETOPS service, introducing Non-stop service between Honolulu and Oakland, Calif.*
27 Jan 2000	*The 737 becomes the first jetliner in history to pass 100 million flight hours.*
18 Feb 2000	*Boeing announces availability of advanced technology "blended" winglets as an option on Next-Generation 737-800.*
14 Apr 2000	*First flight of the 737-700C*
3 Aug 2000	*First flight of the 737-900. Flight-test program begins.*
26 Sep 2000	*First flight of the 737 with blended winglets.*
17 Apr 2001	*The 737-900 earns FAA type certification. JAA follow suit on 19 Apr 2001.*
8 May 2001	*"Blended" winglets make their world debut in revenue service with Hapag-Lloyd Flug.*
17 Sep 2001	*BBJ adds Flight Dynamics' latest head-up guidance system.*
2 Nov 2001	*Boeing delivers first Next-Generation 737-700 Convertible with Quick Change options.*
19 Mar 2002	*Boeing introduces the Technology Demonstrator airplane, a 737-900 outfitted with a suite of new and emerging flight deck technologies to assess their value for enhancing safety, capacity and operational efficiency across the Boeing fleet of airplanes.*
9 Sep 2002	*Boeing Business Jets announced the availability of a lower cabin altitude modification for BBJ operators. The new feature will offer 6,500ft cabin altitude instead of the standard 8,000ft cabin, providing passengers with an improved level of comfort.*
28 Jan 2003	*Boeing delivers a suite of three leading-edge display and flight management software for the 737. The new flight-deck technologies, which include the Vertical Situation Display (VSD), Navigation Performance Scales (NPS) and Integrated Approach Navigation (IAN), promise to reduce flight delays and enhance flight-crew efficiency.*
12 Oct 2003	*Boeing Electronic Flight Bag available for retrofit on BBJ. Boeing is offering an avionics-installed "Class 3" version of the EFB comprising Jeppesen software and data, and electronics and display hardware from Astronautics Corp. of America.*
21 May 2004	*First flight of the 737 AEW&C – "Wedgetail". Flight-test program begins.*
24 May 2004	*A BBJ completes the first North Atlantic flight by a business jet equipped with the advanced Future Air Navigation System (FANS).*
17 Jan 2005	*Final assembly time is cut to 11 days, making it the shortest final assembly time of any large commercial jet.*
3 Feb 2005	*First flight of 737NG without "eyebrow" windows.*
12 May 2005	*737NG First airliner to be certified for Cat I GLS approaches.*
31 Jan 2006	*Boeing launches 737-700ER.*
1 Sep 2006	*First Flight of 737-900ER, with Captains Ray Craig and Van Chaney.*
4 Aug 2008	*737NG carbon brakes certified by the FAA.*
14 Aug 2008	*First BBJ3 completed.*
21 Jul 2011	*First delivery of 737 with CFM56-7BE "Evolution" performance improving engines.*
20 Mar 2013	*Delivery of 7,500th 737.*

737-600

FF 22 Jan 1998, 69 built, 58 remain in service.

The 737-600 was the third of the NG family to be built and originated as the 737-500X with a similar length fuselage, seating up to 132. The launch order came from SAS on 15 March 1995. The fuselage is essentially that of the -700, with two plugs of 1.37m (fwd) and 1.01m (aft) removed giving an overall length of 31.2m (102ft 6in).

Other differences include:
- Engines derated to 18,500lbs.
- Locally increased gauge on wingtip skin panels to avoid flutter.
- Wing-to-body fairing modified to fit aft fuselage contour.

The stubby looking 737-600 series showing the exposed wheels when retracted, a feature common to all 737s. One pilot's description of this series is that "it flies like a wobbly arrow".

737-700

FF 9 Feb 1997, Approx 1050 in service plus 84 on order (exc BBJs).

This was the first of the NGs to fly and the equivalent of the 737-300. The first -700 was retrofitted with winglets on 11 Sept 2001 for Kenya Airways.

A 737-700 showing the double slotted flaps (here at the 30 position) and the slightly longer undercarriage of the NG family

Variants:

737-700C (Convertible)

Fitted with a 3.4 x 2.1m side cargo door, it can carry 18,780kg (41,420lbs) of cargo on eight pallets. The ceiling, sidewalls and overhead bins remain in the interior while the airplane is configured for cargo. It has the strengthened wings of the 737-800 to allow higher zero fuel weights.

737-700QC (Quick Change)

Is a -700C with pallet-mounted seats. This reduces the conversion time from passenger to freighter configuration, and vice-versa, to less than an hour.

737-700ER (Extended Range)

This version is essentially an airline BBJ with the stronger -800 wings and higher MTOW. It has up to nine aux fuel tanks, giving a total capacity of 40,530ltr (10,707USGal) and a range of 5,510nm. It entered service in 2007.

The first 737-700ER at BFI   *Photo Joe Walker*

Notice that in this photograph it does not have winglets - which you would expect on a long-range aircraft. This is because it is a derivative of the BBJ.

BBJ's have never been certified for production winglets and since the 737-700ER is a derivative of the BBJ winglets could not be fitted. So the first task after the aircraft was built & test flown was to retrofit it with APB winglets before delivery! This bizarre situation was resolved in 2008 when the BBJ (and therefore the 737-700ER) was certified for production rather than retrofit winglets.

737-800

FF 31 Jul 1997, Approx 3262 in service plus 947 on order (exc BBJ2s).

The 737-800 was launched on 5 September 1994 when the 737-400X project became the 737-800. The 737-800 is significantly longer at 39.4m (129ft 6in) and seats up to 189 with an MTOW of up to 78,960kg (174,000lbs). The first delivery was to Hapag-Lloyd in April 1998. This has been by far the most successful series of 737 taking 68% of all 737NGs orders.

Differences from the -700 include:
* Fuselage plugs of 3m (9ft 10in) fwd and 2.84m (9ft 4in) aft of wing.
* Additional pair of overwing exits added (similar to -400).
* Tailskid added to section 48 (same as -400).
* Environmental Control System riser ducts added (same as -400).
* Re-gauged skins and stringers to strengthen wing and centre section.
* Strengthened main landing gear structure.
* 44.5in (1.13m) tyres, heavy-duty wheels and brakes.
* Engine thrust increased to 26,400Lbs.

Hapag-Lloyd was the launch customer for the 737-800 and in 2001 became the first operator to fly the 737 with winglets

737-800ERX

This is a heavier (83,500kg / 184,000lbs MTOW), longer range version of the -800 designed to meet the needs of the P-8/MMA and BBJ2. It has various components from the -900ER including its heavier gauge wing, nose & main gear and section 44 (wing-body join section); it also has some parts from the BBJ1. Amongst the unique features is strengthening to the empennage.

737-800SFP

The Short Field Performance improvement package was developed in 2005/6 to allow GOL airlines to operate their 737-800s into the 1,465m (4,800ft) Santos Dumont airport. The modifications enable weight increases of approx 4,700kg (10,000lbs) for landing and 1,700kg (3,750lbs) for take-off from short runways. It includes the following changes:

- Flight spoilers are capable of 60 degree deflection on touchdown by addition of increased stroke actuators. This compares to the current 33/38 degrees and reduces stopping distances by improving braking capability.
- Slats are sealed for take-off to flap position 15 (compared to the current 10) to allow the wing to generate more lift at lower rotation angles.
- Slats only travel to Full Ext when TE flaps are beyond 25. Autoslat function available from flap 1 to 25.
- Flap load relief function active from flap 10 or greater.
- Two-position tailskid that extends an extra 127mm (5ins) for landing protection. This allows greater angles of attack to be safely flown thereby reducing Vref and hence landing distance. This is monitored by a new Supplemental PSEU (SPSEU).
- Main gear camber (splay) reduced by 1 degree to increase uniformity of braking across all MLG tyres.
- Reduction of engine idle-thrust delay time from 5s to 2s to shorten landing roll.
- FMC & FCC software revisions.

737-800SFP prototype. Notice the extended tailskid and trailing static cone.   *Photo: Brian Lockett, Goleta Air & Space Museum*

The SFP package has now become an option on all 737-800s (known as 737-800SFPs) and standard on the 737-900ER. Some of the features may also be fitted to the 600/700 series. The first SFP was delivered 31st June 2006.

737-900

FF 3 Aug 2000, 52 delivered.

Boeing began work on the 737-900 in April 1997 which was stretched to compete with the 185/220 seat, Airbus A321. It featured a 2.4m (7ft 10in) fuselage extension giving it an overall length of 42.1m (138ft 3in), actually 40cm longer than the 707-120. The 900 has 9% more cabin floor space and 18% more cargo space than the -800; however Boeing opted to use the same NG emergency exit layout, with 4 main exit doors and 4 overwing exits, thereby still restricting the maximum passenger load to 189. It has been succeeded by the 737-900ER due to slow sales.

Director of flight testing, John Corrigan, said during the certification program: "The aircraft rotates a little more slowly than an -800, pretty much like a -400, but it flies completely smoothly and also achieves all the expected performance data. As a precaution we have strengthened the tail skid somewhat, but up to now tailstrikes have not been a problem. The changes in flying behaviour tend to be more subtle. We did have a problem with vibration during trimming of the elevators, which meant we had to change the trim tab. But we have already successfully modified a component on the prototype aircraft, which will now need to be certificated for the series."

KLM were one of the few airlines to operate the basic 737-900

737-900ER

FF 1 Sept 2006, 260 in service plus 261 on order (exc BBJ3s).

The -900ER is the successor to the -900 and is what it should have been in the first place. It has the same length fuselage as the -900 but seating has been increased to 215 passengers by adding a pair of Type II doors aft of the wing for passenger evacuation regulations and installing a new flattened aft pressure bulkhead which added an extra fuselage frame (approx 1 row of seats) of cabin space.

The prototype 737-900ER landing at BFI   *Photo: Joe Walker*

Range is increased to 3,200nm with the addition of two 1,970ltr aux fuel tanks (or 2,800nm without aux tanks) and optional winglets. The 900ER has reinforced landing gear legs, wing-box and keel beam structure to handle the increased MTOW of 85,139kg (187,700lbs). Take-off and landing speeds (and hence field length) are reduced by the short field performance improvement package originally developed for the 737-800, this is standard on all 737-900ERs. MZFW is 67,721kg (149,500lbs) & MLW 71,400kg (157,500lbs); the brakes are also upgraded.

Production started in 2006. The flight test program took 7 months and the type gained FAA certification on 26 Apr 2007. The first aircraft was delivered the following day to Lion Air.

BBJ

BBJ1 FF 4 Sep 1998, 118 BBJ1, 21 BBJ2, 6 BBJ3 & 1 BBJC ordered.

The record breaking BBJ1 on the apron at Nice, N737ER

A corporate version of the 737-700 dubbed the Boeing Business Jet (BBJ) was launched on 2nd July 1996 as a joint venture between Boeing and General Electric. It combines the fuselage of a 737-700 with the strengthened wings and undercarriage of the 737-800. Up to 12 fuel tanks, giving 37,712kg (83,000lbs) of fuel can be fitted as a customer option. The BBJ pictured here (N737ER) was designed for medical evacuations and charter operations and flew a record 6,854 nautical miles (12,694 kilometres) from Seattle to Jeddah in 14 hrs 12 minutes. The aircraft still landed with 2,700kg (6,000lbs) of fuel remaining!

Externally BBJs usually differ from standard production 7/8/900s by having various windows blanked to accommodate interior fittings and more antennas for comms equipment; all have winglets for range. All versions of BBJ only require one overwing exit each side because of their low maximum seating capacity.

The BBJ2 has the 737-800 fuselage, wings and undercarriage. It has 25% more cabin space and twice the cargo space or aux fuel tank space of the BBJ1. The BBJ3 is based on the 737-900ER and was available from mid-2008. It has 1120 square feet of cabin space and a range of over 5400nm with 5 aux fuel tanks. Boeing also offers a convertible cargo version of the BBJ, called the BBJC, based on the 737-700C.

So far private individuals have bought 40% of the BBJs. Another 36% have been bought for government heads of state and the rest were sold to corporations and jet charter operators.

The BBJ2 is based on the 737-800

C-40 "Clipper"

FF 14 Apr 2000. Orders: 9 C-40A, 4 C-40B, 6 C-40C.
The C-40 family are the first of the US military versions of the 737. All have the -700 fuselage combined with the stronger -800 wing and landing gear, similar to a BBJ1.

C-40A: US Navy, Fleet logistics support aircraft.
Certified to operate in an all-passenger configuration (121 passengers), an all-cargo variant or a "combi" configuration which can accommodate up to three cargo pallets and 70 passengers on the main deck. This is the only C-40 version without winglets.

USAF C-40B   Photo: Kevin Dawes

C-40B: US Air Force, High-priority personnel transport & communications aircraft.
Modified C-40A to include distinguished visitor compartment for combatant commanders and communications system operator workstation. The C-40B is designed to be an "office in the sky" for senior military and government leaders. Communications are paramount aboard the C-40B which provides broadband data/video transmit and receive capability as well as clear and secure voice and data communication. It gives combatant commanders the ability to conduct business anywhere around the world using on-board Internet and local area network connections, improved telephones, satellites, television monitors, and facsimile and copy machines. The C-40B also has a computer-based passenger data system.

C-40C: Air National Guard & AFRC, High-priority personnel transport aircraft.
Modified C-40A to include convertible cargo area. May be converted for medevac, passenger transport or distinguished visitors such as members of the Cabinet and Congress. The C-40C is not equipped with the advanced communications capability of the C-40B. Unique to the C-40C is the capability to change its configuration to accommodate from 42 to 111 passengers.

E-737 AEW&C

FF 21 May 2004, 14 orders (Australia 6, Turkey 4, South Korea 4)

The first 737 AEW&C "Wedgetail" bristling with antennae   *Photo: Brian Lockett, Goleta Air & Space Museum*

The 737 Airborne Early Warning and Control is designed for countries that don't need (or can't afford) the capability of the much bigger 767 or 707 AWACS. The base plane is essentially a Boeing Business Jet, which has the 737-700 fuselage with the stronger 737-800 wing to support its extra weight and the BBJ aux fuel tanks. The 737 AEW&C costs between $150 million & $190 million, this compares with about $400 million for the 767 AWACS. It carries a mission crew of between 6 and 10 in the forward cabin.

Its main external feature is the "Top Hat" antenna. This is a phased-array, Multi-role Electronically Scanned Array (MESA) radar sensor developed by Northrop Grumman and mounted in a rectangular fairing over the rear fuselage. The antenna alone weighs 2950kg (6500lbs) and is 10.7m long. It provides fore and aft coverage with a low drag profile allowing the system to be installed without a significant impact on aircraft performance. The pylon air intakes are for the liquid cooling system.

Inevitably, the Top Hat reduces the airflow over the fin & rudder which has necessitated the addition of two ventral fins to increase the directional stability for engine-out flight. Also the elevator feel pitot probes have had to be moved higher up the fin away from the disturbed airflow (see photo).

Other modifications include a new section 41 with a cut-out for an air-to-air refuelling receptacle and nose, wingtip and tail mounted counter measure systems. The aircraft will also have chaff and flare dispensers and approx 60 antenna and sensor apertures. The IDGs are uprated to 180kVA and can be seen bulging from the engine cowls. DOW is expected to be just over 50,000kg (110,000lbs). Each AEW&C contains an extra 863 electronic boxes, 300 kilometres of wiring and four million lines of software code more than a standard 737 NG.

The first green aircraft arrived at Wichita in December 2002 for structural modifications. Flight testing of the airframe ran from May 2004 until Jul 2005 with the aircraft logging more than 500 flight hours in 245 flights. According to Boeing "The plane performed superbly in terms of its avionics, structure, systems, flight handling characteristics and performance". This was followed by flight testing of the mission system, including the MESA radar. All appeared to be going well for the project until 2006 when the first of the delays was announced because of "development and integration issues with certain hardware and software components". Deliveries began to Australia in late 2009 and reached full operational capability in November 2012.

The aircraft is known as the "Wedgetail" by the RAAF after the Australian Wedgetail Eagle, which according to the Aussies, "Has extremely acute vision, ranges widely in search of prey, protects its territory without compromise and stays aloft for long periods of time." The Turkish AF call theirs the "Peace Eagle", presumably for similar reasons. Boeing are hoping to sell up to 30 AEW&Cs by 2016.

MMA / P-8 "Poseidon"
FF 25 April 2009. 116 Orders (US Navy 109, India 8, Australia 8)

Jack Zerr, the Multi-mission Maritime Aircraft (MMA) programme manager described the aircraft as "A bit of JSTARS (Joint Surveillance Acquisition Radar System), a little bit of AWACS and a little bit of MC2A (Multirole Command and Control), but with the added ability to go and kill a submarine."

The MMA, US Navy designation "P-8A Poseidon" and Indian Navy designation "P-8I", is based on the 737-800 fuselage and the stronger 737-900 wing, with raked wingtips that have anti-icing along all leading edge slats. A weapons bay aft of the wing, (effectively in the aft hold) carries internal stores such as Mark 54 torpedoes. There are four underwing hardpoints for AGM-84D Harpoon or similar. The fuselage is strengthened for weapons employment and to permit ASW profiles. Up to seven mission consoles and a rotary sonobouy launcher can be fitted in the cabin. Like the AEW&C, the MMA will have 180kVA IDGs as standard. The MMA also has an in-flight refuelling receptacle over the flight deck.

Northrop-Grumman provide the electro-optical/infrared sensor, the directional infrared countermeasures system and the electronic support measures system. Raytheon provide an upgraded APS-137 maritime surveillance radar system and signals intelligence (SIGINT) solutions. Finally, Smiths Aerospace provide the flight management system and the stores management system. The flight management system provides an open architecture along with a growth path for upgrades. The stores management system permits the accommodation of current and future weaponry. The basic open architecture of the MMA is believed to have 1.9 million lines of code!

The P-8 displaying at Farnborough Air Show in 2014 with bomb doors open and dummy AGM-84D Harpoon anti-shipping missiles

Much of the provisioning for the modifications is being done by Boeing during production to save time and cost at the conversion stage. Boeing has built a third production line which is dedicated to the MMA alongside the commercial 737 assembly. After the aircraft are assembled at Renton they are flown over to Boeing Field for mission system installation.

The current P-8As flying are known as "Increment 1" these have the basic tactical system. Increment 2 when it becomes operational in 2016 will have ASW improvements that provide a better performance at high altitude as well as AIS fitted. Increment 3 with its further improvements is scheduled for 2021.

737 MAX

FF Expected 2016, operational by 2017. 2200 Orders

The 737 Replacement Study, project name "Yellowstone 1" (Y-1), started in 2005 but was put on hold in 2008. Boeing had been aiming for a replacement that would give a 20-25% improvement in operating costs and were hoping to use 787 technologies to achieve this. Unfortunately the expected improvements were only around 10%. A Boeing spokesman said in 2008 that "you can't just do a shrink of the 787; it's not as easy as that because of the different missions, higher cycles and shorter range. You can't shrink the 787 because of the systems."

In July 2011, some 8 months after Airbus announced its LEAP-1A re-engined A320 NEO, Boeing followed suit with a CFM-LEAP-1B re-engined 737NG which it called the 737 MAX and claims will produce a 7-percent operating cost advantage over the Airbus A320 NEO. The advantages of a re-engined aircraft over a whole new type are faster certification times so Boeing don't lose market share to Airbus and a common type rating which makes is cheaper to introduce for the airlines. In October 2013 Boeing announced that the overall fuel burn improvement of the 737MAX over the 737NG would be 14%.

The new nomenclature for the 737 MAX family closely resembles that of the 787; Boeing has adopted the names 737 MAX 7, MAX 8 and MAX 9, respectively, to correspond with the 737-700, -800 and -900. The first aircraft is due for first flight in 2016 and delivery in 2017.

In September 2014 Boeing announced the launch of the 737 MAX 200. This is a 200 seat version of the MAX 8 which has an extra pair of Type II doors aft of the wing (as in the 900ER and MAX 9) giving it a maximum certified passenger capacity of 200. The seat pitch will remain at 30 inches with space being gained by slimline seats and removing space from front and rear galleys.

New features will include:

- New 69in (175cm) CFM LEAP-1B fan.
- New CFM LEAP-1B custom core with 11-12% reduction in fuel burn and 7% reduction in operating cost.
- New engine nacelle and pylon will cause engines to project further forward than CFM56-7BE on 737NG.
- Updated EEC software, fuel and pneumatic systems.
- Nose gear extension of 8in (20cm) to give more engine ground clearance.
- Minor changes to nose wheel well to accommodate longer nose gear strut.
- Fly-by-wire spoiler system - to improve production flow, reduce weight and improve stopping distances.
- Hydraulic system redundancy to resemble 757.
- Reshaped tailcone to reduce drag giving a 1% reduction in fuel burn.
- Advanced technology winglets which feature upward and downward-directed composite airfoils
- Widespread structural strengthening.
- Flightdeck, fuselage lengths and door configs all frozen from 737NG.
- Possible eco-improvements currently under consideration include:
    o Mini-split flap
    o Variable area fan nozzle
    o adaptive trailing edges
    o flight-trajectory optimisation
    o Regenerative fuel cells

The 737 MAX is also being offered as a BBJ. The BBJ 2 MAX, based upon the MAX 8, will have a potential range of 6,200nm.

PRODUCTION

The first 271 737s were built in Seattle at Boeing Plant 2, just over the road from Boeing Field, (BFI). However, with the sales of all Boeing models falling and large scale staff layoffs in 1969, it was decided to consolidate production of the 707, 727 and 737 at Renton just 5 miles away. In December 1970 the first 737 built at Renton flew and all 737s have been assembled there ever since.

However not all of the 737 is built at Renton. For example, since 1983 the fuselage including nose and tailcone has been built at Wichita and brought to Renton by train. They appear green due to their protective zinc-chromate coating. Much of the sub-assembly work is also outsourced beyond Boeing (see page 32).

737 fuselages passing BFI en-route to Renton by train from Wichita   Photo: Joe Walker

Production methods have evolved enormously since the first 737 was made in 1966. The main difference is that instead of the aircraft being assembled in one spot they are now on a moving assembly line similar to that used in car production. This has the effect of accelerating production, which not only reduces the order backlog and waiting times for customers but also reduces production costs. The line moves continuously at a rate of 2 inches per minute; stopping only for worker breaks, critical production issues or between shifts. Timelines painted on the floor help workers gauge the progress of manufacturing.

When the fuselage arrives at Renton, it is fitted with wiring looms, pneumatic and air-conditioning ducting and insulation before being lifted onto the moving assembly line. Next, the tailfin is lifted into place by an overhead crane and attached. Floor panels and galleys are then installed and functional testing begins. In a test called the "high blow", the aircraft is pressurised to create a cabin differential pressure equivalent to an altitude of 93,000 feet. This ensures that there are no air leaks and that the structure is sound. In another test, the aircraft is jacked up so that the landing gear retraction & extension systems can be tested. As the aircraft moves closer to the end of the line, the cabin interior is completed – seats, lavatories, luggage bins, ceiling panels, carpets etc. The final stage is to mount the engines. There are approximately 367,000 parts in a 737, held together by approximately 600,000 bolts and rivets.

The present build time is reducing from 11 days (5,500 airplane unit hours of work) towards a future target of 8 days (4,000 airplane unit hours of work). The production rate has increased from 31 aircraft a month in 2005 to 42/month in 2014 and is scheduled to reach 52 aircraft a month by 2018 for the 737MAX.

After construction they make one flight, over to BFI where they are painted and fitted out to customer specifications. It takes about 200ltrs (50USgallons) of paint to paint a 737. This will weigh over 130kg (300lbs) per aircraft, depending on the livery. Any special modifications or conversions (eg for the C40A, AEW&C or MMA) are done at Wichita after final assembly of the green aircraft. Auxiliary fuel tanks, winglets, SATCOM and specialist interiors for BBJs are fitted by PATS at Georgetown, Delaware as these are STC (Supplemental Type Certificate) items.

A "green" 737 seen landing at BFI after its maiden flight   Photo: Joe Walker

Materials

The fuselage is a semi-monocoque structure, ie the loads are carried partly by the frames & stringers and partly by the skin. It is made from various aluminium alloys except for the following parts.

- **Fibreglass**: radome, tailcone, centre & outboard flap track fairings.
- **Kevlar**: Engine fan cowls, inboard track fairing (behind engine), nose gear doors.
- **Graphite/Epoxy**: rudder, elevators, ailerons, spoilers, thrust reverser cowls, dorsal of vertical stab.

Different types of aluminium alloys are used for different areas of the aircraft depending upon the characteristics required. The alloys are mainly of aluminium and zinc, magnesium or copper but also contain traces of silicon, iron, manganese, chromium, titanium, zirconium and probably several other elements which remain trade secrets. The different alloys are mixed with different ingredients to give different properties as shown below:

- Fuselage skin, slats, flaps, stabilizers - areas primarily loaded in tension - **Aluminium alloy 2024** (Aluminium & copper) - Good fatigue performance, fracture toughness and slow propagation rate.
- Frames, stringers, keel & floor beams, wing ribs - **Aluminium alloy 7075** (Aluminium & zinc) - High mechanical properties and improved stress corrosion cracking resistance.
- 737-200 only: Bulkheads, window frames, landing gear beam - **Aluminium alloy 7079** (Aluminium & zinc) - Tempered to minimise residual heat treatment stresses.
- Wing upper skin, spars & beams - **Aluminium alloy 7178** (Aluminium, zinc, magnesium & copper) - High compressive strength to weight ratio.
- Landing gear beam - **Aluminium alloy 7175** (Aluminium, zinc, magnesium & copper) - A very tough, very high tensile strength alloy.
- Wing lower skin - **Aluminium alloy 7055** (Aluminium, zinc, magnesium & copper) - Superior stress corrosion.

Speaking at Farnborough 2006, Alan Mulally former CEO Boeing Commercial Aircraft said that the 737 replacement will have a composite airframe; "composites now being the "material of choice" for airplane design". Mulally said "Composites don't corrode, don't fatigue and are more reliable and easier to maintain....Composites also allow a greatly simplified manufacturing process, and that can significantly drive down costs."

The design life for the 737 was originally set at 20 years, 75,000 cycles or 51,000 hours; this was increased to 130,000 cycles in 1987. Unfortunately one year later an Aloha 737-200 was lost due to structural fatigue, that aircraft was only 19 years old but had 89,193 cycles, the second highest of all 737s.

Outsourcing

Most of the 737 is not built by Boeing but is outsourced to other manufacturers both in the US and increasingly around the world. This may be either for production cost savings, specialist development or as an incentive for that country to buy other Boeing products. Here is a list of some of the outsourced components:

- Nose, fuselage, engine nacelles and pylons - Spirit AeroSystems (formerly Boeing), Wichita.
- Wing-to-body fairing panels and tail cone - BHA Aero Composite Parts Co. Ltd, China.
- Doors - Vought, Stuart, FL.
- Fwd entry door & Overwing exits - Chengdu Aircraft, China.
- Vertical fin - Xi'an Aircraft Industry, China. (also claimed by Spirit)
- Rudder - Bombardier Aerospace, Belfast.
- Horizontal stabiliser - Korea Aerospace Industries.
- Elevator - Fuji, Japan.
- Aileron panels - Asian Composites Manufacturing, Malaysia.
- Aileron units, Krueger flaps - Hawker de Havilland, Australia.
- Spoilers - Ducommun AeroStructures, California.
- Winglets – Kawasaki Heavy Industries, Japan and GKN, UK.
- Wing spars (front & rear), Flaps and slats - Spirit AeroSystems, Tulsa.
- Inboard Flap - Mitsubishi, Japan. Moving production to Vietnam in spring 2009.
- Tail section (aluminium extrusions for) - Alcoa / Shanghai Aircraft Manufacturing, China.
- Main landing gear doors - Aerospace Industrial Development Corp, Taiwan.
- Flightdeck windows – PPG Aerospace, Alabama, USA

Sales

The Boeing 737 is the best-selling commercial aircraft in aviation history with over 12,000 aircraft ordered.

In 2013 Boeing took net orders for 1208 737s; that stands as the most of any model by any major commercial manufacturer in history. By Sep 2014, the total number of 737NGs ordered was over 6,800, 70% of which were for the -800 series.

By Sep 2014, the 737 had an order backlog of over 4000 aircraft, including more than 2,200 orders for the 737 MAX. This is about one-third of the entire order backlog for all large commercial jets built by Boeing and Airbus and will ensure the 737's production until at least 2020.

The table and chart shows orders and deliveries by series and year. Data is correct up to September 2014.

Model	Total Orders	Total Delivered
737-100	30	30
737-200	991	991
737-200C	104	104
737-300	1113	1113
737-400	486	486
737-500	389	389
737-600	69	69
737-700	1215	1107
737-700 C40	20	16
737-700 AEWC	14	14
737-800	4732	3382
737-800 P8	69	37
737-900	52	52
737-900ER	512	273
737-T43A	19	19
BBJ	120	116
BBJ2	21	20
BBJ3	6	6
737 MAX	2295	0
Total:	**12257**	**8224**

Boeing 737 Order and Delivery History

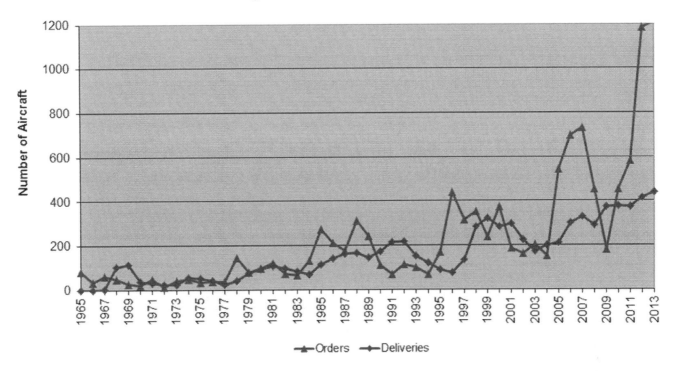

Price

One of the more difficult questions to answer is "How much does a 737 cost?" The answer is not that straightforward. It depends upon series, customer options about avionics and cabin fittings, and most significantly size of order and when it is placed.

The average list prices for 737s as of October 2014 were as follows:

- 737-700 $78.3 million USD
- 737-800 $93.3 million USD
- 737-900ER $99.0 million USD
- 737 MAX 7 $87.7 million USD
- 737 MAX 8 $106.9 million USD
- 737 MAX 9 $113.3 million USD
- 737 P-8 $150.0 million USD

These prices are an average price reflecting a range of available options and configurations for each model. Options include performance capability, interiors, avionics, fuel capacity, etc.

In February 2005, Ryanair placed an order for 70 737-800s with options for a further 70. The public list price was between $61.5 million to $69.5 million per aircraft as given on Boeing's Web site. However because of the size of the order and the fact that it was placed when Boeing had being losing a significant amount of orders to its rival Airbus, it is widely believed that Ryanair paid less than $51 million each; and by the time other concessions such as credit and allowances, support services and free winglets are factored in, the price could have been as low as $29 million - less than half of the list price.

SYSTEMS

AIRCRAFT GENERAL

Lights

From left to right along the panel:

O/B Landing: (Not NG) 3 position switch: OFF - EXTEND (off) - ON. This was modified from a 2 position switch in 1969 to eliminate distracting light reflections during extension in clouds. The lights are located on the outboard flap track fairing. The same modification also introduced the gang bar which operates the inboard & outboard lights.

Retractable Landing: (NG only) Replaces the outboard landing lights on the earlier series. These are located on the fuselage just beneath the ram air intakes. The word is that they may be being moved back to their original position on the flap track fairing due to excessive stone damage.
Note: Use of both of these lights should be avoided at speeds above 250kts due to excessive air loads on their hinges.

I/B Landing: Known as fixed landing lights on the NG. These are located in the wing roots, usually used for all day and night landings for conspicuity.

R/W Turnoff: Also in the wing roots, normally only used at night on poorly lit runways.

Taxi: This 250W light is located on the nose gear, on later models it will switch off automatically with gear retraction. It is common practice to have this on whilst the aircraft is in motion as a warning to other aircraft and vehicles.

Logo: Are on each wingtip or horizontal stabiliser and illuminate the fin. Apart from the advertising value on the ground, they are often used for conspicuity in busy airspace.

Position: Depending upon customer option this can be a three position switch (as illustrated) to combine the strobe. Strobe & Steady / Off / Steady, where steady denotes the red, green & white navigation lights. The three Nav lights are no-go items at night.

Position and logo light

Strobe: (Not illustrated) Off / Auto / On. Auto is activated by a squat switch. They are also in the wing tips and are very brilliant. This gives rise to great debate amongst pilots about when exactly they should be switched on as they can dazzle other pilots nearby. Many people choose to put them on as they enter an active runway for conspicuity against landing traffic.

Anti-Collision: Are the red rotating beacons above and below the fuselage. They are universally used as a signal that the engines are running or are about to be started. They are typically not switched off until N1 has reduced to below 3.5% (or N2 below 20%) when it is considered safe for ground personnel to approach the aircraft.

Wing: These are mounted in the fuselage and shine down the leading edge of the wing for ice or damage inspection at night.

Wheel Well: Illuminates the main and nose wheel wells. Normally only used during the turnaround at night for the pre-flight inspection, but must also be on to see through the gear downlock viewers at night, hence they are a no-go item at night in all but the NGs. There is also a switch for the main wheel well light in the port wheel well.

Eye Position Indicator

This enables pilots to set the correct seat height for optimum inside & outside visibility. When the three balls are in line you are at the recommended height. This device was only fitted to the 1/200 series as the recommended position on the Classics was changed so that you could see under the glareshield panel to the top of the EADI and over the yoke to the bottom of the EHSI.

In practice, most pilots tend to set their seating position by feel.

Eye position indicator on the central windscreen pillar

Water System

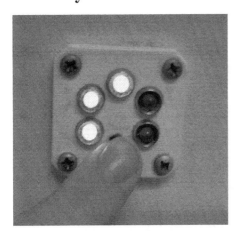

Series -3/4/500 Potable Water Quantity Indicator

There is a 30 US Gal tank (40 US Gal –400 series) behind the aft cargo hold for potable water. This serves the galleys and washbasins, but not the toilets as they use chemicals. Waste water is either drained into the toilet tanks or expelled through heated drain masts.

The tank indicator (-3/4/500 version shown left) is located over the rear service door. Press-to-test indications are clockwise from 7 O'clock: Empty, 1/4, 1/2, 3/4, Full.

The NG attendant panel has an LED panel that is always lit for both potable water and waste tank quantity.

Potable Water Quantity Indicator - NG

Airstairs

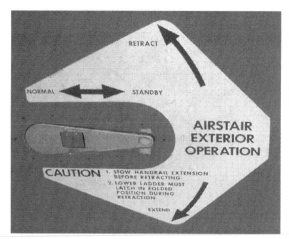

Exterior Airstair Controls

The forward airstairs may be operated from either internal or external panels. The internal panel requires the forward entry door to be at least partially open. Both panels have NORMAL and STANDBY systems. Normal requires AC and DC power, standby only requires DC. External standby system power comes from the battery bus and so does not require the battery switch to be on.

On Classics, if the airstairs will not operate, check the striker pin (see photo right) at the bottom left of the door frame. Move it about and ensure it is vertical; this will often cure the problem. They have a tendency to freeze in position on long flights were the doors have got wet.

Caution: The handrails must be stowed before retraction. The use of the standby system from either panel will bypass the handrail and lower-ladder safety circuits. Note that the NG has a red covered EMERG switch underneath the airstairs for emergency retraction; this also bypasses any safety circuits.

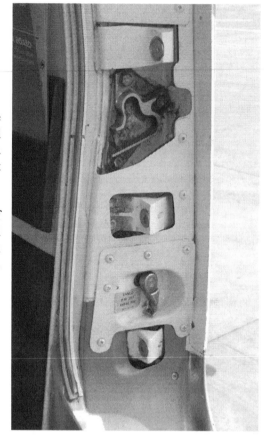

737-3/4/500 Door showing latch fitting (above), striker pin (below) and 3 stop pin fittings

Underneath NG Airstairs - Notice the guarded MAINT switch.

Limitations:

Maximum wind speed for airstair operation: **40kts**.

Maximum wind speed for airstair extended: **60kts**.

Airstairs should not be operated more frequently than 3 consecutive cycles of normal system operation within a 20 minute period.

Aft Airstairs

In the drive for self-sufficiency, these were fitted to about 120 737-200s. They were much more complicated than forward airstairs as they folded in two places and took the door downwards with them. If you have ever considered the forward airstairs to be temperamental then you would not get on with aft airstairs.

There were several reports of inadvertent deployment and even two instances of them extending after take-off. Boeing say that after one of the in-flight deployments the crew landed with little control problem and apart from some scuff marks on the foot plates where they made contact with the runway, they were still in working order after the event!

Doors

200C door panel

Classic door panel

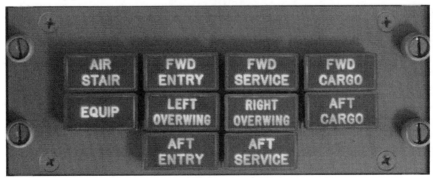

NG door panel

An amber light will illuminate with Master Caution "DOORS" when a door is unlocked. The airstair must be fully stowed, even if fwd entry door is closed. Equip is for E & E bay and Radar bay.

The sequence of door lights is changed in the NGs to accommodate the left and right overwing annunciators. They are located between the fwd & aft entry/service door lights.

On the NG, it is not uncommon to get an overwing caption illuminate for a fraction of a second as you start the take-off run. This is due to the overwing exit automatic locking function being slightly slow.

Limitation - NG: Installation of handle covers on the overwing exits must be verified prior to departure whenever passengers are carried.

737-8/900 Overwing exit doors

Nose, wing tip & tail clearance.

The following table shows the increased radius of turn of the wing and tail relative to the nose during a turn with full nose wheel steering applied. This table shows that turning in a 737-600 is probably most hazardous because the wings and tail turn out much further than the nose.

Radius of turn (ft)			
Series	Nose	Wing	Tail
-300	55	+5	+9
-400	61	+1	+7
-500	50	+9	+10
-600	51	+17	+11
-700	56	+13	+10
-800	66	+6	+9
-900	71	+2	+7

Cargo Systems

The cargo holds are designed to confine a fire without endangering the safety of the aircraft. Series 1-500 before l/n 3079 and NGs before l/n 91 were built as Class D. Subsequent aircraft were built to Class C standard and retrofit kits are available. The differences include a smoke/fire detection and extinguishing system and a pressurisation modification which prevents smoke getting from the hold into the passenger cabin (see page 104).

The holds are sealed and pressurised but have no fresh air circulation. They have no temperature control but are heated by exhausting cabin air around their walls. The forward hold also has additional heating from E & E bay air. Live cargo can be carried on either cargo compartment but the forward hold is preferred because it is warmer.

The optimum distribution for loading the cargo holds to keep the aircraft in balance is as follows:

Series	Fwd	Aft
-300/700	1/3	2/3
-400/800	0	All
-500/600	All	0

Cargo / freighter aircraft have a side cargo door. The systems, controls and annunciators vary with the manufacturer (Pemco, Bedek, AEI etc). The door is normally powered from the hydraulic B system. This can be pressurised either by an additional electric hydraulic pump located in the tail section or, on some aircraft, by the normal EMDP in the wheel well. The door has operating wind limits which vary between 30kts (x-wind) and 60kts if aircraft is parked nose into wind; and whether the door is fully opened (sail position) or partially opened (canopy position). A further AFM limitation prohibits moving the aircraft with the door open or unlatched.

The vent door has an interlock to prevent the cargo door from unlocking unless cabin pressure is released.

The side cargo door from the outside showing the vents

The inside of a 737-200C, the pallet positions are marked above the windows

Note that cargo aircraft must be fitted with either a 9G barrier net or a 9G Bulkhead to prevent loads from shifting into the flightdeck.

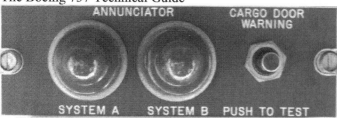

Cargo door not safe annunciator

The controls for opening & closing the door are near the fwd entry door. There is an annunciator on the flight deck to warn if the door becomes unlatched. The push to test button also tells you if the door is not locked, latched, closed or the vent doors are open.

QC & SF versions have optional heat pads near the bottom of the cargo door for passenger comfort since the door does not have as much insulation as the rest of the fuselage.

Smoke detector systems vary; the Systron Donner system shown here divides the cabin into 5 zones and monitors centre (C) and side (S) ports in each zone. The single blue light simply indicates that the system is on.

Some cargo 737s had a pressurisation feature which allowed the crew to pressurise or unpressurise the passenger compartment for smoke clearance (see page 50).

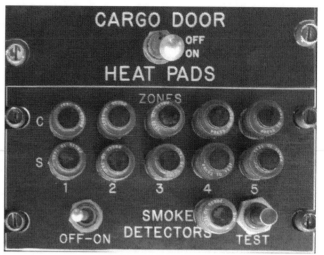

The 737-300QC cargo door heat pads and smoke detector panel

All versions have secondary main gear downlock annunciators because the viewers would be inaccessible when cargo is carried.

Note that side cargo doors are very heavy and move the CofG forward to the extent that when ferrying an empty aircraft ballast may need to be carried to keep in trim.

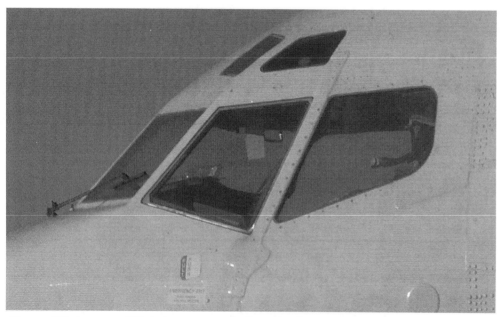

Another subtle difference between cargo versions and all others is that the Captains DV window can also be opened from the outside, not just the F/O's. This is in case the doors are made inaccessible by the shifting of the cargo in the event of an accident.

All cargo 737s have extra DV window access on the Captains side

The pitot-static probes on NG cargo aircraft (eg C-40) are in the same location as the Classics because the side cargo door would interfere with the location of the usual NG static ports.

AIR CONDITIONING & PRESSURISATION

APU or engine 5th and if necessary 9th stage bleed air (hot), is pre-cooled by fan air before entering the pack. Inside the pack, bleed air is cooled by ram air through heat exchangers and an air cycle machine. A water separator collects water condensed by the cooling process to avoid icing.

Series 3/5/6/700 pack output temperatures are independently controlled. Series 4/8/900 packs work to the coldest of the three zones; the two warmer zones are then heated by trim air after the mix manifold. In both series, flight-deck air is taken from before the mix manifold, however series 4/8/900 flight-deck air may also contain trim air.

-3/5/6/700 Air-conditioning Panel

-4/8/900 Air-conditioning Panel

DUCT OVERHEAT means that the duct temperature has exceeded 88C.

ZONE TEMP (4/8/900) means that either the duct temperature has exceeded 88C or the temperature controller has failed.

-1/200 Air-conditioning Panel

An option on some -1/200 series air-conditioning panels (left) was the facility to display the pack temperatures. Otherwise it is identical to the existing panel on the NG.

Some BBJ 1's have the option for separate fwd & aft cabin temperature controls as the cabin is often divided for privacy. This photo (right) shows the panel as fitted by Raytheon. Notice the trim air and zone control buttons.

BBJ1 Air-conditioning Panel

The cabin temperature sensor is behind the small grill located just over the overhead locker on the starboard side at approximately row 3. This is why you get the burning smell of a duct overheat in the cabin when you have the fwd doors open on a cold day and the wind blowing icy air through the cabin. On such a day it is better to use the temperature controls in manual to avoid this. Also if the cabin temperature is not regulating well in flight, you should have an engineer clean the foam air filter behind this grill. The flightdeck temperature sensor is near the dome light.

Cabin temperature sensor

Behind the cabin temperature sensor

Recirculation Fan(s)

The recirculation fan simply re-circulates filtered cabin air back to the cabin to reduce bleed air requirements.

A pack in HIGH flow will produce more cold air than normal, but has a 25% higher bleed air demand. Approximately 25% of the cabin air is recirculated for passenger comfort compared to 50% on the 757/767 and none on the MD80. The recirc fan will switch off if either pack is in high flow, giving a net reduction in the ventilation rate of 15%, so best cooling is achieved with pack(s) AUTO and recirc fan(s) ON; this also reduces pack load, bleed air demand and fuel consumption.

The ventilation rate of the 737-3/500 is 1900 cubic feet per minute (CFM) or about 13 CFM per passenger. When the larger 737-400 was designed, an extra recirculation fan (also on the –4/8/900) was installed to increase the ventilation rate to 2100CFM, and hence comfort levels, for the increased passenger capacity of the larger aircraft. Unfortunately, the second fan is quite loud on the flightdeck - as are the NG fan(s).

Gasper Fan

There was no recirculation fan on the 1/200 series but there was a gasper fan. This was an electric fan designed to increase pressure in the gasper system (the outlets above pax seats), under conditions of low supply pressure or high cold air demand - normally on the ground on a hot day. The gasper fan is still effective even if the packs are off as cabin air is drawn into the distribution ducts, down the risers, into the main manifold and mixing chamber where it is then blown into the cold air risers and ducts and out of the gasper fans.

The extra caption between the two RAM DOOR FULL OPEN lights on the -1/200 series is a blue F OUTFLOW CLOSED light.

-1/200 Pneumatics Panel

Ram Air

Ram-Air Inlet - Classic

Ram-Air Inlet - NG

Ram air is controlled by variable geometry inlet doors, turbofans, and on some aircraft variable exhaust louvers. Maximum ram air is available on the ground and on take-off. In-flight there is less cooling requirement so it is modulated by the doors and louvers which reduces both the ram air flow and also drag. The deflector doors are extended whenever the aircraft is on the ground.

The RAM DOOR FULL OPEN lights will be illuminated on the ground and should extinguish during the climb. If they do not, then either the cabin is still very hot and the packs require full ram airflow for cooling or the heat exchanger may be faulty.

Air-conditioning Packs

A PACK TRIP OFF annunciation means that the pack temp (not output temp) has exceeded its limit.
A PACK annunciation (-4/8/900 only) means that the pack temp has exceeded its limit or failure of pack controls.

On the ground: 1-500s: Use only one pack from the APU because packs are working hardest when delivering cold air. NG APUs are more powerful so you can use both packs for cooling or heating. In fact using both packs causes the APU to burn slightly less fuel than single pack operation.

If one pack fails with the pack switches in AUTO, the other will regulate to high flow (unless flaps are down). Note series 1/200s do not have an AUTO mode, the pack switch is simply ON/OFF.

If you dispatch with one pack inoperative, then max altitude is FL250. If a pack fails when above this level, then you may continue at the higher level. Note if a pack should fail, even at the aircrafts maximum certified altitude, it should be able to maintain cabin pressure. If both packs fail the cabin altitude will climb, probably between 2000 & 4000fpm depending upon the condition of the aircraft seals.

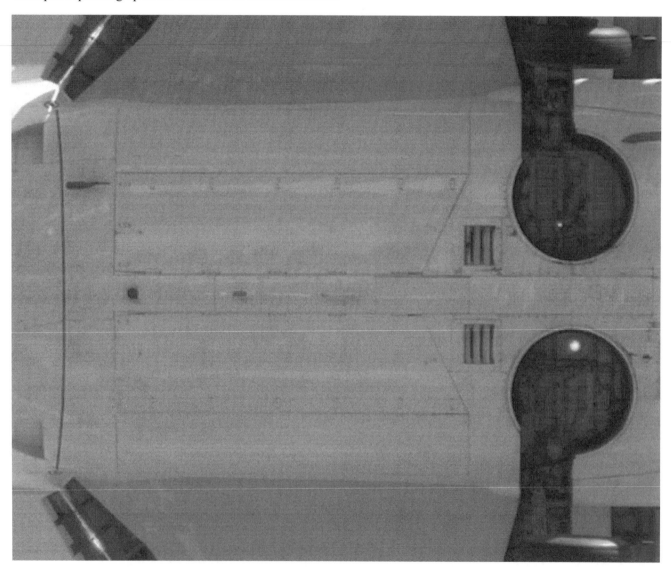

This photo shows the location of the two air conditioning packs underneath the aircraft. Notice that the two ram air inlets have their deflector doors retracted because the aircraft is in-flight. The packs are accessed through the two large access panels between the deflector doors and the wheel well, these panels are hinged inboard. Aft of the access panels are the ram air exit louvers, these will give you a nice warm blast of air on your legs as you stand in the wheel well during your walk-around on a cold winter day!

Equipment Cooling

The flight deck panels, display units, c/b panels and E&E bay are all cooled by replacing the warm air with cool air from the cabin with electric fans. The supply fans push air and the exhaust fans pull air through these units. The second fan was added to EFIS equipped aircraft to cool the CRTs in 1986.

On the ground, the air is then dumped through the flow control valve (Classics) / overboard exhaust valve (NG). In-flight above 2psi cabin differential, the air is used for heating the forward hold. The valve closing can be heard as a sudden hiss on the flightdeck on climb out or during final approach at 2psi.

Note: both cargo holds are also heated by exhausting cabin air around their walls. The forward hold is guaranteed to maintain at least 4.5C (40F) and the aft hold at least 0C (32F) at a distance of 8 inches from the walls.

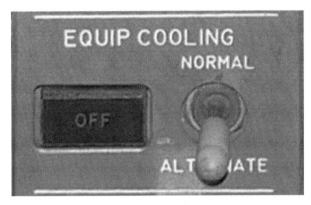

737-1/200 Equipment cooling

737-3/900 Equipment cooling with separate supply & exhaust fans.

Pre-Conditioned Air

This is being used more frequently as airports start restricting the use of the APU on the ground.

Pre-conditioned air is attached here straight into the mix manifold.

Remember that the temperature controls in the flight deck have no effect when using pre-conditioned air.

This is not the same as ground air for engine starting. That is high pressure air put into the right pneumatic manifold and is connected to the aircraft near the right wing root.

Pre-conditioned air flows straight into the mix manifold

Water Separator

Air leaving the air cycle machine is at its coldest and any moisture it contains will condense downstream. This could lead to problems with icing or corrosion so the excess water is removed by the water separator.

Air goes into the water separator and through a polyester coalescer bag fixed on a support which collects water mist from the air. The mist becomes water droplets as more moisture goes through the bag. The support has slots that move the air in a circular motion. The air with the water droplets moves to the collection chamber which is a baffle that causes the water and air to make a sharp bend. This separates the heavier water droplets but allows the air to leave freely. The collected water is then drained overboard; you can see this dripping from the packs on the ground at hot & humid airports. If the ACM output air is below 2C warm air is introduced to prevent icing.

If you have a noisy air conditioning pack, the coalescer bag may be full of dirt and need changing (like a Hoover bag!). When the bag is blocked the air bypasses the water separator.

With the panels open you can see the water separator (in black), air cycle machine and the heat exchangers (silver boxes at bottom right of picture)

Air-conditioning compartment showing the water separator, air cycle machine and primary & secondary heat exchangers

Limitations

Air Conditioning

With engine bleed air switches ON, do not operate the air conditioning packs in HIGH for takeoff, approach or landing.

One pack may be inoperative provided maximum altitude is limited to FL250 (Dispatch limitation only).

Pressurisation

Maximum cabin differential pressure:

Series	Max Diff
1/200	7.5psi - Aircraft before l/n 400 with max certified altitude of 35,000ft
200Adv	7.8psi - Aircraft after l/n 400 with max certified altitude of 37,000ft
Classic	8.65psi
NG	9.1psi
BBJ	9.74psi – Aircraft with reduced cabin altitude option

Maximum differential pressure for takeoff & landing: **0.125psi** (236ft below airport PA)

Maximum takeoff and landing altitude: **8,400ft** or **12,000ft** with high altitude landing option.

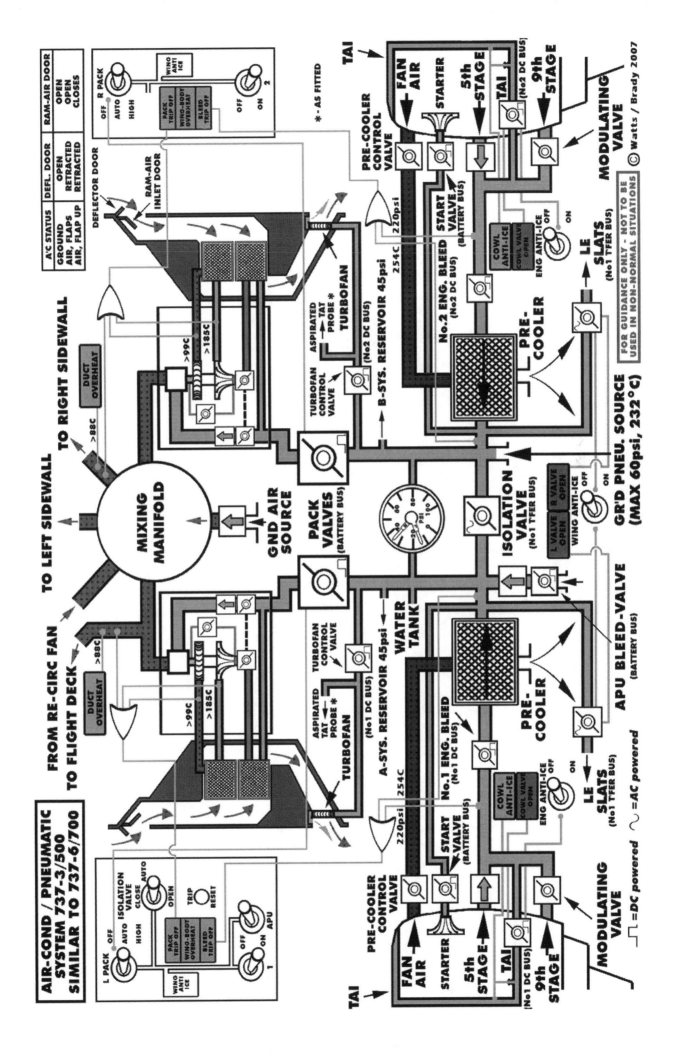

AIR-COND / PNEUMATIC SYSTEM 737-3/500 SIMILAR TO 737-6/700

FROM RE-CIRC FAN

TO LEFT SIDEWALL

TO RIGHT SIDEWALL

TO FLIGHT DECK

© Watts / Brady 2007

FOR GUIDANCE ONLY - NOT TO BE USED IN NON-NORMAL SITUATIONS

A/C STATUS	DEFL. DOOR	RAM-AIR DOOR
GROUND AIR, FLAPS	OPEN	OPEN
AIR, FLAP UP	RETRACTED RETRACTED	CLOSES

WING ANTI ICE

PACK TRIP OFF   WING-BODY OVERHEAT   BLEED TRIP OFF

OFF   AUTO   HIGH

R PACK

OFF   ON   2

* - AS FITTED

DEFLECTOR DOOR

RAM-AIR INLET DOOR

DUCT OVERHEAT

>88C

>99C   >185C

MIXING MANIFOLD

GND AIR SOURCE

PACK VALVES (BATTERY BUS)

ASPIRATED TAT PROBE *   TURBOFAN

TURBOFAN CONTROL VALVE

B-SYS. RESERVOIR 45psi   (No2 DC BUS)

220psi   254C

No.2 ENG. BLEED (No2 DC BUS)

START VALVE (BATTERY BUS)

PRE-COOLER

COWL ANTI-ICE   COWL VALVE OPEN

ENG ANTI-ICE   OFF   ON

LE SLATS (No1 TFER BUS)

MODULATING VALVE

TAI   FAN AIR   STARTER   5th STAGE

TAI (No2 DC BUS)   9th STAGE

PRE-COOLER CONTROL VALVE

TAI   TAI

WATER TANK

ISOLATION VALVE (No1 TFER BUS)

L VALVE OPEN   R VALVE OPEN   WING ANTI-ICE   OFF   ON

GR'D PNEU. SOURCE (MAX 60psi, 232°C)

APU BLEED-VALVE (BATTERY BUS)

PRE-COOLER

No.1 ENG. BLEED (No1 DC BUS)

A-SYS. RESERVOIR 45psi (No1 DC BUS)

254C   220psi

START VALVE (BATTERY BUS)

COWL ANTI-ICE   COWL VALVE OPEN   ENG ANTI-ICE   OFF   ON

LE SLATS (No1 TFER BUS)

PRE-COOLER CONTROL VALVE

FAN AIR   STARTER   5th STAGE

TAI   9th STAGE (No1 DC BUS)

MODULATING VALVE

TAI   TAI

ASPIRATED TAT PROBE *   TURBOFAN

TURBOFAN CONTROL VALVE (No1 DC BUS)

DUCT OVERHEAT   >88C

>99C   >185C

L PACK   OFF   AUTO   HIGH

ISOLATION VALVE   CLOSE   AUTO   OPEN

TRIP   RESET

PACK TRIP OFF   WING-BODY OVERHEAT   BLEED TRIP OFF

OFF   ON   APU   1

WING ANTI ICE

= DC powered   ∿ = AC powered

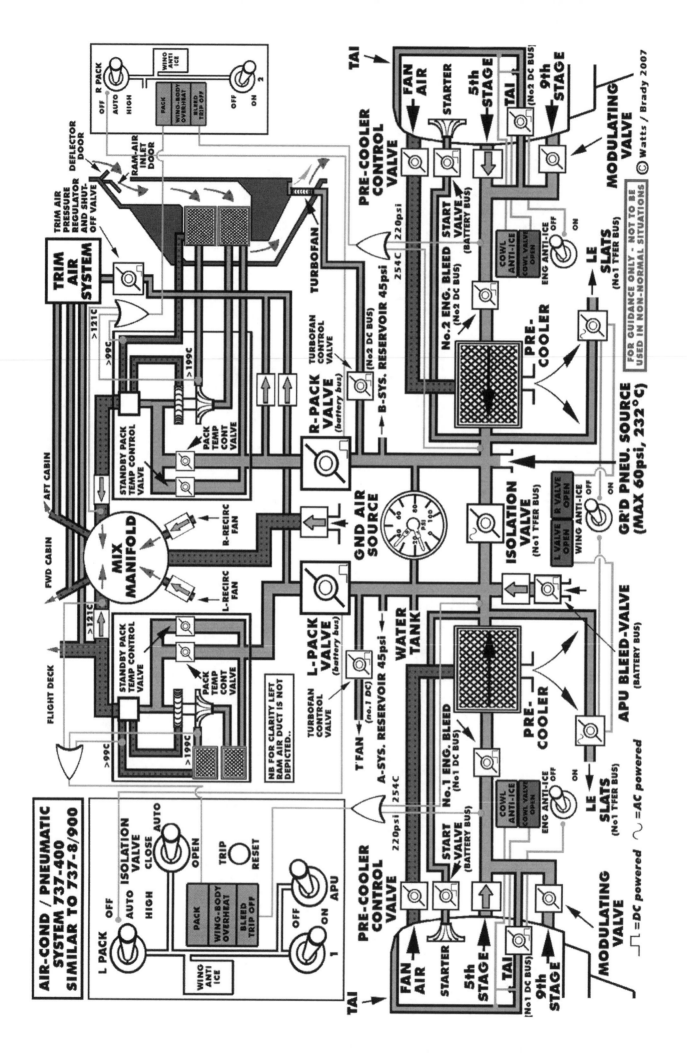

Pressurisation

Digital Cabin Pressure Control System (DCPCS)

The aircraft is pressurised by bleed air supplied to the packs and controlled by outflow valves.

Digital pressurisation controllers have two automatic systems (AUTO & ALTN) instead of a standby system, these alternate every flight. If the auto system fails, the standby / alternate system will automatically take over. The AUTO FAIL light will remain illuminated until the mode selector is moved to STBY / ALTN (tidy but not necessary). On CPCS panels, the cabin rate selector, for use in standby mode, adjusts cabin rate of change of altitude between 50 and 2000fpm, the index is approx 300fpm. The normal (AUTO) scheduled rate of climb is 500 sea level fpm and 300 slfpm for descent.

In manual mode, you drive the outflow valve directly. The sense of the spring-loaded switch can be remembered by: "Moving the switch towards the centre of the aircraft keeps the air inside."

If you have to return to your departure airfield, you do not need to adjust the pressurisation panel. You will get the OFF SCHD DESC light, but the controller will program the cabin to land at the take-off field elevation. If the flight alt selector is pressed, this facility will be lost.

The 737NG pressurization schedule is designed to meet FAR requirements as well as maximize cabin structure service life. The pressurization system uses a variable cabin pressure differential schedule based on airplane cruise altitude to meet these design requirements. At cruise altitudes at or below FL 280, the max differential is 7.45psi which will result in a cabin altitude of 8000ft at FL280. At cruise altitudes above FL280 but below FL370, the max differential is 7.80psi which will result in a cabin altitude of 8000ft at FL370. At cruise altitudes above FL 370, the max differential is 8.35psi which will result in a cabin altitude of 8000ft at FL410. This functionality is different from other Boeing models which generally use a fixed max differential schedule thus can maintain lower cabin altitudes at cruise altitudes below the maximum certified altitude.

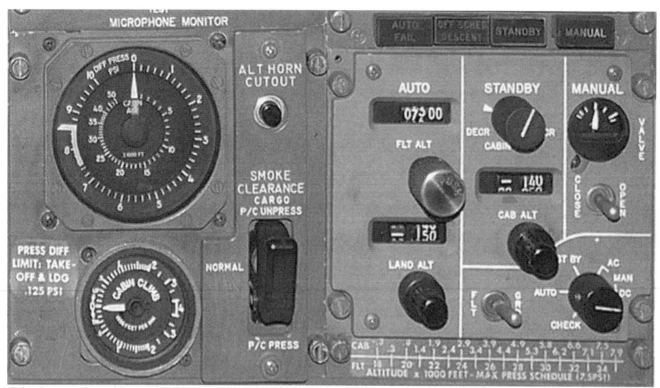

Cabin Pressure Control System (CPCS). Notice the extra SMOKE CLEARANCE controls in this cargo version.

Reduced Cabin Altitude System

The pressurisation system of all series of 737 ensures that the cabin altitude does not climb above approx 8,000ft in normal operation. However in 2005 the BBJ was certified to operate with a reduced cabin altitude of 6,500ft at 41,000ft (ΔP of 8.99psid above 37,000ft) to increase passenger comfort. The payback for this is a 20% reduction in airframe life cycles, ie from the standard 75,000 down to 60,000 cycles. This is not a problem for a low utilisation business jet but would be unacceptable in airline operation where some aircraft are operating 10 sectors a day.

High Altitude Airport Operations System

Is a customer option for operations into airfield elevations of up to 10,000ft (all NG), or 13,500/14,500ft (-6/700 series only). Changes include: Addition of high altitude landing selector switch to transfer to the high altitude mode, cabin altitude at warning horn activation is increased, an extra hour of emergency oxygen, 10 minute take off thrust if engine inop; winglets and carbon brakes are recommended.

Cabin Altitude Warning Horn

The cabin altitude warning horn will sound when the cabin altitude exceeds 10,000ft. It is an intermittent horn which sounds like the take-off config warning horn. It can be inhibited by pressing the ALT HORN CUTOUT button. Note the pax oxygen masks will not drop until 14,000ft cabin altitude although they can be dropped manually at any time. Following the Helios accident (page 323) where the crew did not identify the cabin altitude warning horn, new red "CABIN ALTITUDE" and "TAKEOFF CONFIG" warning lights have been

fitted to the P1 & P3 panels to supplement the existing aural warning system.

AUTOMATIC FLIGHT

Autopilot

The autopilot is just one part of the Flight Control System (FCS) / Digital Flight Control System (DFCS). The flight director, autothrottle, Mode Control Panel (MCP) & Flight Control Computer (FCC) are separate but interrelated. There have only been four different autopilots in the history of the 737.

Autopilot	Aircraft fitted	Dates
SP-77	737-1/200/200Adv	1967-1980
SP-177	737-200Adv	1980-1988
SP-300	737 Classic / NG	1984-2003
EDFCS-730	737 NG	2003 onwards

Flight Control Computer (FCC)

The FCC, located in the E&E bay, is the brains of the DFCS and like any other computer, its hardware and software is being improved (and debugged!) all the time. There are two identical FCCs in each aircraft and either one is capable of managing all of the DFCS functions; however, both are required for Cat III autoland and autopilot go-around operation.

This chapter gives every 737 FCC hardware and software update and their features. On the NG series you can find out which FCC and software update that your aircraft has through the MAINT pages of the FMC. In this example the aircraft has a Collins EDFCS with rudder channel and software update -140.

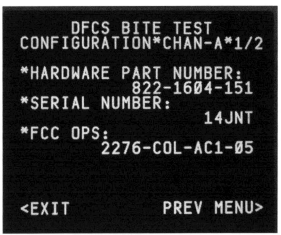

The FCC configuration can be shown on the FMC

Mode Control Panel (MCP)

The MCP is the main interface with the autopilot and autothrottle but much of the control can also be done with the PDCS/FMC. The MCP is located on the glareshield panel, (a.k.a. P7 panel) which also contains the master caution lights & annunciations, fire warning lights and, on the NGs, the EFIS control panels. Whilst the MCP has evolved with the autopilot, the master caution system and fire warning lights have remained unchanged through to the present day which is testimony to their good initial design.

The P7 panel contains the MCP, fire & master caution warning lights and EFIS control panel

Sperry SP-77

The autopilot in the first 737s was the Sperry SP-77, recognisable by the old MCP. The fire panel was moved to the throttle quadrant to allow space for the master caution, flight director and autopilot controls.

The SP-77 MCP with two Collins flight directors

The MCP comprised three panels, the autopilot (centre) and one for each flight director (sides). All three were independent so any mode to be used with the autopilot (eg ALT HOLD or VOR/LOC) had to be selected three times. HDG SEL and VOR/LOC could be coupled to the autopilot but would only be driven by the heading bug set on the Captains compass and the course set on the Captains HSI.

The SP-77 autopilot consisted of a Pitch Control Computer and a Roll Computer. For a dual-channel configuration, there were two Pitch Computers. Airplanes with this configuration had separate Flight Director controllers for the Collins FD-108, FD-109, or FD-110 system, whichever was installed. The FD controllers were either built into the ADIs and HSIs, or were of various shapes and sizes with different combinations of switches and position nomenclatures. Modes available were GA, OFF, HDG, VOR LOC, AUTO APP & MAN GS, which was mainly used for capturing the glideslope from above!. There is also a PITCH CMD knob which has now become vertical speed.

The centre panel is for autopilot selections and has two paddles to engage/disengage the ailerons and elevators "AIL" & "ELEV" for roll and pitch modes and could be engaged separately or together. The LH knob has the roll modes of HDG, VOR LOC, AUTO APP & MAN G/S and has a HDG OFF / (HDG HOLD) / HDG SEL switch to its right (see para 2). The RH knob has pitch modes of ALT HOLD or TURB. The TURB mode was controlled by the vertical gyro to allow smoother movements to regain altitude during turbulence. Some also had an ALT SEL mode. The small knob at the top, left of centre (labelled INOP/INOP in this photo) is the hydraulic system selector source for the autopilot and should read "A", "B", "AB". The early SP-77 only used hydraulic system B PCUs.

The early SP-77s had no ALT Select (ie no altitude capture) or Speed Select and were flown most of the time in Control Wheel Steering (CWS) - it was used like a sort of sophisticated "wing leveller". The A/P was then "Flown" via the normal controls.

The SP-77 MCP with F/D heading & course windows

This asymmetric looking version of the MCP was the first to have heading and course windows. Although everything looks biased to the Captain it is more First Officer friendly because he can set a heading (centre window) or course (the two outside windows) on this panel rather than having to rely on what the Captain sets on his compass or HSI. The usual F/D MODE SEL, ALT HOLD & PITCH CMD controls are all in this single panel and are repeated to both the Capt's and F/O's F/Ds. The autopilot panel is displaced to the right.

Sperry SP-177

The SP-177 MCP

In 1980 the 737-200Adv was fitted with a Sperry SP-177 autopilot with integrated PDCS/FMC and autothrottle. After several developments (see table) it became Cat IIIa autoland capable in 1982. This MCP has remained virtually unchanged through to the NGs being delivered today.

Sperry SP177 Autopilot FCC Summary of changes		
FCC P/N 10-61984-x	**Date**	**Summary of changes**
-1	Jan 1980	The original FCC for the SP177
-7	?	?
-5	Jun 1980	Replaced -1 & -7 FCC. Disengages A/P servos whenever a power interrupt is sensed to prevent an abrupt pitch manoeuvre.
-8	Jul 1980	Provides A/P CMD to CWS amber warning lights.
-110	Jan 1981	"Reduce in-service problems and improve operational characteristics"
-2	Jul 1981	Interim measure toward fully operational AFCS
-6	Sep 1981	Cat II & auto G/A capable
-111	Apr 1982	Replaced -2 & -6 FCC to be Cat IIIa capable
-113	?	?
-114	Jul 1983	Improved G/A, LVL CHG and F/D & PDC interfaces
-214	Nov 1986	Fixes non-selected MCP changes
-314	Nov 1987	Selectable bank angle limit (20/25/30) with LNAV.

Sperry (Honeywell) SP-300

All Classics and NGs until 2003 were fitted with the Sperry (who later became Honeywell) SP-300 autopilot-flight director system (AFDS). This early 737-300 MCP has paddles to engage & disengage autopilots and CWS. The two small grey panels either side of the MCP are to select the source of navigation information. The options are FMC (normal mode), ANS-L or ANS-R if the alternate navigation system (IRS based) is required.

An SP-300 MCP with A/P engage paddles

The differences between this SP-300 MCP and the SP-177 MCP are:
- EPR button became N1
- VNAV/PDC button became VNAV
- ALT & SPD INTV buttons (optional).
- LCD windows instead of Number tapes.

On pre-1991 MCPs, turning the altitude knob changed MCP altitude in 1000ft increments, when pressed in it changed to 100ft increments.

Differences between various SP-300 MCPs are:
- The A/P CMD/CWS/Disengage paddles changing to select buttons and bar or a mixture of both.
- ALT & SPD INTV buttons for FMC hard speed and alt restraints, (shown blanked off below).

An SP-300 MCP with A/P engage buttons and disengage bar

FCC P/N 10–62038–1 (known as -1 FCC) - Feb 1984

Original FCC fitted to the first (non-EFIS) 737-300s.

FCC P/N 10–62038–2 (known as -2 FCC) - Nov 1986

Developed for EFIS equipped 737-300s and non-EFIS aircraft after line number 1275 (Sept 1986).
1. The FCC is now EFIS compatible.
2. The F/D take-off mode is changed to allow entry for up to 150 seconds after take-off and less than 2000ft rad alt. This makes F/D take-off guidance available for windshear encounters even though the take-off mode was not initiated before take-off.
3. An option is added to allow entry into TOGA mode even if one or both F/D switches are in the OFF position.
4. The F/D TOGA control laws are modified to enhance flyability in windshear.
5. Control laws modified to give smoother roll entries and reduced bank angles during close in localiser captures.
6. Capability is added to transfer the LNAV source between the FMC and an Alternate Navigation CDU.
7. Single channel autopilot approach now allowable when only the associated radio altimeter is valid.
8. Selectable option added to inhibit glideslope capture before localiser capture.
9. MCP/FCC interface monitor added.
10. MCP selected altitude display validity monitor added.
11. Filtering added to the neutral shift feedback input signal to increase autopilot pitch stability during approach.
12. Mach trim logic modified to minimise failure nuisance warnings.
13. F/D comparator thresholds adjusted to minimise nuisance F/D command display retractions during approach.
14. Several BITE modifications also made.

FCC P/N 10–62038–3 (known as -3 FCC) - Nov 1988

Basic FCC for 737-400 and 737-300 after Nov 1988.
1. Speed trim failure will be indicated if "in air" logic from the air/ground sensing system is received when airspeed is less than 80 knots and N1 from either engine is greater than 20%. This will cause master caution and speed trim fail annunciations which alert the flight crew that a system failure exists prior to initiation of take-off.
2. The FD command bars are no longer biased out of view during TOGA mode when the radio altimeters are invalid.
3. Occasional flight director mistracking in altitude capture and hold modes due to differences between L & R ADC computed altitudes are eliminated.
4. Loss of valid flight path acceleration inputs to the autopilot, when modes are selected requiring the inputs, will result in an autopilot disconnect and/or loss of flight director commands.
5. The localiser-on-course mode is automatically initiated as a default at 135 seconds, rather than 45 seconds after localiser capture. This ensures completion of the localiser capture manoeuvre before bank angle limiting is reduced to eight degrees.
6. Speed trim warning logic is revised to eliminate nuisance warnings.

When modifications to existing autopilot option selection configurations are accomplished, the following autopilot functional changes are also available:

1. ALT HOLD mode is always initiated and maintained at the point where the MCP ALT HOLD button is pressed.
2. MCP ALT knob changes altitude by 1000 if rotated by one detent provided knob is not depressed when rotated.
3. If a max continuous N1 power limit value has been selected and an engine out condition exists, the autopilot will not change this limit except at initiation of TOGA.
4. When in HDG SEL mode, the aircraft will turn in the same direction as the HDG SEL knob is rotated.
5. Roll rate limits are set as a function of manually selected bank angle limits. ie Lower selected bank angle limits will result in reduced roll rate limits.
6. When airspeed is reduced in VNAV mode to a point that the angle of attack limit (alpha floor) is reached, the autopilot will no longer revert to level change mode. The VNAV mode will be maintained and the aircraft will be controlled to the alpha floor speed until a higher speed is commanded.
7. The minimum climb rate in LVL CHG mode in engine-out operation is reduced from 1000fpm to 60fpm. This allows for increased acceleration to reach desired airspeeds more quickly.
8. Speed trim calculations use IRU v/s instead of ADC v/s to improve system performance.
9. If an engine fails after take-off with the A/P engaged, a continuous N1 limit may be selected and will be retained. VNAV cruise mode as well as automatic engagement to CLB, CRZ or DESC N1 limits is inhibited.
10. If a TRUE/MAG heading switch is installed in the aircraft, utilisation of either true heading and track or magnetic heading and track is possible.

FCC P/N 10–62038-4 (known as -4 FCC) - Dec 1991

Boeing developed and certified the -4 FCC as the basic autopilot computer for the 737–500 airplane. This FCC contains functional improvements and additional selectable options, as well as control law changes required for the 737–500 airplane. The most significant new capability is electronic aileron force limiting. Use of which allows deletion of the mechanical aileron force limiter assembly on 737–500 airplanes. Since FCCs –1 through –3 do not have electronic aileron force limiting incorporated, they cannot be used on the 737–500.

FCC P/N 10–62038-5 (known as -5 FCC) - Initial 737-700s - Feb 1997

Installed on the first 64 737-700s. Came with built in software. I can only imagine that the "-5" designator had been allocated to the NG program before the requirement for "-6".

FCC P/N 10–62038-6 (known as -6 FCC) - March 1995

Introduced to replace FCC-3 on 737-3/400s with mechanical aileron force limiters. FCC-6 provides improved high altitude alpha floor detection between the FMS as in the FCC-4, without activating all of the features present in the FCC-4. The improved alpha floor mode is a function of altitude which provides more margin between the normal operating envelope and detection of the alpha floor, and also prevents the airplane from accelerating above the speed that existed when alpha floor was detected. This follows an incident when the alpha floor speed mode was prematurely triggered in the cruise causing an overspeed.

FCC P/N 10–62038-7 (known as -7 FCC) - Initial 737-800s - Jul 1997

This was the first FCC that could be loaded with the Operational Program Software (OPS) on board the airplane. This allowed updates to be made without changing the FCC thereby saving time and money.

Honeywell FCC software P/N 2215-HNP-03B-02 (known as -702 software) - Apr 98

1. Added 737-800 model dependent constants.
2. Added improvement to altitude hold and altitude capture control laws to reduce overshoots.
3. Updated Alpha V2 tables for the 737-700.
4. Redesigned the glideslope and localizer beam anomaly detectors. They are now active at glideslope engage plus 2 seconds and LOC on course, respectively.
5. Revised speed trim to: a) Inhibit for bank angles greater than 40 degrees. b) Synchronize when the flaps are up and the gear are down and airspeed is greater than a value that is model dependent.

6. Reduced flight director bar steps when transitioning between cruise modes.
7. Increased the threshold of the trim sensors monitor when above Mach 0.82.
8. Reduced control wheel oscillations in CWS roll.
9. Changed when the flap and gear placard limiting and warning occurs to reduce nuisance trips.
10. Revised maintenance monitor for mach trim faults to start recording with engine start.
11. Revised nominal values and limits in a number of BITE tests.
12. Replaced uncorrected baro altitude with inertial altitude in cruise modes that provide altitude control.
13. Improved "speed on elevator" operation when operating near Vmo/Mmo.
14. Reduced the limit on the flare gust compensation path.
15. Eliminated a TOGA initialization problem which can occur when transitioning to TOGA from some pitch cruise modes.
16. Inhibited selection of roll cruise modes below 400 feet during LNAV takeoffs.

The basic 737NG MCP is virtually unchanged from the SP-177 MCP but the EFIS control panels have been moved into the glareshield from the aft electronics panel in an arrangement similar to the 747-400.

Honeywell FCC software P/N 2213-HNP-03B-04 (known as -704 software) - Oct 98

1. Updated the control laws for the 737-600 and 737-700 IGW.
2. Added model dependent table values for the 737-600 and 737-700 IGW.
3. Revised speed trim to include changes associated with JAA stall identification requirements.
4. Updated the V2 speeds for the 737-800.
5. Revised the level change mode to reduce vertical speed variations.
6. Revised the pitch single channel camout detector threshold and enabled this function for all 737NG models.
7. Delayed localizer on course engagement by 2 seconds to filter out LOC beam noise.
8. Made further changes to the CWS roll to improve performance.
9. Latched the wheel spin touchdown logic.
10. Provided wings level roll command at touchdown.
11. Added a roll gain change for speed brake deployment.
12. Improved the CWS pitch flight characteristics.
13. Revised the approach control laws to improve performance during turbulence.
14. Changed the 737-700 IGW go-around target speeds.

FCC P/N 10–62038-8 (known as FCC-8) - Aug 2000

Hardware update which provides provision for Integrated Approach Navigation. Note that the capability is added in the -709 software. This airplane option is active when other airplane system interfaces have been installed and are enabled. This allows the autopilot of flight directors to fly ILS look-alike signals from the FMC, via the approach push button, down to CAT I minimums.

Honeywell FCC software P/N 2212-HNP-03B-05 (known as -705 software) - March 2000

1) Prevents one RA failure from causing an FCC to default to the 8 degree bank angle limit during an LNAV takeoff.
2) Corrected a 1-2 Hz oscillation in the pitch axis that could occur at high altitude in certain light weight configurations on the 737-700BBJ. Also, the pitch inner loop gains for the 737-700, 737-700IGW, and 737-800 airplanes were updated.
3) Revised the Mach Trim operation when the autopilot is controlling "speed on elevator" to reduce interaction. The mach trim command is held fixed to the selected mach when the autopilot is in the speed mode.
4) 737-700C added to the allowable configurations. This addition includes changes to the power up configuration test, airplane model gain and table switching, BITE self test messages and verification tests, and sensor values airplane identification.
5) Changes were made to the "speed on elevator" control laws to reduce the .1 Hz oscillations.
6) Changed the G/S Beam Anomaly Detector in one of the processors to eliminate a bug that could result in an erroneous trip.
7) To support flight data recording requests, new CWS Pitch and Roll inputs were added. When this software is installed with a future FCC hardware version in development, the scaling of the digital signals to the DFDAU will be increased and the values recorded will be scaled as pounds at the sensors.

8) Limited autopilot operation will be allowed on a single IRU when the IRU transfer switch is selected to "BOTH ON X".
9) Modified the speed on elevator control law to help keep the airplane below clacker speed when crossing the mach change-over altitude and Mach is being flown.
10) Changed the processing of the cross channel data to better insure that data necessary for synchronous entry by both FCCs into certain modes (e.g. F/D TOGA) is captured.

Honeywell FCC software P/N 221F-HNP-03B-08 (known as -708 software) - May 2001

1) Added changes specific to the aerodynamic characteristics of the 737-900.
2) Revised software interlocks with the Nav Transfer switch to cover the case where the airplane is configured to switch both the FCCs and displays.
3) The Bank Angle limits between 200 and 400 feet R/A are changed from +/-15 degrees to +/- 30 degrees for LNAV T/O.
4) The Roll F/D Comparator was enabled for LNAV T/O.
5) Increased the time delay associated with failures of the CDUs to reduce Approach and TOGA mode resets.
6) Added interlock such that if both AOA vanes are invalid or no computed data, the A/P will disconnect if in go-around.
7) Previously stab trim thresholds were a function of elevator position error for flaps down and elevator command for flaps up. For this part number, the use of elevator command was extended (instead of elevator position) down to flaps 15 if the autopilot is not G/S engaged.
8) Changed F/D G/A logic to also allow entry above 2000 feet RA if G/S is engaged or if the flaps are down.
9) Added logic to reset the F/D Approach mode on the ground, after sufficient time delay.
10) Changed the stab trim monitoring to increase detection of erroneous stab trim motion.
11) Revised autoland control laws on previous models to improve performance toward the end of the C of G, airport altitude, and landing speed ranges. These changes are activated via a configuration program pin.
12) Revised logic to re-synchronize Speed Trim references whenever a flap change is made.
13) Revised FCC logic to insure disconnect of both autopilots if one of the CWS pushbuttons is pushed after both autopilots have been engaged in a dual approach.

Honeywell FCC software P/N 221E-HNP-03B-09 (known as -709 software) - Jun 2002

1) Added provisions for FMC IRNAV capability (hardware FMC IRNAV capability was previously added in the –8). This airplane option is active when other airplane system interfaces have been installed and are enabled. This allows the autopilot or flight directors to fly ILS look-alike signals from the FMC, via the approach pushbutton, down to CAT I minimums.
2) Automatic entry into LNAV from Roll Go-around is allowed when the flight crew has entered a valid LNAV go-around path into the FMC. This feature is an airplane option requiring properly configured FMCs and a CDS capable of displaying LNAV ARM.
3) In order to preclude A/T underspeeds from causing a reversion to Level Change from Vertical Speed or VNAV Vertical Speed during VNAV/LNAV or VOR approaches, these reversions are inhibited for flaps > 12.5 degrees.
4) Improved the detection and monitoring of both the localizer and glideslope beam anomaly detectors.
5) Revised the GLS ARINC 429 input provisions to use the MMR input receivers rather than dedicated receivers.
6) Corrected a problem associated with the display of MACH/IAS on the MCP when switching from FMC elevator speed control to FMC A/T speed control, or from FMC speed control to non-FMC speed control.
7) Enabled Limited Loc Beam averaging for dual channel GLS provisions.

Honeywell FCC software P/N 2216-HNP-03B-10 (known as -710 software) - Aug 2007

1. Added the capability to arm VNAV prior to selection of Takeoff when compatible CDS and FMC part numbers are also installed. When armed, pitch takeoff will transition to VNAV engage at 400 feet. Note, due to inconsistencies associated with the arming of VNAV prior to takeoff, Boeing released the reference a) ops manual bulletin instructing flight crews to not attempt arming VNAV on the ground prior to takeoff.
2. Added the capability to arm LNAV prior to selection of Takeoff when compatible CDS and FMC part numbers are installed. When armed, the takeoff roll mode will transition to LNAV engage at 50 feet.
3. Revised design of "flight director only" LNAV ARM to LNAV ENG in roll go around to allow auto engagement of LNAV from Track Hold down to 50 feet whether the flight director switches are on or off. Note: This function is available as an option. This option is activated by incorporation of a negotiated Boeing Service Bulletin that will specify the correct FCC software, FMC software, CDS OPS and OPC software.
4. Added logic to reduce false altitude acquires due to erroneous but un-flagged altitude inputs.

Collins EDFCS-730

The Rockwell Collins Enhanced Digital Flight Control System (EDFCS) was introduced in Feb 2003 (l/n 1278). It is a fully digital, fail operational AFDS. The autothrottle computer has been eliminated by integrating it into the FCC and it has an optional rudder servo to enable Cat IIIb autoland and rudder controlled rollout.

The Collins MCP has been designed to be similar to the Honeywell MCP. The main difference to the pilots are that the buttons are larger and have the caption printed on them and the angle of bank selector is different.

These are Boeings comments upon how the new MCP was designed:
"Collins provided a preliminary MCP design to Boeing in 2000 for Boeing pilots and airline pilots to evaluate in the simulator and comment on. Based upon those comments, a revised MCP was installed on a test 737NG in November 2001. Boeing test pilots evaluated that design for approximately 4 months. Based upon that evaluation, changes were made in the tactile, lighting and thermal characteristics to increase the similarity of the Collins and Honeywell MCPs. The goal during this evaluation was to make the Collins MCP operationally transparent to the flight crew when compared with the Honeywell MCP. Recent certification and service-ready testing has indicated that the latest Collins MCP has obtained a high level of crew transparency."

Collins EDFCS FCC, P/N 822-1604-101 (-101 / -151 FCC or "P1.0" software) - Feb 2003

The Collins system, known as a "**-101 FCC**", is recognisable by the new style MCP. The system was offered from the outset with an optional rudder servo card to make Cat IIIb available for the first time on the 737, they are known as "**-151 FCC**". Both FCCs came with **P/N 2271-COL-AC1-02** software known as "**-111 software**".

Collins FCC software P/N 2270-COL-AC1-03 (known as -120 or "P2.0" software) - Jan 2004

This software is known as "-120" from the diskette part number, P/N 831-5854-120.

1) Added provisions for IAN capability to the FCC software. This airplane option is active when other airplane system interfaces have been installed and are enabled. This allows the autopilot or flight directors to fly ILS look-alike signals from the FMC, via the approach pushbutton, down to CAT I minimums.
2) Corrected a condition that may occur when selecting a new altitude when the FCC is in the final transition between ALT ACQ and ALT HLD (within a 200 millisecond window), where an altitude somewhere between the previously selected MCP altitude and the new MCP altitude will be stored as the reference ALT HLD altitude.
3) Corrected an error where, in certain scenarios, the FCC software may leave the glideslope engage logic set internally in either FCC without a corresponding G/S flight mode annunciation (on the FMA) on that side. This will cause the flight director takeoff mode and some autopilot or F/D pitch cruise modes to be inhibited. This condition may also occur when an autopilot or F/D approach is disconnected using means other than autopilot or flight director go-around.
4) Revised the FCC ARINC 429 flap output to be that of the internal FCC calculation from synchro position, including flat spots representing the flap detents, instead of retransmitting the SMYD input values. With the previous software version, the FMC may display incorrect target speeds at certain flap setting, and may incorrectly keep the autothrottle in the FMC Speed mode during descents.
5) Added autothrottle engine out capability.
6) Changed fault response for failures of the TOGA switch.
7) Corrected operation of the Flight Director Master light logic and Approach pushbutton light logic that may result in incorrect annunciation if one FCC is not powered.
8) Increased localizer signal useable range.
9) Corrected the DME logic to ensure that the data used is from the frequency selected on the navigation control panel.

10) Implemented improvements to CAT IIIb annunciation, initial localizer tracking, engine out go-around performance, and localizer beam anomaly detector during rollout.
11) Added logic to reset the F/D Approach mode on the ground.

Collins FCC software P/N 2277-COL-AC1-04 (known as -130 or "P3.0" software) - Feb 2005

1) Corrected logic sequencing when capturing an altitude in VNAV Altitude Acquire.
2) Improved filter initialization to prevent mistracking of runway course and selected course. Note: This is only applicable to single channel approaches on airplanes where the CAT IIIb option is selected.
3) Improved repeatability of entry into go-around from LNAV ENG when an FMC is invalid.
4) Revised design of "flight director only" LNAV ARM to LNAV ENG in roll go-around to allow auto engagement down to 50 feet whether the flight director switches are on or off. Note: This function is available as an option.
5) Improved the operation of the lateral beam anomaly detector when coupled to final approach course.
6) Increased filtering on the angle of attack input signals from the stall management yaw damper computer.
7) Reduced the airspeed devs which can occur when operating near the changeover altitude in LVL CHG or VNAV Speed.
8) Revised FCC logic to set NCD on the course output signals when the buses transmitting those signals are invalid.
9) Corrected the Flight Director Anomaly where, during takeoff, the roll mode could be reset to "no mode" if the TOGA switches were reselected after lift-off.
10) Corrected takeoff mode logic to ensure that if "engine out" is reset, the target speed transitions back to V2+20 instead of remaining at V2.
11) Corrected an implementation error which prevented autopilot engagement with a failed IRU when the Single IRU Option pin has been enabled.

Collins FCC software P/N 2276-COL-AC1-05 (known as -140 or "P4.0" software) - Aug 2006

1) Corrected flap placard overspeeds which can occur in flight director takeoff when target speeds are dialled to large values with flaps remaining at their original setting.
2) Added logic to reduce false altitude acquires due to erroneous but unflagged altitude inputs.
3) Added logic to reduce pitch manoeuvres on approach due to erroneous but unflagged radio altitude inputs.
4) Corrected a failure condition which could result in different Flight Director roll GA modes on each side.
5) CAT IIIb Config Only: Corrects an erroneous NO AUTOLAND which can occur if both runway courses are not selected by 2000 feet.
6) Revised enabling of lateral acceleration input at dual channel engage to insure signal transients are below second up channels synchronization capability. This corrects instances of autopilot disconnects which could occur near dual engage.
7) Added the anti-ice bias to the autopilot calculation of go-around alpha floor.
8) Prevent the A/T disconnect warning lights from flashing if the autothrottle is in the ARM mode upon landing.
9) Added logic for the A/T to look to the SMYD on whether or not to apply the speed brake bias to its min manoeuvre speed.
10) CAT IIIb Config Only: Increased roll gain during rollout to account for loss of spoiler effectiveness when speed brakes are fully deployed.
11) Improved glideslope control laws when the beam anomaly detector is tripped at large beam angles.
12) Added changes in the application program associated with selection of the Short Field Performance (SFP) Option.
13) Added autothrottle logic to mitigate throttle retards on approach due to erroneous but unflagged radio altitude inputs.
14) CAT IIIb Config Only: Fixed a nuisance NO LAND 3 Latch message which can occur if an approach is done after an autopilot GA.
15) CAT IIIb Config Only: Corrected course monitor initialization errors which erroneously set NO AUTOLAND messages when the course was near +- 180 degrees.
16) CAT IIIb Config Only: Desensitized monitor on FCC to MCP engage solenoid output to prevent nuisance NO AUTOLAND faults.
17) Added the capability to arm LNAV and/or VNAV prior to flight director takeoff. Enabled when compatible FMC and CDS are installed.

Collins FCC software P/N 2275-COL-AC1-06 (known as "-150" or "P5.0" software) - Jun 2007

1) Added changes in the application program associated with selection of the 737-900ER model pin (including a variable flare height).
2) Removed a "pitch bump" which can occur when glide slope is captured from above.

3) Eliminated the cause of a disconnect which may occur when flying dual channel autopilot go-arounds on the Alpha submode.
4) Changed the ratcheted Radar Altimeter rate limit (used for ILS gain programming) from 21 to 50 feet per second to allow faster updating of rapid rises in approach terrain.
5) Insured that an invalid minimum speed input disengages the VNAV mode or prevents entry into the VNAV mode.

Collins FCC software P/N 2275-COL-AC1-07 (known as "P6.0" software) – 2012

1. Improved Autothrottle Radio Altimeter comparator.
2. Improved the interface between the Autothrottle Servo Motor and the FCC to reduce Autothrottle disconnects

NASA 515 – TSRV

Photo Bob Bogash

Although not a production MCP I have included this for interest and completeness. This MCP is unique, it is the final MCP of the prototype 737-100, N515NA, formerly N73700 which was used by NASA until 1997 as the Transport Systems Research vehicle (see page 4).

This aircraft was often flown (and landed!) from a research flightdeck located in the cabin. This "Control and Command" panel was used by the safety pilots at the front to monitor the rear pilots' vital actions and allow them to take control from the rear cockpit if the situation required.

Autopilot / Flight Director System – AFM Limitations

1/200:

- Use of autopilot not authorised for takeoff or landing.
- Do not use the autopilot roll channel above 30,000ft with yaw damper inoperative.
- Do not use autopilot pitch channel above 0.81M with hydraulic system A or B depressurised.
- Do not use ALT HOLD mode when Captain's alternate static source is selected.

3/500:

- Use of aileron trim with autopilot engaged is prohibited.
- Do not engage the autopilot for takeoff below 1000ft AGL.
- For single channel operation, the autopilot shall not be engaged below 50ft AGL.
- Maximum allowable wind speeds, when conducting a dual channel Cat II or Cat III landing predicated on autoland operations, are:
  - Headwind 25 knots
  - Crosswind 15 knots
  - Tailwind: 10 knots

- Maximum and minimum glideslope angles are 3.25 degrees and 2.5 degrees respectively.
- Autoland capability may only be used with flaps 30 or 40 and both engines operative.

6/900:

- Use of aileron trim with autopilot engaged is prohibited.
- Do not engage the autopilot for takeoff below 400ft AGL.
- For single channel operation during approach, the autopilot shall not remain engaged below 50ft AGL.
- The autopilot must be disengaged before the airplane descends more than 50 feet below the minimum descent altitude (MDA) unless it is coupled to an ILS glide slope and localizer or in the go-around mode. (JAA Rule).
- Maximum allowable wind speeds, when conducting a dual channel Cat II or Cat III landing predicated on autoland operations, are:
  - Headwind 25 knots
  - Crosswind 20 knots
  - Tailwind: Varies between 0 and 15kts depending upon field elevation and flap setting.
- Maximum and minimum glideslope angles are 3.25 degrees and 2.5 degrees respectively.
- Cat IIIA autoland capability may only be used with flaps 30 or 40 and both engines operative.
- Cat IIIB autoland capability may only be used with flaps 30 with one engine operative and only for DH at or above 50ft.
- Autoland capability may only be used to runways at or below 8,400ft pressure altitude.

AUXILIARY POWER UNIT

The APU is a source of bleed air and AC electrics for the aircraft. This gives independence during turnarounds, electrical backup in the event of engine failure and provides air conditioning & pressurisation during an engine bleeds off take-off. The power source for the APU starter is the battery, or on the NG, transfer bus 1 if available. Some Classics have an extra, dedicated APU battery to preserve main battery usage.

There are many different APUs available for the 737. The most common is the Garrett GTCP (Gas Turbine Compressor [air] Power unit [electrics]) 85-129. This was standard for the series 1/200 but when the -300 was

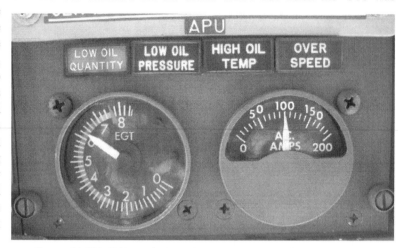

introduced it was found that two to three times the energy was needed to start the larger CFM56 engines. Garrett produced the 85-129[E] which had a stretched compressor, ie the impellers were lengthened and the tip diameters increased. When the 737-400 was introduced, even more output was required and Garrett produced the 85-129[H]. This has an Electronic Temperature Control which limits hot section temperatures depending upon demand and ambient temperatures. By 1989 the 85-129[H] was the most standard APU in all 737 models, although there are actually 14 different models of the 85-129 in service with 737s.

Garrett 85-129 APU panel – Note that NG APU panels do not have an ammeter

Sundstrand APU panel

Other APUs available are the Garrett GTCP 36-280(B) and the Sundstrand APS 2000 on the 3/4/500; and the Allied Signal GTCP 131-9B for the NGs. The main difference between them is that the Garrett is hydro-mechanical whereas Sundstrand and Allied Signal are FADEC controlled. I am told by engineers that whilst the Garrett is more robust, the Sundstrand and Allied Signal APUs are easier to work on. On the 3/4/500s, we pilots prefer the Sundstrand because it has no EGT limits and faster restart wait times. The easiest way to tell which is fitted is to look at the EGT gauge limits; the GTCP 85-129 has an 850C limit

and also runs at 415Hz, the GTCP 36-280 has an 1100C limit if no EGT limits are marked you have a Sundstrand. Later aircraft have MAINT instead of LOW OIL QUANTITY and FAULT instead of HIGH OIL TEMP warning lights.

Some aircraft have APU timers fitted on the aft overhead panel, since APU running time cannot be measured by aircraft logbook time.

The AlliedSignal APU has a 41,000ft start capability and incorporates a starter/generator, thus eliminating a DC starter and clutch. In practice this means that it can be started either by battery or AC transfer bus 1 (the Classics are battery start only). It has an educter oil cooling system and therefore has no need for a cooling fan. It is rated at 90KVA up to 32,000ft and 66KVA up to 41,000ft. The Garrett and Sundstrand APUs are only rated to 55KVA.

The fuel source is normally from the No 1 main tank and it is recommended that at least one pump in the supplying tank be on during the start sequence (and whenever operating) to provide positive fuel pressure and preserve the service life of the APU fuel control unit. Boeing responded to this need by installing an extra DC operated APU fuel boost pump in the No 1 tank on newer series 500 aircraft which automatically operates during APU start and shuts off when it reaches governed speed. You can quickly tell if this is installed by looking for the APU BAT position on the metering panel and the APU BAT OVHT light on the aft overhead panel.

It is recommended that the APU be operated for one full minute with no pneumatic load prior to shutdown. This cooling period is to extend the life of the turbine wheel of the APU. NGs have a 1 minute shutdown timer for this.

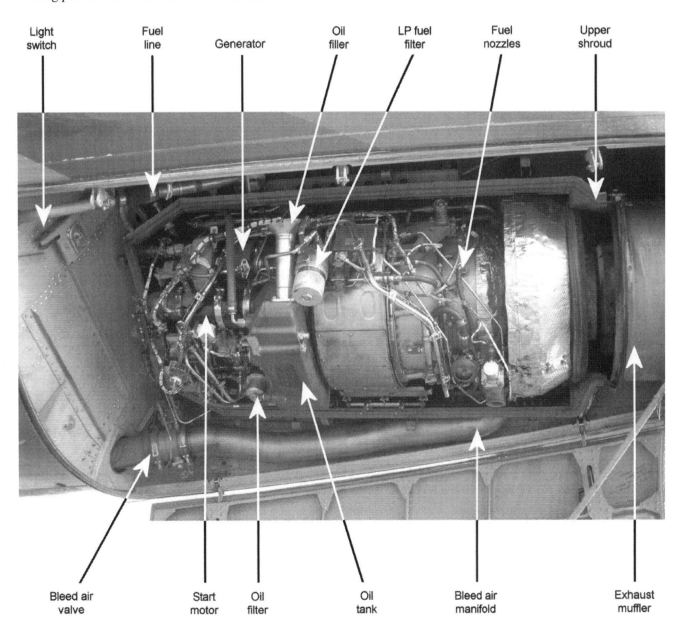

Sundstrand APS 2000 Components

The APU will auto-shutdown for the following reasons:
- Fire
- Low oil pressure
- High oil temperature / Fault
- Overspeed

FAULT – Although these malfunctions will cause the APU to auto-shutdown, additional restarts may be attempted. Faults can include any of the following:
- Fuel shutoff valve not in commanded position
- Loss of dc power to ECU
- ECU (Electronic Control Unit) failure
- APU fire (detected or APU fire handle pulled)
- Inlet door not in command position
- APU inlet overheat (compressor inlet temp >350F/180C for 3 secs)
- Loss of both EGT signals
- Loss of both speed sensing circuits
- No acceleration (acceleration is less than 0.2% per second for 12.5 secs)
- No APU rotation (if speed is less than 7% 20 seconds after the ECU gives the start command)
- No flame (EGT <300F/149C 20 secs after fuel introduced)
- Starter-generator filter clogged.
- High oil temperature (> 290F/143C for 10 seconds)
- Over-temperature (TIT >2200F/1204C or EGT > ECU calculated limit)
- Reverse flow (load compressor airflow zero for 6 secs)
- Sensor failure (Oil temperature or inlet air temperature failure on ground only)
- Underspeed. (Accel <0.5%/sec or speed <85% or APU is not starting)

LOW OIL QTY/MAINT (Blue) – When illuminated, you may continue to operate the APU for up to 30 hrs. Note: this light is only armed when APU switch is ON.

The OVERSPEED light may illuminate for any of the following reasons:
- An aborted start (overspeed signal given to shutdown). – Further restart may be attempted.
- A real overspeed while running. – Do not restart.
- On shutdown (failed test of the overspeed circuit). – Do not restart.

Note that on Classics any amber caption will extinguish when the APU is switched off, on NGs any amber caption will remain illuminated for 5 minutes after the APU is switched off.

There is no CSD in the APU because it is a constant speed engine.

If the APU appears to have started but no APU GEN OFF BUS light is observed then you may have a hung start.

The current limit is 125A -air and 150A -ground, due to better airflow cooling on the ground. The galley power will automatically be load shed if the APU load reaches 165A. Because of these limits, the APU may only power one bus in the air. However, if you should accidentally take-off with the APU on the buses then it will continue to power both buses. If the APU EGT reaches 620-650°C, the bleed air valve will modulate toward closed. (This can lead to an aborted engine start if the electrics do not load shed first.)

Max recommended start altitude – 25,000ft Classics; No limit NGs.

Each start attempt uses approx 7mins of battery life.

Classic: Switching the battery off will shutdown the APU on the ground only.
NG: Switching the battery off will shutdown the APU in the air or on the ground.

The APU is enclosed within a fireproof, sound reducing shroud which must be removed before access can be gained to its components.

There are two drain masts. The one just aft of the port wheel-well is shared with the hydraulic reservoir vent and is a shrouded line enclosing the APU fuel supply line, this collects any leakage of fuel into the shroud which can be drained when a stop cork is pushed up in the wheel-well. If fuel drains when the stop cork is pushed, it indicates a leak in the APU fuel line.

The APU shroud drain and hydraulic reservoir vent

The APU shroud (centre), fuel supply line (left), bleed air duct (right) and cooling air vent. Note this metal shroud is replaced by thermal fire protection blanket on the NG

APU Oil Panel

Only fitted to older aircraft, this small panel is located underneath the rear fuselage and has a "FULL" and an "ADD" light for the APU oil tank. The engineers tell me the indicator lights are not particularly reliable.

On the 737NG the oil level can be checked from the CDU with a BITE check.

```
         APU BITE TEST
 OIL QUANTITY REPORT   1/1

      OIL LEVEL : FULL
```

NG Eductor cooling system

The 737 NG APU is immediately recognisable by the new "eductor" cooling air inlet above the exhaust. This and the new silencer make the NG APU 12dB quieter than the Classics.

The eductor works by using the high speed flow of the APU exhaust which forms a low pressure area. The low pressure pulls outside air through the eductor inlet duct to the APU compartment. The cooling air then goes through the oil cooler and out of the APU exhaust duct below, eliminating the need for a separate cooling air vent or fan.

Vortex Generator

The protrusion on the lower right hand side of the photo is the vortex generator on the APU air inlet door. This was added in 1977 to fix poor in-flight start performance. According to Boeing it "reversed the former negative pressure difference across the APU so that aerodynamic cranking could assist the starter motor during in-flight starts". This was further modified by adding the trailing edge articulated flap which folds inboard 25 degrees when the APU is started and running.

One of the easiest ways to tell a 737 NG apart from a Classic is to count the APU holes

Pack Operation and Fuel Flow

Single pack operation is not recommended with the Allied Signal 131-9 APU. The following 737-700 CDU BITE pages show the reason why:

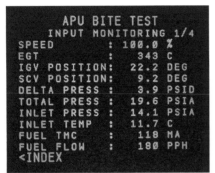

```
     APU BITE TEST
   INPUT MONITORING 1/4
SPEED        : 100.0 %
EGT          :   343 C
IGV POSITION:  22.2 DEG
SCV POSITION:   9.2 DEG
DELTA PRESS :   3.9 PSID
TOTAL PRESS :  19.6 PSIA
INLET PRESS :  14.1 PSIA
INLET TEMP  :  11.7 C
FUEL TMC    :   118 MA
FUEL FLOW   :   180 PPH
<INDEX
```
No packs on

```
     APU BITE TEST
   INPUT MONITORING 1/4
SPEED        : 100.0 %
EGT          :   452 C
IGV POSITION:  77.8 DEG
SCV POSITION:  47.0 DEG
DELTA PRESS :   6.1 PSID
TOTAL PRESS :  58.7 PSIA
INLET PRESS :  14.0 PSIA
INLET TEMP  :  11.6 C
FUEL TMC    :   154 MA
FUEL FLOW   :   240 PPH
<INDEX
```
One pack on

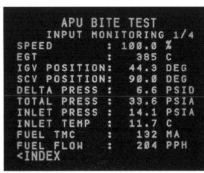

```
     APU BITE TEST
   INPUT MONITORING 1/4
SPEED        : 100.0 %
EGT          :   385 C
IGV POSITION:  44.3 DEG
SCV POSITION:  90.0 DEG
DELTA PRESS :   6.6 PSID
TOTAL PRESS :  33.6 PSIA
INLET PRESS :  14.1 PSIA
INLET TEMP  :  11.7 C
FUEL TMC    :   132 MA
FUEL FLOW   :   204 PPH
<INDEX
```
Both packs on

A single pack must work harder than two packs to cool the cabin to a given temperature. Hence the APU must supply higher bleed air pressures to assure proper environmental control system operation. This higher pressure requires a greater Inlet Guide Vane (IGV) open position than that required for 2–pack operation. Since there is less airflow required to operate 1–pack than is needed, a significant amount of unused bleed air is exhausted through the Surge Control Valve (SCV). This higher IGV open position and large quantity of unused air translates into higher APU fuel burn and higher EGTs during 1–pack operation. Also, the high airflow levels exhausting through the surge control valve increases the overall APU generated noise by 2dbA. With 2 packs supplying the cabin cooling requirements the pressure requirement is lower, resulting in lower turbine inlet temperatures, EGTs and far less unused air being discharged through the surge valve.

The Future

Boeing is believed to have started flight tests of a Solid Oxide Fuel Cell (SOFC) APU. The SOFC uses jet fuel as the reformer in the proton exchange membrane to give a 440kW APU that is 75% efficient compared to the conventional 40-45% efficient APUs. This would give a typical fuel saving of 1,360t for a 737 over a year. It is actually a hybrid gas turbine / fuel cell due to the sudden surges in demand eg engine starts and gear retraction etc. The SOFC will use air from a compressor passed through a heat exchanger for its gas turbine section.

A potential drawback is that it has a 40min start-up time, so it would have to remain on for the whole day. Depending upon its noise levels this could be a problem at airports which require the APU to be shutdown during the turnaround.

The technology for the SOFC APU to replace the current APU is not likely to be available until at least 2020.

Limitations & Operating Techniques

APU life can be shortened by incorrect operating techniques. This can be helped by allowing the correct warm-up & cool-down times and bleed configuration for each type of APU. They all differ slightly due to engine core and design differences, but the manifestation of the failure is usually a turbine wheel rotor and/or blade separation. The following table is based on manufacturer's recommendations.

APU Type:	Garrett 85-129	Sundstrand APS 2000	Garrett 36-280	Allied Signal 131-9(B)
A/C Series:	737-1/2/3/4/500s	737-3/4/500	737-3/4/500	737-NG
EGT Gauge Markings	850C gauge with colour bands	850C gauge (old FADEC) / 1100C gauge (new FADEC)	1100C gauge	1100C gauge
EGT Limits	Max start 760C Max cont 649C	No limits	Max start 760C Max cont 710C	No limits
Starter Limits	2nd – No wait 3rd – 5 mins 4th – 1 hour	1st – 3rd – No wait 4th – 30 mins	2nd – No wait 3rd – 5 mins 4th – 1 hour	No limits
Max alt Bleed & Electrics	10,000ft	10,000ft	10,000ft	10,000ft
Max alt Bleeds	17,000ft	17,000ft	17,000ft	17,000ft
Max alt Electrics	37,000ft	37,000ft	37,000ft	41,000ft
Warm up period	3 min	3 min	3 min	3 min
Bleed Pack Operation	1 pack	1 pack	2 pack	2 pack
MES to APU shutdown	1 min (unloaded)	Immediately	Immediately*	Immediately*
APU shutdown	1 min (unloaded)	Immediately	Immediately*	Immediately*

*Initiates automatic cool down cycle.

Warm up period: The minimum time to run the APU before a pneumatic load is applied. This allows the turbine wheel temperature to stabilise before a load is applied. Whilst 3 minutes is the recommended figure, 1 minute should be the absolute minimum. Note that an electrical load may be used with no warm up period.

Bleed Pack Operation: The number of packs to use on the ground. APUs which should run both packs have load compressors to supply bleed air. For these APUs, two pack operation gives both cooler turbine wheel temperatures and a lower fuel burn.

Main Engine Start (MES) to APU shutdown: The cool-down time to allow after main engine start.
Note, also that there should be a minimum amount of time between turning off the pack(s) and starting the first engine. Additionally, minimum delay should occur between starting the first & second engine. This prevents the turbine wheel temperature from decreasing and then significantly increasing when the second engine is started.

APU shutdown: The cool-down time to allow after flight, after the packs have been switched off. It is important to allow the APU to complete its shutdown sequence before the battery is switched off.
Ref: Flt Ops Tech Bulletin 99-1

Further AFM limitations:
    APU bleed valve must be closed when:
- Ground air connected and isolation valve open
- Engine no. 1 bleed valve open
- Isolation and engine no. 2 bleed valves open.
    APU bleed valve may be open during engine start, but avoid engine power above idle.

COMMUNICATIONS

Audio Control Panel

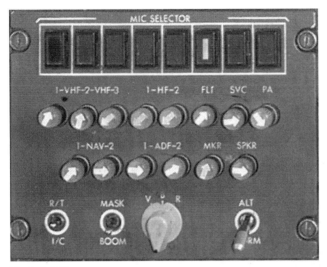

Typical ACP

Also known as the Audio Selector Panel (ASP), the ACP allows you to select which radio / interphone you transmit on, which microphone to use, which radio / interphone / beacon to listen to and at what volume.

This type of ACP has cylindrical button volume controls, others have sliders.

R/T / I/C works in the same way as the rocker switch on the control column. ie the I/C position bypasses the mic selector to transmit on the flight interphone.

Mask/Boom simply selects either mask or boom microphone. Check this if nobody can hear you transmit - especially after your oxygen mask mike check!

The filter switch, Voice-Both-Range, allows better reception of either voice or Morse identifiers on NAV & ADF radios. Check that this switch has not been left in the V position if you can not hear a beacon ident.

Alt/Norm in the ALT position puts the ACP into degraded mode. If the Captain's ACP is in degraded mode, he can only transmit on VHF1 through mask or boom and can only receive VHF1 at a preset level. The F/O's ACP in degraded mode is the same but uses VHF2. Note that aural warnings will still be heard from the aural warning module (not the overhead speaker).

ACP with sliders

Typical 737-200 ACP

VHF Radio

Most 737s have three VHF radios and at least one HF radio. This unit allows for selection of any of those at this station. The TEST button is a squelch and is used to hear faint stations which do not have the strength to breakthrough. If you switch the panel OFF at the same time as TEST is applied this holds the test condition thereby allowing you to hear faint stations without having to hold down the test button - very useful for copying distant weather!

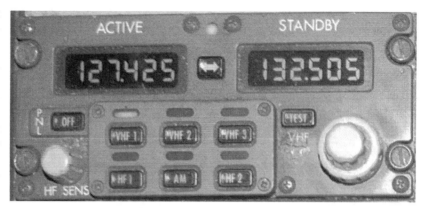

This particular box has an "offside tuning" feature whereby any of the VHF or HF radios can be tuned from this box by selecting it on this panel. The light (here VHF 1) indicates which radio is in use.

HF Radio

HF is still the most common method of long range communications, but will probably be superseded by ACARS in the future. Compare this box with those on the B707 overhead panel (page 245) and you can see how little has changed!

If you can't hear anything try switching between USB and LSB (upper/lower side band), turning the sensitivity up or choosing a more appropriate frequency for the time of day. As a general rule try the highest frequencies when the sun is

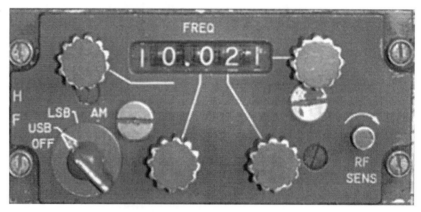

between you and the station and the lowest at night. The higher the frequency you can get through on, the better the signal will be. AM is used to receive commercial radio stations, perhaps for those all important football scores.

Selective Calling - SELCAL

The SELCAL light will illuminate and a two-tone chime sounds if the aircraft is being SELCAL'd on either HF or VHF. This particular panel is a very old unit and most operators have had to improvise the method of radio connections to it. Typically, in the past, diodes would be used to "OR" the VHFs together to illuminate one of the lights.

Over the last 15 years, most SELCAL panels are provided with a light for each of the radios, VHF-1, VHF-2, VHF-3, HF-1, HF-2 and in some cases, include the Attendant call, and SATCOM call.

Transponder

Mode S transponders have been mandatory in Europe since March 2007. Note that with the mode switch in the STBY position, modes A and C will not function, but mode S will continue to operate and is unaffected by switch position.

Marker Beacons

The marker lights are pre-tuned to their frequency (75MHz) and flash when overflown. The marker tone can also be heard if selected on the ACP.

Marker type	Colour	Ident	Tone Freq
Outer	Blue	Continuous dashes	400 Hz
Middle	Orange	Alternate dots and dashes	1300 Hz
Inner	White	Continuous dots	3000 Hz
Backcourse	White	Continuous paired dots	3000 Hz
Airways	White	Station identifier	3000 Hz

ELT

An ELT is a customer option. It contains an accelerometer trigger which will automatically activate on impact - if the switch is in the ARM position. It transmits on 121.5, 243 and 406MHz through an antenna at the top rear of the fuselage. The 406MHz transmitter shuts down after 24 hours to conserve battery life.

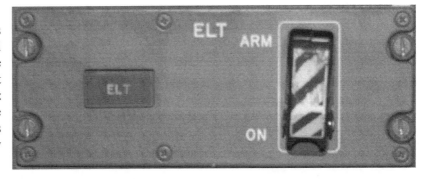

The ELT has its own battery pack which will last for at least 72 hours after activation. It consists of 4 size D lithium manganese dioxide cells.

Cockpit Voice Recorder

The CVR records the headset and microphone of all 3 ASPs and the ambient cockpit sounds all on separate channels. The recordings start with the first rise in engine oil pressure and go onto a 120 or 30min continuous loop tape until 5mins after last engine shutdown. In the event of an incident crews are advised to pull the CVR c/b after final stop to avoid automatic erasure. It is illegal to stop the CVR in flight. The CVR is located in the aft cargo hold.

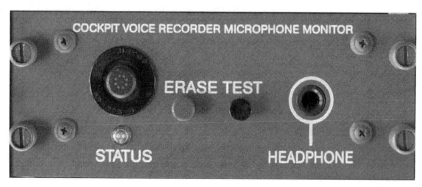

Some aircraft may be fitted with a CVR AUTO/ON switch, located on the aft overhead panel. This enables the CVR to be manually switched on before engine start, allowing the cockpit preparation to be recorded. The switch is solenoid held and moves back to the AUTO position during the first engine start.

Interphone

The **EXTERNAL POWER HATCH** is located beneath the F/O's DV window. It is used by groundcrew to connect the Ground Power Unit and headset for pushback communications with flight interphone.

The service interphone is used by engineers to communicate with the service interphone stations inside the aircraft. Note that the Service Interphone switch on the aft overhead panel must be switched ON for this use.

There is also a pilot call button and a nose-wheelwell light switch to assist the groundcrew to insert the steering bypass pin.

The service interphone switch on the aft overhead panel activates the many external jacks around the aircraft to the service interphone system. Normal internal service interphone operation is unaffected by switch position.

External Power Hatch

Data Loader

Data Loader Control Panel - Used to upload software for the FMC updates & database, DEU options, Flight Control Computer (FCC), Aircraft Condition Monitoring System (ACMS), Digital Flight Data Acquisition Unit (DFDAU), EGPWS terrain database, EEC & APU engine control software etc. You also use this unit to select data for download, eg flight parameters from the DFDAU or fault history from the DEUs.

Airborne Data Loader – is where the disk or PCMCIA card is inserted. This is the AlliedSignal (Sundstrand) version, the decode for the status lights is given on the door.

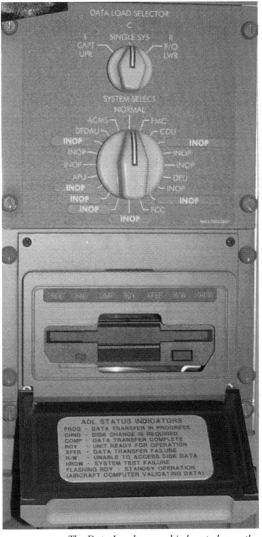

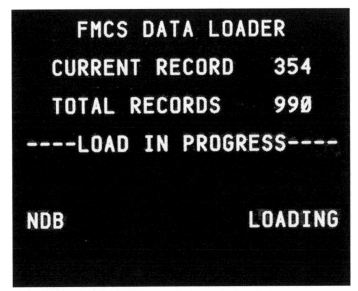

The CDU during data load

The Data Loader panel is located near the supernumerary oxygen mask on the NG

Satcom

Satcom is used for mobile phone and internet access. It used to be the preserve of BBJ's and government aircraft but are now starting to appear on business class aircraft such as the 737-700ER. The systems can be either Ku-band or L-band. Ku-band antennae are larger & heavier and may be fuselage or fin mounted, but give a faster connection speed (0.5Mbps at time of printing). L-band antenna can be made as phased arrays which have the advantage of a low profile and look like a patch approx 3ft x 1.5ft x 2inches on the side of the fuselage. L-band is catching up on the connection speeds of Ku-band systems but at the moment they are only about half as fast.

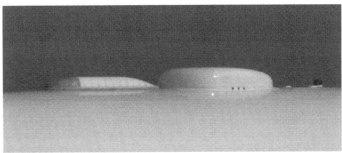

Ku-band Satcom antenna

L-band Satcom antenna on the upper side fuselage

Antennas

The antennas (called aerials in UK English) can reveal a lot about the avionics of the aircraft.

This aircraft has an old style HF wire which identifies it as an early 737-200, yet it has TCAS, now required by law, so it has been in service for some considerable time. The ELT is an option usually only taken by operators in remote areas.

As a general rule the longer the wavelength, the larger the antenna must be. Exceptions to this are phased antennas, such as those in Satcom or the top hat of the AEW&C.

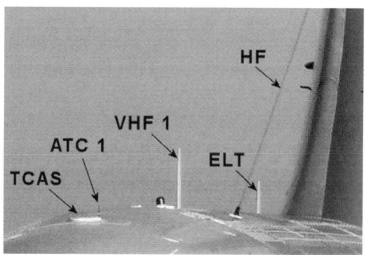

A selection of antennas on an early 737-200

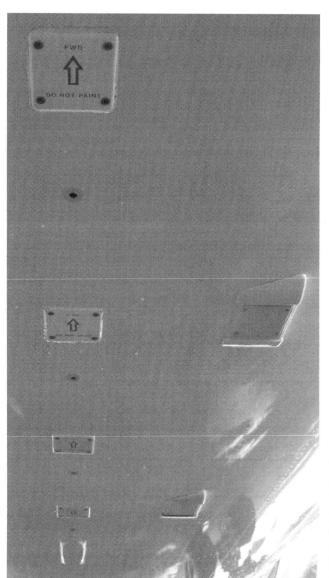

Radio Altimeter Tx & Rx (2 pairs plus 1 provision)

The following selection of antennas shows their wide variety of sizes & shapes.

Marker Beacon (75MHz)

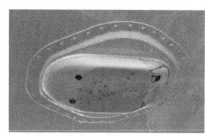

TCAS (1030 MHz)

Transponder (1030 MHz)

The radio altimeter aerials (left) are located aft of the nosewheel. They are in pairs as they have a separate transmit and receive antenna and have a different polarisation to avoid interference between the two altimeters. This particular aircraft has provision for a third pair of rad-alt aerials. It is believed that these may have been in anticipation of Cat IIIB capability but were deemed unnecessary.

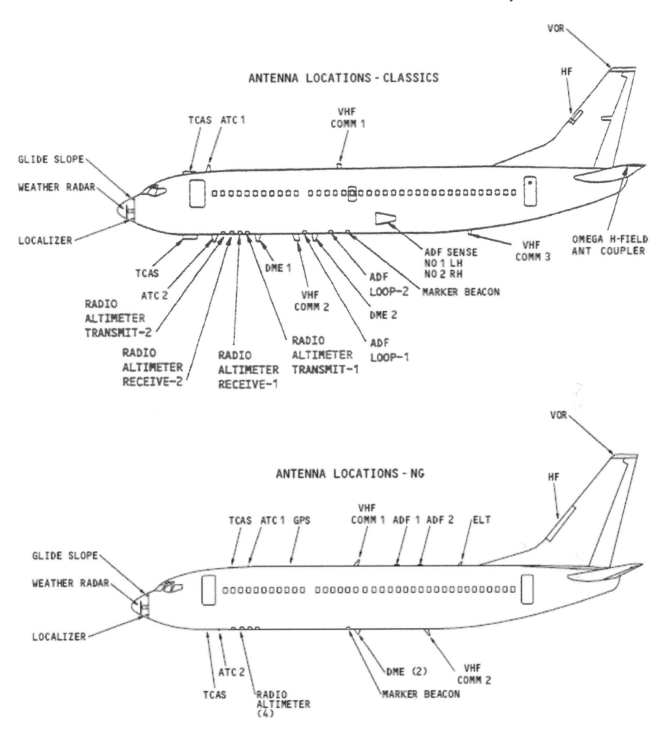

The antenna locations for the 1/200s are basically the same as for the Classics, except that in early aircraft the HF antenna was strung between the mid fuselage and the fin, and normally only one radio altimeter was fitted. I suspect that the Omega antenna was not installed on many Classics. Note that these location diagrams are not definitive as there are many avionics options available. This is particularly so with BBJs and military variants which may be fitted with an array of exotic comms equipment such at Satcom.

Antennas and static discharge wicks should be inspected carefully on the walkaround for integrity and burns, especially if lightning or St Elmo's fire has been observed.

ELECTRICAL

AC & DC Metering panel

This panel is slightly unusual because it contains the optional APU BAT position on the DC side. Most Classics do not have this second battery.

The **RESIDUAL VOLTS** button (not installed on the NG) may be used to test a generator that has dropped off a bus. When pressed, if a voltage is seen then the generator is still turning, therefore a generator showing zero residual volts has failed and will not reconnect. Residual volts is the only selection to use the 30V scale on the AC voltmeter; for this reason residual volts should never be pressed when a generator is connected to a bus (will try to display 115V on a 30V scale and possibly damage the meter).

The TEST positions are used in conjunction with the Power System Test panel (see page 84). This test information is all contained within the metering panel on the NG.

AC & DC Metering panel – 2 battery Classic

The purists may like to know that the DC voltages are measured at the following points:

DC Selector switch	Voltage measurement point	Typical Voltage	Typical Current
STBY PWR	DC Standby bus	24-30	N/A
BAT BUS	Battery bus	24-30	N/A
BATT	Hot battery bus	22-30*	0
TR1	DC bus 1	24-30	20-25
TR2	DC bus 2	24-30	20-25
TR3	TR 3	24-30	10-15
TEST	Power system test module	See table on page 84	

*May be up to 33V during pulse mode charging.

Do not leave the DC meter selector at BAT on a dead aircraft, because indicating draws some current and will eventually drain the battery.

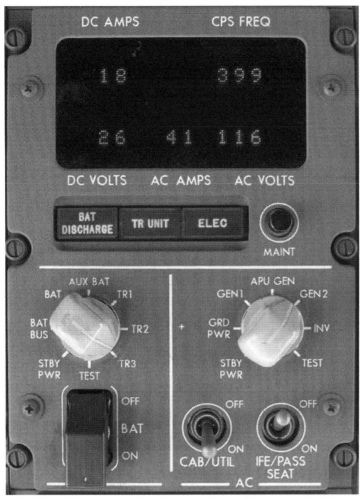

AC & DC Metering panel - NG

The NG metering panel has replaced the meters with digital displays and added an AC amps display.

Captions are now included for battery discharge and failed TRUs. The ELEC caption will only illuminate on the ground and indicates a fault in the DC or standby power systems.

Notice the new CAB/UTIL & IFE/PASS SEAT switches that replace the GALLEY switch. These control the following services:

CAB/UTIL	IFE/PASS SEAT
Recirculation fan(s)	115V AC audio IFE
Door area heaters	115V AC video IFE
Drain mast heaters	28V DC video IFE
Lavatory water heaters	Airphone equipment
All galley buses	Pax seat elec outlets
Shaver outlets	
Logo lights	
Potable water comp	

The TEST positions are used in conjunction with the MAINT button to check for faults.

The residual volts button has been removed.

TRUs

The Transformer Rectifier Units (TRUs or TRs) convert 115V AC into 28V DC. The check for TR serviceability is current, not voltage, because the TR voltage indicates that of the associated DC buses (for TRs 1 & 2). TRs should be checked before committing to an autoland because the TR3 disconnect relay/cross bus tie relay opens at glideslope capture and this will leave DC Bus 1 unpowered if TR1 had previously failed.

The primary function of TR3 is to power the battery bus. Note that on Classics, TR3 is powered by main bus 2, whereas on NGs it is powered by transfer bus 2 backed up by transfer bus 1.

NGs have a TR UNIT light which illuminates if either TR1 or TR2 and TR3 fail in flight or if any TRs fail on the ground. The TRs are unregulated and output rated to 50 Amps (Classics) / 75 Amps (NGs).

Limitations:

TR voltage range: 24-30V

Battery voltage range: 22-30V (May be up to 33V during pulse mode charging)

Gen Drive & Standby Power panel

Gen Drive & Standby Power panel – Classic

Gen Drive & Standby Power panel - NG

The STANDBY PWR OFF caption will illuminate if the AC standby bus is unpowered. On the NG series it will also illuminate if either the DC standby bus or battery bus are unpowered. Note that with the standby power switch in AUTO, for NGs this allows automatic switching of power in the air and on the ground; for Classics automatic switching will only occur in the air.

NG: LOW OIL PRESSURE and HIGH OIL TEMP cautions are replaced by a single DRIVE caption. This will illuminate for an IDG low oil pressure (<165psi), since the IDGs will auto disconnect for a high oil temperature (>182C).

NG: The IDGs will auto-disconnect (by opening the GCR and GCB) for a frequency deviation if the following limits are exceeded:
Over-frequency: 425 Hz for more than 1.5 seconds or 435 Hz for 35 milliseconds.
Under-frequency: 375 Hz for more than 1.5 seconds or 355 Hz for 150 milliseconds.

Classics: A higher than normal rise (ie above 20C) indicates excessive generator load or poor condition of drive. These temperature gauges were deemed to be redundant and have been removed from the NG.

Limitations:

Max generator drive rise:	20°C
Max generator drive oil temp:	157°C
Max generator load (1/200 only)	111 Amps

Any aircraft fitted with a VSCF must operate within **45mins** of a suitable aerodrome.

For more details about the different types of generators (CSD, VSCF, IDG) fitted, see page 254.

Generator Bus panel

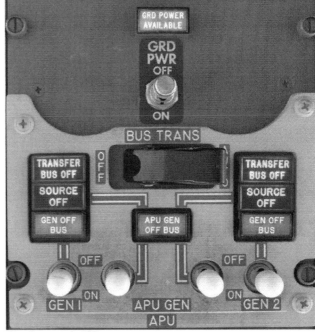

Gen Bus panel – Classic   *Gen Bus panel - NG*

The blue GND POWER AVAILABLE light, on Classics only, means that the GPU is physically plugged in to the aircraft and gives no indication about the quality of the power. You may not be able to connect the ground power to the buses even if the light is illuminated. On NGs the quality is checked and the light will only illuminate when external AC power is connected and the quality is good.

The amber TRANSFER BUS OFF light comes on when the respective AC transfer bus does not have power.

The amber BUS OFF light (1-500s) indicates that the respective AC generator bus is not energized.

The amber SOURCE OFF light (NGs) indicates that the respective AC transfer bus is not energized by the source you last selected.

The engine and APU generator OFF BUS lights illuminate when the respective generator is running and of the correct quality.

There are three golden rules of 737 electrics:
1) There is no paralleling of AC power.
2) The source of AC power being connected to a generator bus takes priority and automatically disconnects the existing source.
3) A source of AC power does not enter the system automatically (when it reaches proper voltage & freq). It must be manually switched on. NB this rule has been relaxed on the NG with the "automatic generator on-line" feature. This will automatically connect the engine generators if the aircraft has taken off with the APU still powering the busses and it subsequently fails or is switched off.

Buses

A bus (full name bus-bar) is a conductor that distributes electrical power on to services or other smaller buses.

AC Buses - Classics

- Gen buses – Point of connection for the power sources (engines/APU/GPU). Used for heavy, important loads, eg hydraulic pumps. Effectively now renamed transfer buses on the NG.
- Main buses – Fed from the respective gen bus. Used for heavy non-essential loads, eg fuel boost pumps.
- Transfer buses – Normally powered by respective gen bus. If these fail, will feed from other gen bus if BUS TXFR switch is in AUTO. Used for essential loads, eg trim.
- AC standby bus – Powered by transfer bus 1 or the battery via an inverter. Used for essential loads, eg ATC 1.
- Ground service bus – Powered by gen bus 1 or external AC bus if GPU available. Used for battery charger, service outlets, cargo hold lights etc.

AC Buses - NGs

- Transfer Buses - Point of connection for the power sources (engines/APU/GPU). Used for heavy, essential loads, eg hydraulic pumps.
- Main Buses - Fed from respective transfer bus. Used for non-essential loads, eg recirc fans. The main buses are next to be load shed after the galley buses.
- Galley Buses - First in line to be load shed.
- AC standby bus – Powered by transfer bus 1 or the battery via an inverter. Used for essential loads, eg ATC 1.
- Ground service buses – Powered by respective transfer buses or from GPU if available. Used for battery charger, service outlets, cargo hold lights etc.

DC Buses

- DC buses – Powered by the respective transfer buses via a TRU.
- DC standby bus – Powered by DC bus 1 (Classics) / TRs (NGs) or battery bus (Classics) / battery (NGs).
- Battery bus – Normally powered by TR3, alt power is battery. Powered when the battery switch is ON or the standby power switch is BAT.
- Hot battery bus – Always live, used for fire extinguishing & Captains clock.
- Switched hot battery bus – Only powered when the battery switch is on.

Standby Buses

- Are for essential AC & DC loads and are guaranteed for 30mins from the battery (single battery a/c).
- SBY AC bus – Is powered from AC transfer bus 1 or the battery via an inverter.
- SBY DC bus – Is powered from DC bus 1 or the battery via the battery bus.
- Bus Transfer switch - when off will completely isolate left & right sides of the electrics.

Batteries

The battery is 20 cell, 24V, nickel-cadmium and is located in the E&E bay. It sits on the external DC power receptacle which is used to start the APU if the battery voltage is insufficient. Note the DC power receptacle is not fitted to the NGs because if the battery voltage becomes low it must be replaced. The battery has been upgraded several times in the aircrafts history and many aircraft have two batteries.

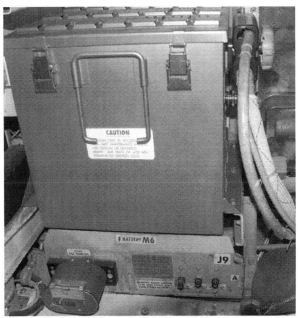

Battery & External DC receptacle (1-500 only) in the E & E bay

Series	Rating (Ampere-hours)	Standby power time (mins)
Originals	22	20 early aircraft / 30 late aircraft
Classics	36 / 38 / 40	30 single / 45 with APU battery
NGs	36 / 48	30 single / 60 dual battery

APU Battery - This is a customer option that I have only seen on some Series 500 aircraft. It is primarily used for starting the APU, but also works in parallel with the Main Battery to provide 45mins of standby power. One of its best functions is to keep power on the Captain's EFIS after the loss of all generators.

Aux Battery - This is a reserve battery on the NG which is normally isolated unless the Main Battery is powering the standby system when it operates in parallel with the Main Battery. The Aux Battery combined with the Main Battery will provide 60 minutes of standby power.

The NG also has 2 small, extra dedicated batteries for the engine and APU fuel shut off valves and the ISFD (150mins capacity).

BAT OVHT & APU BAT OVHT lights are a customer option on Classics. They are located on the aft overhead panel and no crew action is required if they should illuminate.

Battery captions on the aft overhead panel

Normal battery voltage range is 22-30 volts.

The BAT DISCHARGE light on the NG series will illuminate if the current draw from either battery is more than either:

- 5 amps for 95 seconds or
- 15 amps for 25 seconds or
- 100 amps for 1.2 seconds.

The Generator Diagnostics (Annunciator) Panel

Series -1/2/3/4/500 only

The Generator Diagnostics Panel (left) and the Power System Test panel (right)

The Generator Diagnostics panel, a.k.a. the M238 panel, is easily missed as it is tucked away on the right hand side wall as you enter the flight deck. It is used as an indication of whether or not individual AC & DC buses are powered, and provides reasons in the form of malfunction lights, why a generator has tripped.

It is arranged into three areas:

DC Bus lights:
The first 3 lights on the top 2 rows. Hold switch to INDICATE to see which DC buses are powered.

AC Bus lights:
These lights show which AC buses are powered. They are behind the shield to avoid distracting the crew. The top row is phase A, the bottom row phase C. Phase B is checked on the AC meter panel on the overhead panel.

Malfunction lights:
The last 6 lights on the top 2 rows will illuminate immediately when a fault occurs on either engine or APU generator.

If any lights are illuminated not covered by the shield, something may be wrong, make a note of the light and report to an engineer. If the fault is on either Gen 1 or 2 and you have VSCFs fitted you can confirm the fault by the light test on the VSCF unit. A list of malfunction lights and their possible causes is given below.

Possible causes of Generator Diagnostic Panel lights are as follows:

Annunciator Panel Troubleshooting Guide	
Malfunction Light	**Possible Causes**
FF - Feeder Fault. **Light comes on followed by GCR, GB tripping:**	• **Defective CT** • **Defective GCU** • **Over-current condition**
MT - Manual Trip. **Light comes on**	• **Defective circuit to manual trip** • **Generator switched off** • **CSD disconnect**
HV - High Voltage. **Light comes on at 130+/-3V:**	• **Defective Generator Control Unit (GCU)**
LV - Low Voltage. **Light comes on at 100+/-3V:**	• **Defective generator** • **Damaged CSD shaft or spline** • **Defective GCU** • **Fire handle pulled**

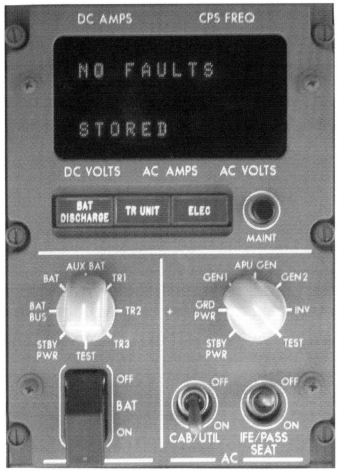

The NG metering panel during BITE test

NG Series

The functions of the Generator Diagnostics panel and Power System Test panel are all contained within the AC & DC Metering panel on the NG. A BITE test can be initiated by pressing the MAINT button with the selectors in the TEST position. This function is inhibited in-flight. Fault messages can include the following:

- PANEL FAILURE (soft, display operates)
- INTERFACE FAILURE
- BAT CHGR INOP
- STAT INV INOP
- SPCU INOP.

If the fault message has the letter I after the fault, it means that the fault is intermittent and is not present at this time.

Pressing the MAINT button a second time will show any subsequent messages and holding the MAINT button for 3 seconds will clear the faults from memory.

The M400 panel space is now usually occupied by the Data Load Panel on the NG series.

The Power System Test Panel

The Power System Test Panel, a.k.a. the M400 panel, is used to check the individual phases of a generator or bus. To use it you must first select TEST on both the AC and DC metering panels. Now, by selecting the switches on this panel to combinations of A to H and 1 to 8 in accordance with the table below, the voltages & frequencies can be read on the metering panel. Eg to check the external power bus, select the power system test switches to D and 3 for external power bus phase A and read from the AC Freq and AC Volts meters.

There are also AC & DC measuring points at the bottom of the panel for use with a more accurate hand held meter.

The individual phase currents of each generator (Eng 1, Eng 2 or APU) can be read from the appropriate generator ammeter by selecting switch position A, B or C.

Note: S2 (Left switch) is normally left in position B. This connects all 3 generator ammeters to phase B and leaves the M400 selector relays relaxed.

	A	B (Default posn)	C	D	E	F	Metering Panel
1	No1 Gen field No1 Main bus ØA	No2 Gen field No1 Main bus ØB	APU Gen field No1 Main bus ØC	No1 Trans bus ØA	No1 Trans bus ØB	No1 Trans bus ØC	DC Volts AC Volts & Freq
2	No1 GCU DC No2 Main bus ØA	No2 GCU DC No2 Main bus ØB	APU GCU DC No2 Main bus ØC	No2 Trans bus ØA	No2 Trans bus ØB	No2 Trans bus ØC	DC Volts AC Volts & Freq
3	Eng GB1 Close coil Gnd Serv bus ØA	Eng GB2 Close coil Gnd Serv bus ØB	APU GB1 Close coil Gnd Serv bus ØC	Ext Pwr bus ØA	Ext Pwr bus ØB	Ext Pwr bus ØC	DC Volts AC Volts & Freq
4	-	-	APU GB2 Close coil	-	-	-	DC Volts
5	-	-	EPC 1 Close coil	-	-	-	DC Volts
6	-	-	EPC 2 Close coil	-	-	-	DC Volts
7	-	-	APU 95% switch	-	-	-	DC Volts
8	-	-	-	-	-	-	-

The Power System Test panel (M400) – Switch Decode

Circuit Breakers

From the QRH CI.2.3

"Flight crew reset of a tripped circuit breaker in flight is not recommended. Unless specifically directed to do so in a non-normal checklist. However, a tripped circuit breaker may be reset once, after a short cooling period (approximately 2 minutes), if in the judgement of the Captain, the situation resulting from the circuit breaker trip has a significant adverse effect on safety. A ground reset of a tripped circuit breaker by the flight crew should only be accomplished after maintenance has determined it is safe to reset the circuit breaker.

Flight crew cycling (pulling and resetting) of circuit breakers to clear non-normal conditions is not recommended."

According to Boeing, there is approximately 46 miles of wire on the 737 Classic, but only 42 miles on the 737 NG!

Behind the P6 panel

Just to prove that electrics is not the exact science that engineers would have you believe, check out this story from Suzanna Darcy, a Boeing flight test pilot for 18 years:

"Systems that seem fine alone can interfere with one another", she recalled testing a 737 (NG). When she switched the power on, she heard the toilet in the lavatory flush. After confirming that no one was in the lavatory, she switched the power on again. This time, all the toilets on board flushed. The reason: interference between electrical systems.

Limitations

The use of Flight Deck Auxiliary Power outlets in the flight deck requires operational regulatory approval.

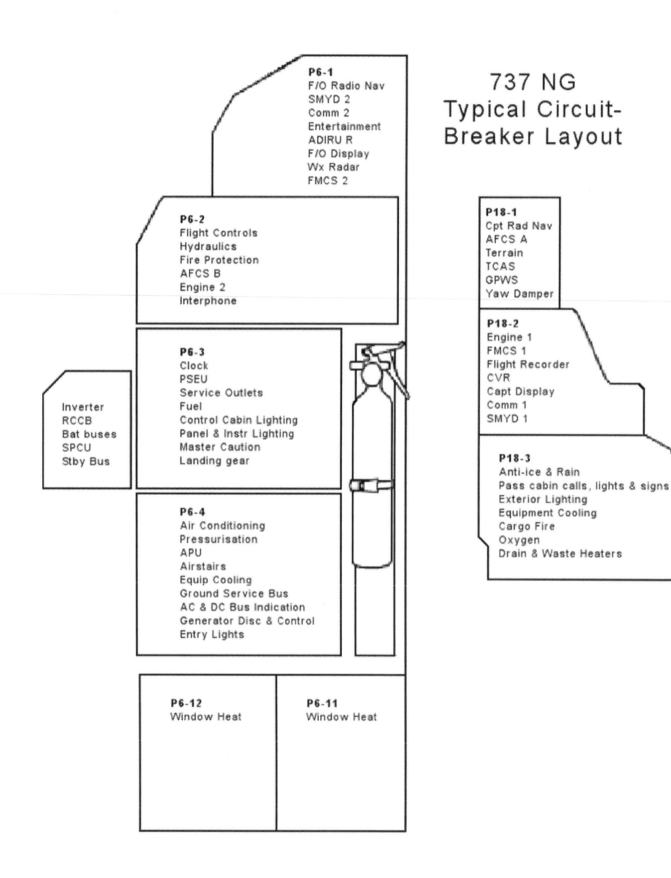

737 NG
Typical Circuit-
Breaker Layout

P6-1
F/O Radio Nav
SMYD 2
Comm 2
Entertainment
ADIRU R
F/O Display
Wx Radar
FMCS 2

P6-2
Flight Controls
Hydraulics
Fire Protection
AFCS B
Engine 2
Interphone

P18-1
Cpt Rad Nav
AFCS A
Terrain
TCAS
GPWS
Yaw Damper

P18-2
Engine 1
FMCS 1
Flight Recorder
CVR
Capt Display
Comm 1
SMYD 1

P6-3
Clock
PSEU
Service Outlets
Fuel
Control Cabin Lighting
Panel & Instr Lighting
Master Caution
Landing gear

Inverter
RCCB
Bat buses
SPCU
Stby Bus

P18-3
Anti-ice & Rain
Pass cabin calls, lights & signs
Exterior Lighting
Equipment Cooling
Cargo Fire
Oxygen
Drain & Waste Heaters

P6-4
Air Conditioning
Pressurisation
APU
Airstairs
Equip Cooling
Ground Service Bus
AC & DC Bus Indication
Generator Disc & Control
Entry Lights

P6-12
Window Heat

P6-11
Window Heat

NG Electrical Buses Services

Given below is a list of all the electrical buses and the services that are taken from each one of them. Please note that this list is not applicable to any other series of 737 and furthermore, due to the many different electrical configurations of individual aircraft, it may not be the same as your aircraft. It is purely intended as a guide to the many services and the loading and redundancy of the various buses.

115V AC TRANSFER BUS 1

Normally powered by a generator
Alternate power is Transfer bus 2 via the BTBs

- **28V AC XFR BUS 1**
- APU START CONVERTER
- B SYS ELEC HYDR PUMP
- BOOST PUMP TANK 1 FWD
- BOOST PUMP TANK 2 AFT
- EQUIP COOL SUPPLY FAN ALT
- **GALLEY BUS C & D**
- **GROUND SERVICE BUS 1**
- **MAIN BUS 1**
- MIDU
- TEMP CONT 35 DEG L
- TEMP CONT L
- TEMP CONT MAN
- TRU 1
- TRU 3 ALT
- VACUUM WASTE BLOWER
- **XFR BUS 1 SEC 1**
- **XFR BUS 1 SEC 2**

115V AC TRANSFER BUS 2

Normally powered by a generator
Alternate power is Transfer bus 1 via the BTBs

- **28V AC XFR BUS 2**
- A SYS ELEC HYDR PUMP
- AIRCO PACK CONT RIGHT AC
- BOOST PUMP CTR TANK RIGHT
- BOOST PUMP TANK 1 AFT
- BOOST PUMP TANK 2 FWD
- EQUIP COOL EXH FAN ALT
- EQUIP COOL SPLY FAN NORM
- **GALLEY BUS A & B**
- **GROUND SERVICE BUS 2**
- **MAIN BUS 2**
- STBY HYDR PUMP
- TEMP CONT VALVE CLOSE R
- TEMP VALVE/FAN CONT FLT DECK
- TEMP VALVE/FAN CONT FWD CABIN
- TRU 2
- TRU 3
- **XFR BUS 2 SEC 1**
- **XFR BUS 2 SEC 2**

115V AC TRANSFER BUS 1 – Sect 1

Powered by Transfer bus 1

- **28V AC XFR BUS 1 SEC 1**
- AFCS A SNSR EXC AC
- ALPHA VANE HEATER LEFT
- ANTI COLLISION LIGHT RED
- ANTI COLLISION LIGHT WHITE
- BOOST PUMP CTR TANK LEFT
- CAPT PITOT HEATER
- DATA LOADER
- ELEVATOR PITOT LEFT
- ENG 1 LEFT IGNITION
- ENG 1 ALT POWER CHAN B
- ENG 1 ALT PWR CHAN A
- FLIGHT REC AC
- GROUND PROX WARNING
- HF 1
- L FIXED LANDING LIGHT
- MACH TRIM A AC
- R RETRACTABLE LANDING LIGHT
- RADIO ALT 1
- STAB TRIM AFCS
- TCAS
- TEMP PROBE HEATER
- WINDOW HEAT CONTROL L FRONT AC
- WINDOW HEAT CONTROL L SIDE AC
- WING ANTI ICE VALVE
- YAW DAMPER INDICATION

115V AC TRANSFER BUS 2 – Sect 1

Powered by Transfer bus 2

- ALPHA VANE HEATER RIGHT
- AUX PITOT HEATER
- ELEVATOR PITOT RIGHT
- FO'S PITOT HEATER
- HEAD UP GUIDANCE DEU ALTN AC
- L RETRACTABLE LANDING LIGHT
- R FIXED LANDING LIGHT
- VOICE RECORDER
- WINDOW HEAT CONTROL L SIDE AC
- WINDOW HEAT CONTROL R FRONT AC

115V AC TRANSFER BUS 1 – Sect 2

Powered by Transfer bus 1

- **AC STBY BUS PWR**
- ACARS MU AC
- AIRCO PACK CONT LEFT AC
- BLEED AIR ISLN VALVE

115V AC TRANSFER BUS 2 – Sect 2

Powered by Transfer bus 2

- 115V AC XFR BUS 2 INDICATION
- **28V AC XFR BUS 2 SEC 2**
- ADF 2
- ADIRU R AC

- CMU 1 AC
- PRINTER
- RAM AIR MOD LEFT
- TEMP CONT VALVE CLOSE L
- TEMP VALVE/FAN CONT AFT CABIN
- WINDOW HEAT POWER L1, L3, L4 & L5
- WINDOW HEAT POWER R SIDE

- AFCS B SNSR EXC AC
- AIL TRIM CONTROL
- ATC 2
- BLEED AIR PRESS INDICATION
- DME 2
- ENG 2 LEFT IGNITION
- ENG 2 ALT POWER CHAN B
- ENG 2 ALT PWR CHAN A
- ENG VIB MON
- FLOOD LIGHT CONTROL
- FMC COMP 2
- FWD AIRSTAIR ACTUATOR
- HF 2
- MACH TRIM B AC
- MCDU 2
- MMR 2
- NAV CONTR PANEL 2
- NOSE GEAR STEERING
- RADIO ALT 2
- RAM AIR MOD RIGHT
- RUDDER TRIM CONTROL
- STAB TRIM ACTUATOR
- TAPE REC AC
- TE FLAP ALT DRIVE
- TEMP CONT 35 DEG L
- TEMP CONT AUTO RIGHT
- VOR 2
- WEATHER RADAR RT
- WHEEL WELL & WING BODY O'HEAT
- WINDOW HEAT POWER L SIDE
- WINDOW HEAT POWER R1, R3, R4 & R5

28V AC TRANSFER BUS 1

Powered by 115V AC TRANSFER BUS 1 via a transformer
- CAPT INSTRUMENT LIGHTS 28V PRI
- DRAIN MAST HEATERS GND
- FASTEN SEAT BELT SIGN LEFT
- FASTEN SEAT BELT SIGN RIGHT
- FLIGHT REC POS SENSORS
- LAV OCCUPIED SIGN
- NO SMOKING SIGN LEFT
- NO SMOKING SIGN RIGHT
- READING LIGHTS LH
- RUNWAY TURN OFF LIGHT LEFT
- TAXI LIGHT
- YAW DAMPER AC

28V AC TRANSFER BUS 2

Powered by 115V AC TRANSFER BUS 2 via a transformer
- ADIRU R EXC
- CAPT AFT ELEX PNL LIGHT
- CONTROL STAND LIGHT
- ELEX PANEL LIGHT FWD
- ELEX PANEL LT
- FO'S AFT ELEX PNL LIGHT
- FO'S INSTRUMENT LIGHTS 28V PRI
- FUEL TEMP INDICATOR
- INSTR & CB PNL LIGHTS
- MAP & KIT LIGHT
- OBSERVERS READING LIGHT
- O'HEAD PNL LIGHT 5V SECONDARY
- O'HEAD PNL LIGHT 28V PRIMARY
- O'HEAD PNL LIGHT 28V SECONDARY
- POS/SKEW SENSOR & IND RIGHT
- PRESS CONTROL LCD LIGHTING
- READING LIGHTS RH
- RUDDER TRIM INDICATOR
- RUNWAY TURN OFF LIGHT RIGHT
- SMYD 2 SNSR EXC AC
- TE FLAP POS/SKEW SENSOR & IND L

115V AC MAIN BUS 1

Powered by AC TRANSFER BUS 1
- DOOR AREA HTR AFT
- DOOR AREA HTR FWD
- DRAIN HEATERS
- DRAIN MAST HEATER AIR
- HOSE HEATERS
- LAV A, D & E WATER HEATER
- LOGO LIGHT
- RECIRC FAN LEFT
- WATER COMPRESSOR

115V AC MAIN BUS 2

Powered by AC TRANSFER BUS 2
- ENT AUDIO
- MUX
- RECIRC FAN R
- VIDEO 1, 2, 3 & 4
- VIDEO CONT CENTER AC
- ZONAL DRYER

115V AC STANDBY BUS

Normally powered by AC TRANSFER BUS 1, alternate power is Hot BB
- 115V AC STBY BUS INDICATION
- **28V AC STBY BUS**
- ADF 1
- ADIRU L AC
- ATC 1
- DME 1
- ENG 1 IGNITION RIGHT
- ENG 2 IGNITION RIGHT
- FLOOD LIGHT STBY
- FMC 1
- MCDU 1
- MMR 1
- NAV CONTR PANEL 1
- RMI
- SCU FAN POWER
- VOR/MKR BCN 1

28V AC STANDBY BUS

Normally powered by AC TRANSFER BUS 1, alternate power Hot BB via inverter
- ADIRU L EXC
- EMERGENCY PANEL LIGHT (via 5V AC transformer)
- SMYD 1 SNSR EXC AC

115V AC GROUND SERVICE BUS 1

Normally powered by AC TRANSFER BUS 1
or external power when available
- **28V AC GND SERVICE BUS 1**
- AFT CARGO LOADER DRIVE
- AUX BATTERY CHARGER
- CEILING LIGHT LEFT
- CEILING LIGHT RIGHT
- ENTRY LIGHT BRIGHT
- EQUIP COOL EXH FAN NORM
- FWD CARGO LOADER DRIVE
- **GND SVCE BUS 1 SEC 1**
- LAVATORY MIRROR LIGHT
- POSITION LIGHT LEFT
- POSITION LIGHT RIGHT
- WING LIGHT

115V AC GROUND SERVICE BUS 2

Normally powered by AC TRANSFER BUS 2
or external power when available
- **28V AC GND SERVICE BUS 2**
- BATTERY CHARGER
- GALLEY LIGHT
- SERVICE OUTLET
- VAC OUTLET AFT
- VAC OUTLET FWD
- WINDOW LIGHT LEFT
- WINDOW LIGHT RIGHT

28V AC GROUND SERVICE BUS 1

Powered by 115V AC GND SERVICE BUS 1
- COMPARTMENT ACCESS LIGHT
- LIGHTS FWD ELECTRONIC RACK
- NIGHT LIGHT CONTROL
- THRESHOLD LIGHT
- WORK & THRESHOLD LIGHT

28V AC GROUND SERVICE BUS 2

Powered by 115V AC GND SERVICE BUS 2
- AFT CARGO COMP LIGHT
- AIRCO COMP LIGHT
- FWD CARGO COMP LIGHT
- WHEEL WELL LIGHT
- WINDOW LTS BRT CONT

28V DC GROUND SERVICE BUS

Normally powered by DC BUS 1, alternate power Hot BB
- AFT CARGO LOADER CONT
- CABIN UTIL RLY PWR
- ENTRY LIGHTS AUTO DIM
- EXHAUST FAN CONT NORMAL
- FWD CARGO LOADER CONT
- PSEU PRIMARY
- VACUUM WASTE
- WASTE WATER LINE HEATERS

28V DC BUS 1

Normally powered by TR1, altn power TR2 or TR3

- AFCS A RUDDER DC
- APU FIRE SWITCH
- B SYS ELEC HYDR PUMP CONT
- BUS PWR CONTROL UNIT
- **DC BUS 1 SEC 1**
- **DC BUS 1 SEC 2**
- DC BUS 1 XFR
- **DC STBY BUS SEC 1**
- **DC STBY BUS SEC 2**
- ELEX PNL AFT CENTER
- GENERATOR 1 LOAD SHED
- MASTER DIM SEC 7
- MASTER DIM SECTION 1, 2, 3 & 4
- PACK CONT VALVE L ALT
- RAM AIR MOD CONT L
- SPCU NORMAL
- VACUUM WASTE CONT
- WATER DRAIN VALVE
- WATER QTY INDICATION

28V DC BUS 2

Normally powered by TR2, altn power TR3 or TR1

- A SYS ELEC HYDR PUMP CONT
- AFCS B RUDDER DC
- AFCS INTLK 2
- APU GENERATOR CONTROL UNIT
- CAPT AUDIO
- CARGO LIGHT CONTROL
- **DC BUS 2 SEC 1**
- **DC BUS 2 SEC 2**
- DC BUS 2 XFR
- ENG 1 SPAR VALVE
- ENG 2 SPAR VALVE
- ENT AUDIO DC
- F/O'S AUDIO
- FORCE FIGHT MONITOR
- GENERATOR 2 LOAD SHED
- INTERPHONE AND WARNING
- MASTER DIM SECTION 5, 6, 7 & 8
- MIX VALVE POS IND
- OBERSERVERS AUDIO
- PACK CONT VALVE R ALT
- RAM AIR MOD CONT R
- RECIRC FAN CONT
- ZONAL DRYER CONT

28V DC BUS 1 – Sect 1

Normally powered by DC Bus 1

- AFCS A ENGAGE INTLK
- AFCS A RUDDER DC (Cat IIIb)
- AUTOTHROTTLE DC 1
- CABIN CREW CALL
- CARGO FIRE DETECTOR AFT A
- CARGO FIRE DETECTOR FWD A
- CHARGER L AFT
- CHARGER R AFT
- DOOR AREA HEAT CONT
- EMER EXIT LT CHARGER FWD
- ENG 1 COWL ANTI ICE VALVE
- ENG 1 RUN/PWR
- ENG 1 THRUST REVERSER IND
- FCC A DC
- FLIGHT RECORDER DC
- FMCS BITE DC
- HEAD UP GUIDANCE CMPTR DC
- HEAD UP GUIDANCE CONT PNL DC
- LAV SMOKE
- LIGHT CONTROL
- LOW FLOW DETECT EXHAUST
- MACH TRIM A DC
- MCP DC 1
- PASSENGER SIGN CONTROL
- PAX CALL LEFT
- RECIRC FAN L CONT
- SELCAL
- TERRAIN DISPLAY
- WINDSHIELD WIPER LEFT
- YAW DAMPER 1 & 2 DC

28V DC BUS 2 – Sect 1

Normally powered by DC Bus 2

- AUTOTHROTTLE DC 2
- CARGO FIRE DETECTOR AFT B
- CARGO FIRE DETECTOR FWD B
- EMERG FLASHLIGHT CHARGER
- ENG 2 ANTI ICE CONTROL
- ENG 2 COWL ANTI ICE VALVE
- LOW FLOW DETECT SUPPLY
- PAX CALL RIGHT
- VHF COMM 3
- VOICE REC RELAY
- WINDSHIELD WIPER RIGHT

28V DC BUS 1 – Sect 2

Normally powered by DC Bus 1

- **28V DC GND SVCE BUS 1**
- 28V DC SERVICE OUTLETS
- AIR/GND SYS 1
- AIRCO PACK CONT LEFT DC
- ANTI SKID OUTBOARD
- AUTOBRAKE BITE CONT 1

28V DC BUS 2 – Sect 2

Normally powered by DC Bus 2

- AFCS B ENGAGE INTERLOCK
- AFCS B RUDDER DC (Cat IIIb)
- AIR/GND RELAY
- AIRCO PACK CONT RIGHT DC
- AIRCO TEMP INDICATION
- ATC ANT SWITCH

- AUTOSLAT DC 1
- BLEED AIR VALVE LEFT
- BLEED AIR XDCR LEFT
- CAPT PROBE IND
- CMU / ACARS DC
- DC BUS 1 INDICATION
- ELT
- ENG 2 HYDR PUMP DEPRESS VALVE
- EQUIP COOL SPLY FAN CONT ALT
- FLAP LOAD RELIEF
- FLAP SHUT OFF VALVE
- FLIGHT CONT SHUT OFF VALVE CONT
- FSEU DC 1
- FUEL CTR TNK L AUTO SHUTOFF VLVS
- FUEL QUANTITY TANK 1
- LDG XFR VALVE PIMARY
- LDG XFR VALVE SECONDARY
- MASTER CAUTION ANN BUS 1
- MASTER DIM BUS 1
- O'WING EXIT FLIGHT LOCK LEFT
- PRESS CONTROL AUTO 1
- PSEU PRIMARY
- PTU VALVE CONT 1
- PTU VALVE CONT 2
- RAM AIR CONT LEFT

- AUTO SPEED BRAKE
- AUTOBRAKE BITE CONT 2
- AUTOSLAT DC 2
- BLEED AIR VALVE RIGHT
- BLEED AIR XDCR RIGHT
- CAPT INTPH POWER DC2
- COCKPIT DOOR LOCK
- LOWER DISPLAY UNIT
- DC BUS 2 INDICATION
- DEU 2 PRIMARY
- ELEVATOR TABE VALVE RIGHT
- ENG 1 HYDR PUMP DEPRESS VALVE
- ENG 2 RUN/PWR
- ENG 2 THRUST REVERSER IND
- EQUIP COOL SPLY FAN CONT NORM
- EXHAUST FAN CONT ALTERNATE
- F/O'S EFIS CONT PANEL
- F/O'S INBOARD DISPLAY
- F/O'S OUTBOARD DISPLAY
- FCC B DC
- FLIGHT CONT SHUT OFF VALVE CONT
- FMC XFR
- FO'S INTERPHONE PWR DC2
- FO'S PROBE IND
- FORCE FIGHT MONITOR
- FSEU DC 2
- FUEL CTR TNK R AUTO SHUTOFF VLVS
- FUEL SHUTOFF VALVE PWR PACK
- FWD AIRSTAIR DOOR
- FWD AIRSTAIR CONTROL NORMAL
- MACH TRIM B DC
- MACH WARNING SYSTEM 2
- MASTER DIM BUS 2
- NAV SENSOR DC 2
- O'WING EXIT FLIGHT LOCK RIGHT
- PRESS CONTROL AUTO 2
- RAM AIR MOD CONT R
- RECIRC FAN R CONT
- RUDDER AUTHORITY LIMITER
- SMYD COMP 2 DC
- SPOILER SHUT OFF VALVE
- STAB TRIM CONTROL
- STICK SHAKER RIGHT
- TRIM AIR PRESS
- VHF COMM 2
- VIDEO CONT CENTER DC

28V DC - STANDBY BUS – Sect 1

Normally powered by TR1, alternate power is the Hot BB
- CAPT EFIS CONT PANEL
- CAPT INBOARD DISPLAY
- CAPT OUTBOARD DISPLAY
- UPPER DISPLAY UNIT
- DEU 1 PRIMARY
- ENG 1 THRUST REVERSER CONTROL
- ENG 1 THRUST REVERSER INTERLOCK
- ENG 1 THRUST REVERSER SYNC LOCK
- INSTR XFR
- MACH WARNING SYS 1
- NAV SENSOR DC 1
- SMYD COMP 1 DC
- STBY ALT/MASI VIBRATOR
- STICK SHAKER LEFT
- VHF COMM 1

28V DC - STANDBY BUS – Sect 2

Normally powered by TR1, alternate power is the Hot BB
- DC INDICATION STBY BUS
- GENERATOR CONTROL UNIT 2

28V DC - BAT BUS

Normally powered by TR3, alternate power is the battery
- AIR CONDITION O'HEAT
- **BATT BUS SEC 1**
- **BATT BUS SEC 2**
- **BATT BUS SEC 3**
- ENT PA SYS BAT
- EXHAUST VALVE RECONFIG CONT
- ISFD

28V DC - BAT BUS – Sect 1

Powered by the battery bus
- ENG 1 AND WING ANTI ICE CONTROL
- ENG 1 START LEVER CHAN A
- ENG 1 START LEVER CHAN B
- ENG 1 START VALVE
- OXYGEN INDICATION
- OXYGEN MANUAL CONTROL
- PAX OXYGEN LEFT
- PAX OXYGEN RIGHT
- STBY ATT IND

28V DC - BAT BUS – Sect 2

Powered by the battery bus
- APU FIRE DETECTION
- BATT BUS INDICATION
- BRAKE PRESS IND
- CAPT INTPH POWER BAT
- DUCT O'HEAT AFT CABIN
- DUCT O'HEAT FLT DECK
- DUCT O'HEAT FWD CABIN
- ENG 1 FIRE DETECTION
- ENG 1 HYDRAULIC SYSTEM SHUT OFF VALVE
- ENG 2 FIRE DETECTION
- ENG 2 HYDRAULIC SYSTEM SHUT OFF VALVE
- ENG 2 START VALVE
- ENG 2 THRUST REVERSER CONTROL
- ENG 2 THRUST REVERSER INTERLOCK
- ENG 2 THRUST REVERSER SYNC LOCK
- ENG 2 START LEVER CHAN A
- ENG 2 START LEVER CHAN B
- EXHAUST VALVE CONT
- FO'S INTERPHONE PWR BAT
- GENERATOR 1 DISC
- GENERATOR 2 DISC
- GENERATOR CONTROL UNIT 1
- L AIRCO PACK VALVE CONT
- R AIRCO PACK VALVE CONT
- MASTER FIRE WARNING & CONTROL
- MCP DC 2
- PA AMP BAT
- PACK/BLEED AIR O'HEAT L & R
- PRESS CONTROL INDICATION
- PRESS CONTROL MANUAL
- STBY RUDDER SHUT OFF VALVE

28V DC - BAT BUS – Sect 3

Powered by the battery bus
- AIR/GND SYS 2
- ALT GEAR EXTEND SOLENOID
- ANTI SKID INBOARD
- AURAL WARNING
- CLOCK DISPLAY
- CROSS FEED VALVE
- DOME LIGHT WHITE
- ENG 1 HPSOV CON

- ENG 1 HPSOV IND
- ENG 2 HPSOV CON
- ENG 2 HPSOV IND
- FUEL QUANTITY TANK 2
- LANDING GEAR LATCH & PRESS WNG
- MASTER CAUTION ANN BAT
- MASTER CAUTION ANNUN CONT 1, 2, 3 & 4
- MASTER DIM & TEST CONTROL
- MASTER DIM BAT
- STBY COMPASS LIGHTING

28V DC - SW HOT BAT BUS

Only powered when the battery switch is on
- ADIRU L DC
- ADIRU R DC
- AFCS A WARN LT (BAT)
- AFCS B WARN LT (BAT)
- APU CONTROL
- APU GENERATOR CONTROL UNIT
- FWD AIRSTAIR CONTROL STANDBY
- FWD AIRSTAIR STANDBY DOOR ACTUATOR
- GENERATOR BUS PWR CONTROL UNIT
- PSEU ALTN
- SPAR VALVE INDICATION
- SWITCHED HOT BATTERY BUS INDICATION

28V DC - HOT BAT BUS

Always powered; either by battery or battery charger
- BAT/STBY SW POS IND
- CARGO FIRE EXT 1
- CARGO FIRE EXT 2
- CLOCK
- DC HOT BAT BUS INDICATION
- DEU 1 HOLDUP
- DEU 2 HOLDUP
- DUAL BATTERY REMOTE CONTROL
- ENTRY LIGHTS DIM
- EVAC SIGNAL
- FIRE EXTINGUISHER APU
- FIRE EXTINGUISHER L ALT
- FIRE EXTINGUISHER LEFT
- FIRE EXTINGUISHER R ALT
- FIRE EXTINGUISHER RIGHT
- FUEL SHUTOFF VALVES BUS
- FUELING CONTROL
- FUELING INDICATORS
- GND SERVICE CONTROL
- GND SVCE DC HOT BATTERY
- PARKING BRAKE
- SPCU STBY
- TR3 XFR RELAY CONT

Electrical Schematics

The following electrical schematics are included to give the reader an overview of the basic electrical configurations of the various series of 737. Please note that although these contain more information than FCOM Vol 2, they are still a great simplification of the full system (particularly in the way I have represented the standby power switch relays / standby power control unit). Furthermore there have been many different configurations over the years for different customers, so please do not assume that your particular aircraft match any of the following.

All schematics have been drawn in the normal in-flight condition; ie both engine generators on, all electrical switches in their normal positions and glideslope not captured. Note that the RCCBs in these schematics refer to Remote Control Circuit Breakers and not Residual Current Circuit Breakers that you may find in your house. Their operation is very different!

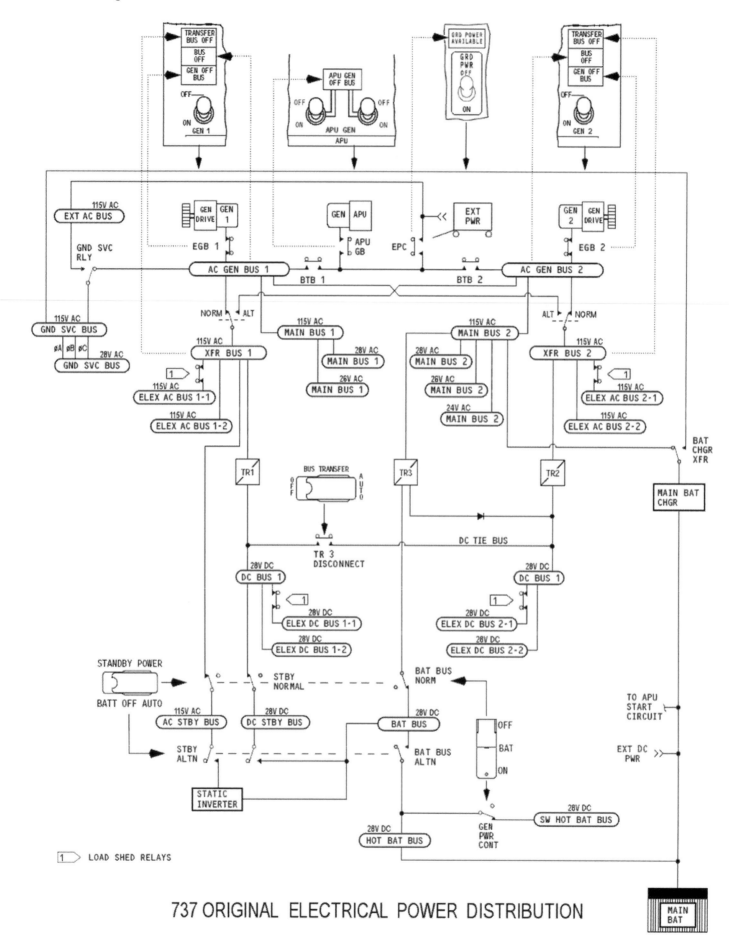

737 ORIGINAL ELECTRICAL POWER DISTRIBUTION

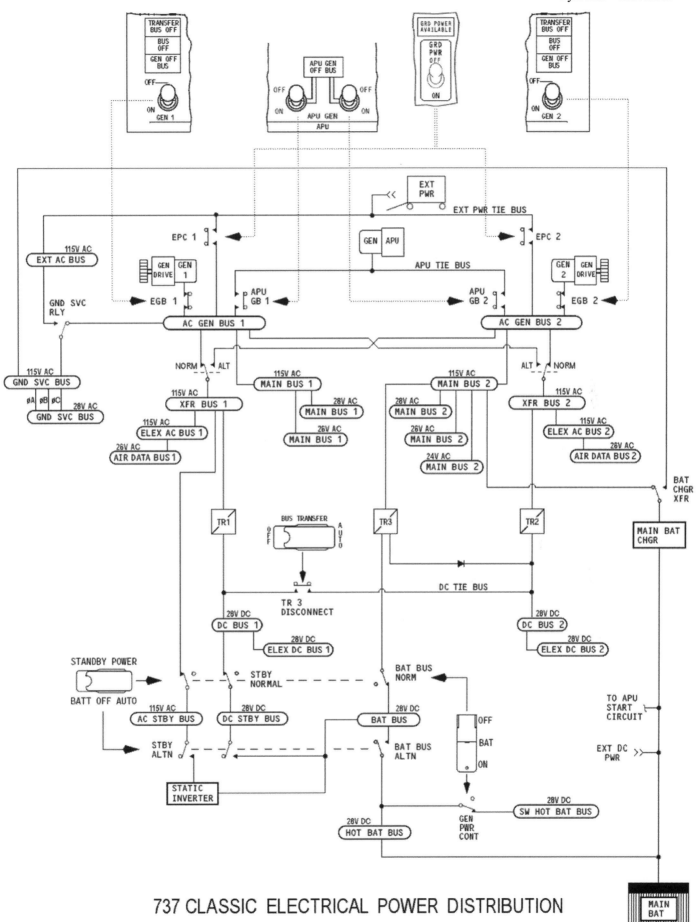

737 CLASSIC ELECTRICAL POWER DISTRIBUTION
(SINGLE BATTERY INSTALLATION)

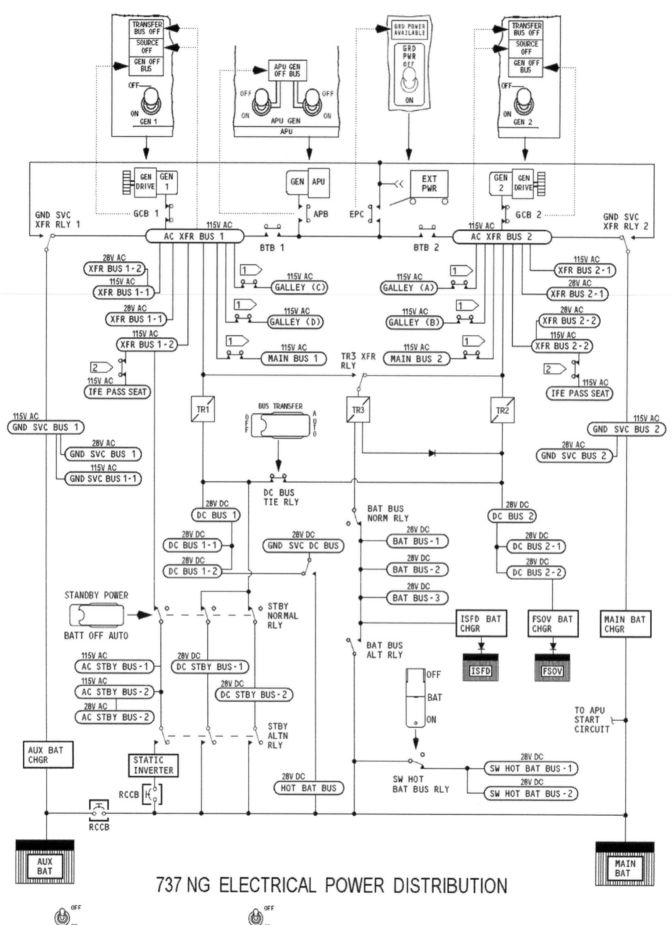

737 NG ELECTRICAL POWER DISTRIBUTION

EMERGENCY EQUIPMENT

Flight Crew Oxygen

When conducting the oxygen mask flow & intercom check, monitor the crew oxygen pressure gauge to ensure a steady flow as any fluctuations may be due to an obstruction in the system. Give a long check of the flow on the first flight of the day in case the crew oxygen shut off valve has been closed. A short check may sound OK, but you may be hearing the residual oxygen left in the lines, rather than fresh oxygen from the bottle.

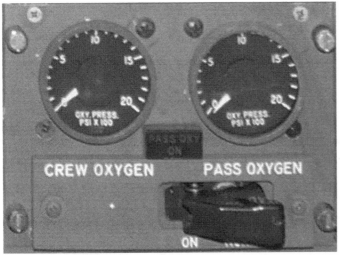

Oxygen Panel -1/200

Oxygen Panel -300+

Crew Oxygen Shutoff Valve (Not installed on NGs)

On the 1-500 series, the F/O should ensure that the crew oxygen shutoff valve, located at the bottom outside of the P6 panel, is open (anticlockwise) and ideally backed off by half a turn to avoid damage to the seal. This should be done during the cockpit preparation, particularly in airlines where it is the practice to close this valve overnight.

Crew oxygen pressure on aft overhead panel should be checked against MEL 35-1 or FPPM 2.2.14. The minimum despatch quantity varies with size of bottle, bottle temp and number of flightdeck crew.

The minimum amount of oxygen is based upon one hour of normal flight at a cabin altitude of 8000ft for one pilot with the diluter set to NORMAL (76 cu ft bottle).

97

Minimum Crew Oxygen Dispatch Pressure				
Oxygen Bottle Size	Temp (C)	Number of Crew		
		2	3	4
39 cu ft	0	1130	1645	
	15	1190	1735	
	30	1250	1825	
76 cu ft	0	620	890	1155
	15	655	940	1220
	30	690	990	1280
114 cu ft	0	445	620	800
	15	470	655	840
	30	495	690	885

Crew oxygen is stored in a bottle in the forward hold. On older aircraft (pre 1990) there is a servicing point on the outside however on most access is gained through the forward hold.

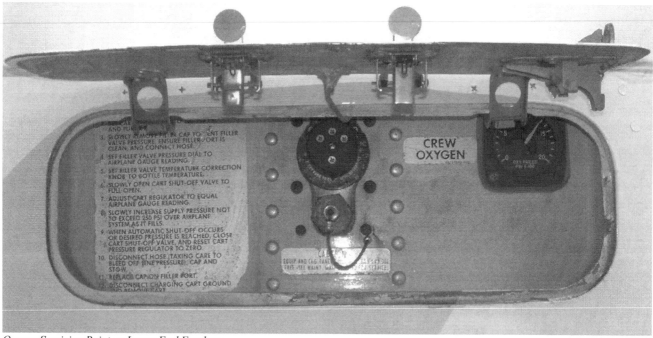

Oxygen Servicing Point on Lower Fwd Fuselage

All aircraft have a green discharge disc on the outside to warn crews if the bottle has discharged from overpressure. This should be checked on every walkaround.

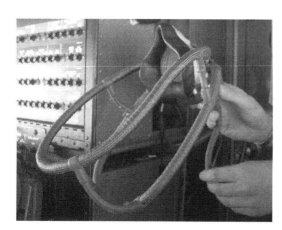

Flight Crew Oxygen Mask

The oxygen regulator has three modes:

- **Normal**: Red latch on left is up - Gives air/oxygen mix on demand. Use if no fumes are present, eg decompression.
- **100%**: Red latch on left is pushed down - Gives 100% oxygen on demand. Use if smoke or fumes are present.
- **Emergency**: Red knob is rotated clockwise - Gives 100% oxygen under pressure. Use to clear mask & goggles of fumes or when at very high altitude.

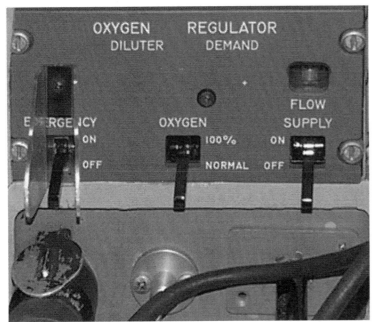

737-200 crew oxygen panel

The 737-200 crew oxygen panel, officially called the "Diluter Demand Regulator" is located above each pilot, near the escape rope. The EMERGENCY and OXYGEN controls are on each mask from the -300 series onwards.

Passenger Oxygen

Classics & NGs: The masks will deploy automatically above 14,000ft cabin alt or when switched on from the aft overhead panel. No oxygen will flow in a PSU until a mask in that PSU has been pulled. Passenger oxygen masks will be ineffective as smoke hoods because the air in the mask is a mixture of oxygen and cabin air. Also the oxygen will increase the fire hazard in the cabin.

There is 12 minutes supply of oxygen in each PSU, this is based upon:
- 0.3 min delay at 37,000ft
- 3.1 mins descent to 14,000ft
- 7.6 mins hold at 14,000ft
- 1 min descent to 10,000ft

Passenger oxygen on 737-1/200s is supplied by two oxygen bottles in the forward hold. The capacity varies with operator but is typically 76.5 cu ft each. Oxygen bottle pressure is indicated on the aft overhead panel.

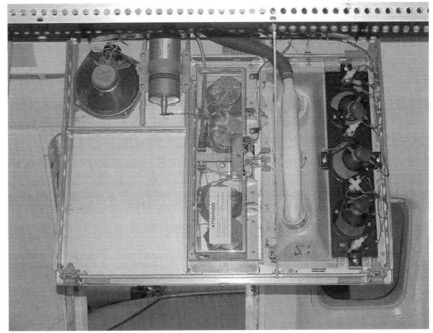

Behind a Passenger Service Unit

Emergency Exit Lights

The emergency exit lights, when armed, will illuminate if power is lost to DC bus 1. They can also be switched on from the aft flight attendant panel. Whenever these lights are on, they are being powered from their own individual Ni-Cad batteries and will last for a minimum of 10mins.

Emergency exit light controls

Aft Attendant panel

Smoke Hood (Drager)

After pulling the toggle, the oxygen generator will operate for less than 30 secs. Don't worry, the oxygen remains in a closed loop system within the mask and filter to prevent contamination from the outside air. It is filtered twice, on inhalation and again on exhalation, and is breathable for approximately 20mins.

Life Jacket

Do not inflate this until you are outside the aircraft. When it has inflated it will be very bulky and will slow you down and you could puncture it on the way out!

Cockpit Fire Extinguisher

Is BCF and works by removing oxygen from the fire triangle of oxygen - heat - fuel. As it does not directly cool the fire, when oxygen returns, so could the fire. To operate, remove ring and press down on top lever. Hold upright and beware, BCF fumes are toxic.

Slides

Serviceability check includes the pressure gauge. Tip: Be extremely careful to remember to disarm any door slides that you may have armed on flights without cabin crew, eg ferry flights or airtests.

Note that the slides are not certified as emergency floatation equipment. Boeing say that an inflated slide could be buoyant, and useful as a floatation device. There are handgrips positioned along the sides of the slide.

FIRE PROTECTION

Engine & APU fire detection – Battery bus
Engine, APU & Cargo fire extinguishing – Hot battery bus.
Cargo fire detection – DC Bus 1 & 2

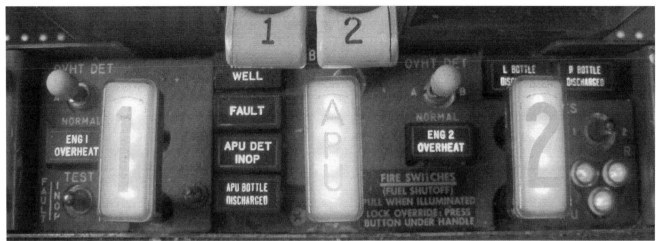

Overheat / Fire Protection Panel 3-900 series

Overheat / Fire Protection Panel -200C series. Notice FWD & AFT CARGO SMOKE detectors to the left of the bell cut-out.

There are two fire detection loops in each engine. Failure of both loops in one engine will illuminate the FAULT light. The individual loops can be checked by selecting A or B on the OVHT DET switches.

Fire switches will unlock in the following situations:
1) Overheat detected
2) Fire detected
3) During an OVHT/FIRE test
4) Pressing manual override buttons

Pulling a fire switch will do the following:
1) Arm firing circuits
2) Allow fire switch to be rotated for discharge
3) Close engine fuel shut-off valve (all series) and spar fuel shut-off valve (NG)
4) Trip the associated GCR (i.e. switches off the generator)
5) Close hydraulic supply to EDP & disarms its LP light (Not if APU)
6) Close engine bleed air valve (If APU will also close air inlet door)
7) Close thrust reverser isolation valve (Not if APU)

The engine fire bottles (NG)

Wheel-Well

There is a wheel-well fire detection system but although the engine fire bottles are located in the wheel-well, there is no extinguishing system for a wheel-well fire. (Suggest extend gear & land ASAP).

Lavatory Smoke Detection (Optional)

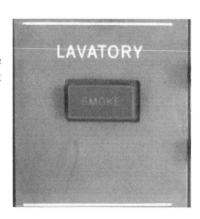

Some 737s have a warning light on the flight deck to warn of smoke in the lavatory. If the smoker is in the forward lav you can usually smell it on the flight deck within seconds without a warning light.

APU

There is only one APU fire extinguisher bottle, despite the fact that the handle can be turned in either direction! It is filled with Freon (the extinguishant) and Nitrogen (the propellant) at about 800psi. When the fire handle is turned, the squib is fired which breaks the diaphragm on the bottle, the pressure of the nitrogen then forces the Freon into the APU compartment which suffocates the fire.

The APU fire extinguisher bottle indicators, not fitted to NGs, comprise of one yellow disc to show if the squib has been fired and one red disc to show if the bottle has over-temperatured (130C) or over-pressured (1800psi). The sight glass to the bottle pressure gauge in this photo is optional. Note: Sight glass and bottle indicators are not fitted to NGs.

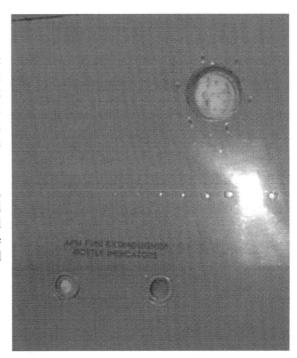

APU bottle indicators and sight glass

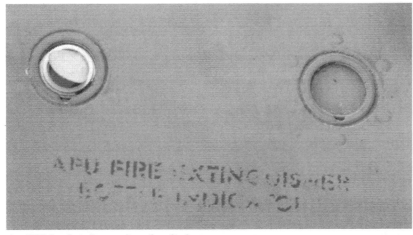

On the only occasion that I have had to discharge an APU fire bottle, the yellow disc (on the left) did not blow clear, it just opened slightly. This could easily be missed on an external inspection.

An APU bottle indicator disc after discharge

There is a second set of APU fire controls on the aft wall of the wheel well.

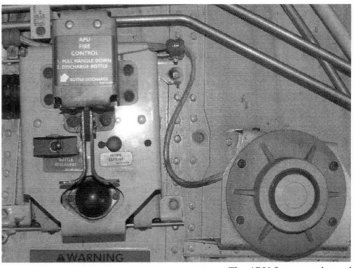

The APU fire control panel

Cargo Compartment

Cargo Fire Panel

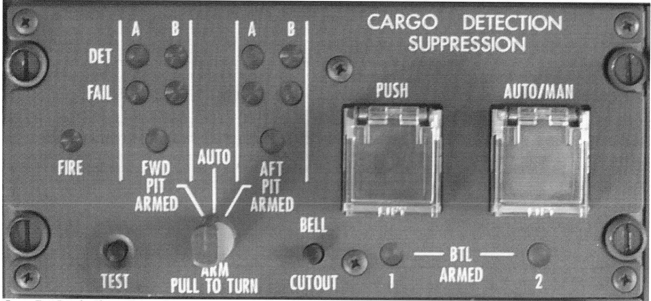

Cargo Fire Panel - Alternative version

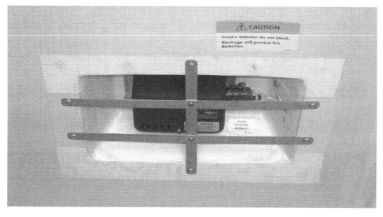

Cargo Hold Smoke Detector

The cargo holds have dual-loop smoke detectors powered by DC bus 1 & 2. There is only one cargo fire bottle, it is powered by the hot battery bus and can be discharged into either the fwd or aft hold. On later 737NGs the cargo fire smoke detector sends a signal to the cabin pressure control system. This triggers the cabin pressure to descend at 750 fpm which helps prevent smoke penetration into the passenger cabin from the lower lobe. This function is inhibited on the ground.

FLIGHT CONTROLS

Roll

Ailerons are powered by hydraulic systems A and/or B. If both hydraulics should fail, manual reversion is available from both control wheels. There are balance tabs and balance panels to assist aileron movement in the event of manual reversion. If the aileron system jams, the co-pilots wheel can be used to move the spoilers (hydraulically).

Aileron trim moves the neutral position via the feel & centering mechanism. There are two aileron trim switches to prevent spurious electrical signals from applying trim. The fwd switch is for direction, the aft switch is simply an earth return.

Aileron trim controls

Use of aileron trim with the autopilot engaged is prohibited because of the potential for excessive roll when the autopilot is disconnected.

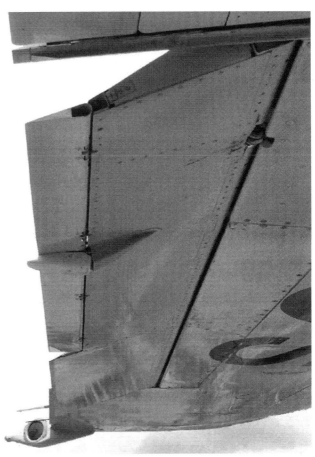

Aileron and balance tab

Speedbrake / Spoilers

Series 1-500 aircraft have 2 flight spoilers and 3 ground spoilers on each wing, NGs have 4 flight and 2 ground spoilers on each wing. They deploy asymmetrically to augment the ailerons and/or symmetrically as speedbrakes.

Half of the flight spoilers are powered by hydraulic system A and half by system B. All ground spoilers are powered by system A.

Most aircraft have no restriction on the speed at which the speedbrakes may be used (on airtests they are deployed at 360KIAS/M0.84); however the resultant position is a function of airspeed due to blowdown.

The 737-300W/500W/700W and 737-900ER have a **Speedbrake Load Alleviation System** which retracts the spoilers and speedbrake lever to the 50% position when above 320kts and a certain gross weight. The weight depends upon the series of 737, eg 64,860kg / 143,000lbs for a 737-700W.

737-700W speedbrake lever with 50% position

The Boeing 737 Technical Guide

The SPEED BRAKE DO NOT ARM light in-flight does not preclude its use on the ground.

On landing, if armed, all spoilers will deploy when the thrust levers are at idle and any two wheels have spun up or right gear is compressed. If not armed, the speedbrakes will deploy when reverse thrust is selected.

737-500 on landing roll with flaps & ground spoilers deployed

The above photo shows how the ground spoilers move more than the flight spoilers. On the 737-NG (below) there is an extra trapezoid shaped outboard ground spoiler panel.

737-800 on landing roll with flaps & ground spoilers deployed. The longer span of the NG can be clearly seen in this photograph.

The SPEED BRAKE TEST buttons are only found on series -1/200s and old, non-EIS -300s. They are located at the top of the centre instrument panel and are used in conjunction with the speedbrake lever and antiskid system to illuminate the SPEED BRAKE ARMED and/or SPEED BRAKE DO NOT ARM lights.

The 737NG (BP99 onwards) has an optional Flight Control Surface Position Indicator which displays the positions of the ailerons, elevator, rudder and flight spoilers on the lower DU when the MFD SYS button is pushed. The DEUs receive the position information from the digital flight data acquisition unit (DFDAU).

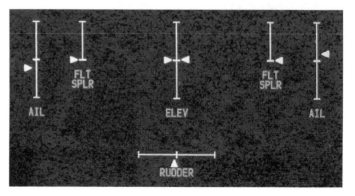

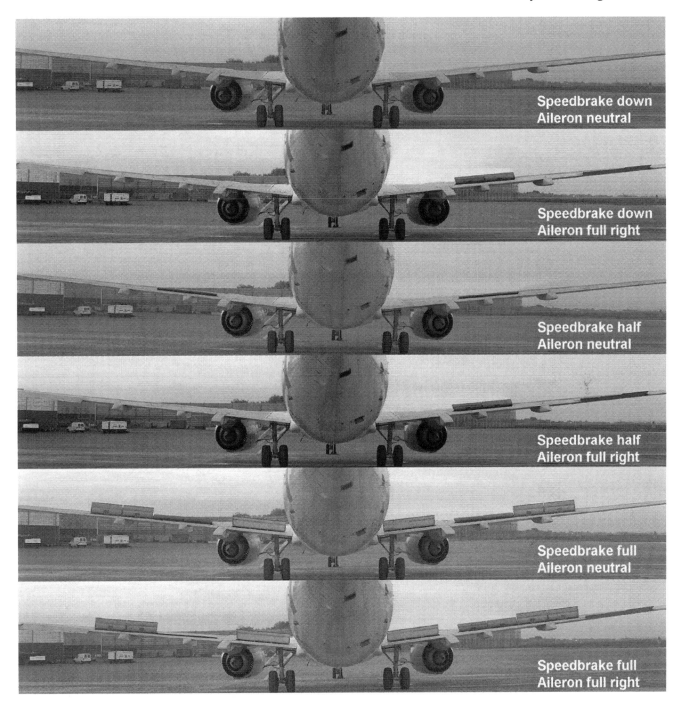

The above series of photographs (737-300) show how the flight spoilers move with various combinations of aileron and speed brake. With speedbrake down, the spoiler simply rises on the down-going wing with aileron. With speedbrake applied, not only do the spoilers on the down-going wing rise but also the spoilers on the up-going wing fall. Notice that even with full speedbrake applied the spoilers still rise on the down-going wing.

This property of the spoilers on both wings to respond to roll inputs is known as differential spoilers. It only occurs when speedbrake is used which is why the roll rate increases with speedbrake extended.

NB In the bottom two photographs the speedbrake lever was only at the flight detent position, but because the aircraft was on the ground, the ground spoilers deployed. This is why if you have any sort of technical problem which might be due to a faulty air-ground sensor, eg QRH "Gear Lever Will Not Move Up After Takeoff", you must not use speedbrake in case the ground spoilers deploy in-flight. The series of photographs illustrates just how much extra drag ground spoilers will give over flight spoilers.

Yaw

The rudder is moved by a Power Control Unit (PCU) powered by hydraulic system A and/or B. If A and/or B fails a standby rudder PCU is powered from the standby hydraulic system. Unlike the ailerons and elevator, there is no manual reversion for the rudder; hence it does not have a tab.

Basic 737 rudder with 3 mass balances

737-200Adv rudder with 2 mass balances

737-200Adv rudder with 1 mass balance

The rudder is a single piece and has a horn balance at the top, it also has mass balances forward of the hinge line. The first 737s had three until 1970 when the lower balance was dropped to save 40lbs in weight. Later 737-200s and all subsequent series only had the upper balance. The reduction in number is probably because the rudders have become lighter as their construction changed from fibreglass to carbon fibre materials.

Rudder Trim

The rudder trim knobs have evolved over the years. The -1/200 series had the large knob from 707 days when there was more hand flying and asymmetric thrust before the days of autothrottle.

737-1/200 Rudder trim knob

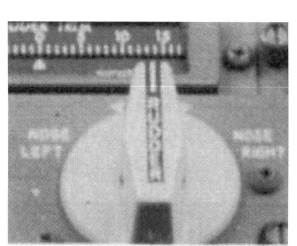

737-3/400 old blade shaped rudder trim knob

The 737 Classics initially had a smaller blade shaped knob but this was changed by FAA AD 90-14-02 in 1990 to the present round fluted knob following the crash of a US Air 737-400 which overran at La Guardia with full left rudder trim set (NTSB/AAR-90/03). It was believed that a jump seat occupant may have inadvertently moved the knob with his foot when it was resting on the aft electronic panel. A guard rail was also fitted around the aft end of this panel to prevent inadvertent movement of the trim controls.

You will sometimes find that a particular aircraft requires an adjustment of the rudder trim in the climb or descent. There are two theories for this: The first is that the out of trim condition arises due to different thermal expansion rates in the components as the aircraft cools / heats up; the second is that whatever slight out of trim condition there may be, it will be magnified at higher IAS, this of course changes during the climb & descent. Either way, the acceptable limits of rudder & aileron trim are as follows:

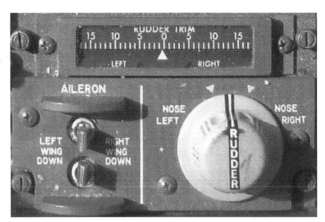

Rudder and aileron trim controls

Configuration	Aileron Trim	Rudder Trim
M0.74 above 30,000ft	¾ Unit	½ Unit
250 kts, Flaps UP	¾ Unit	¾ Unit
190 kts, Flaps 1	1 Unit *	1 Unit
150 kts, Flaps 15	1 Unit *	1 Unit
130 kts, Flaps 40	1 Unit *	1 Unit

*Aileron trim changes should not exceed 1 unit difference between adjacent flap positions after flaps are extended.

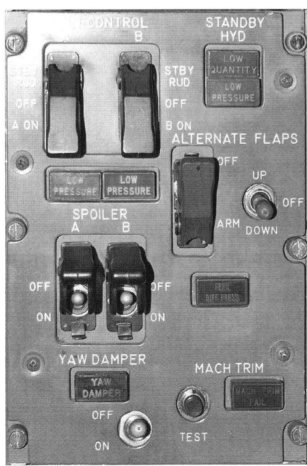

737-1/200 flight controls panel

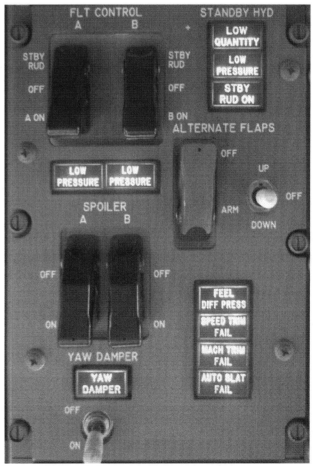

The new flight controls panel with STBY RUD ON light

From Jan 2003 (l/n 1268) all 737s were delivered with the Rudder System Enhancement Program (RSEP) installed. This can be recognised in the flight deck by a new STBY RUD ON light in the STANDBY HYD column. When illuminated, it indicates that the standby rudder PCU is pressurised. It may be pressurised either automatically (loss of system A or B etc) or manually (through the FLT CONTROL switches). See Rudder section (page 217) for more details.

Yaw Damper

737-200 Dual Yaw Damper

The 737 is positively damped in combined lateral-directional oscillations, which in plain English means that if you set up a Dutch Roll the aircraft will gradually stop oscillating. So the yaw damper is not required for dispatch; however, it is fitted for passenger comfort. It is powered by hydraulic system B.

The 1/200 series had a dual yaw damper system because at the design stage, from experience of the 707 and 727, it was expected that the 737 would not be so naturally, positively yaw damped, so two were fitted to allow dispatch in case one failed. As it happened, none were required. The 1/200s have a yaw damper test switch to the right of the indicator. Note that for aircraft with the RSEP installed, the yaw damper test switch is inoperative.

The NGs saw a return to a dual system by having a standby yaw damper. Note that only the main yaw damper inputs are shown on the indicator.

The yaw damper system can move the rudder a maximum of 2 degrees (-1/200), 3 deg (-3/4/500), 2 deg (NG flap up), 3 deg (NG flap down), either side of the trimmed position. Yaw damper inputs are not fed back into the rudder pedals, which is why there is an indicator. However the indicator was withdrawn from new aircraft from April 2010.

Pitch

The control column moves the elevators using hydraulic system A and/or B. If both hyd systems should fail, manual reversion is available from either control column. There are mechanically operated balance tabs, linked to the elevators to assist with pitch, particularly whilst in manual reversion. The tabs always operate even when the hydraulics are powered. Pitch trim is applied to the stabiliser (a.k.a. all moving tailplane) this reduces drag by allowing the elevators to remain fared with the stabiliser.

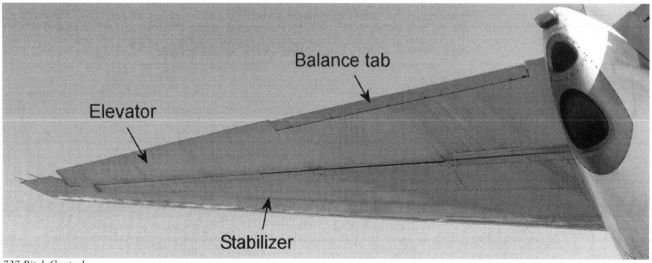

737 Pitch Controls

Pitch trim is applied to the stabilizer. Trim can be applied by electric trim switches, autopilot or a manual trim wheel. Electric and autopilot trim may be disengaged by cut-out switches on the control stand in the event of a runaway or other malfunction.

Moving the control column in the opposite direction to electric trim will stop the trim, unless the STAB TRIM switch (on the aft electronics panel) is set to OVERRIDE. This function could be used to control the pitch of the aircraft with trim, say in the event of a jammed elevator.

The trim authority varies according to aircraft series and method of trim. The full range is only available with the manual trim wheel, but if at an extreme setting, electric trim can be used to return to the normal range. There are two electric trim switches on each control column, the right is for the direction and the left is an earth return for protection against spurious electrical signals.

The STAB TRIM light was only fitted to the 1/200 series and simply tells you that the main electric trim motor is operating.

Trivia: On the 737-100 to 500, the stab trim control wheels should be mounted with their white marks 90 +/- 15 degrees apart from the other wheel, so that in the event that the trim wheel handles need to be used, one handle will be in an accessible position. Surprisingly this is not a requirement for the NG.

737-1/200 Stab trim switches & light

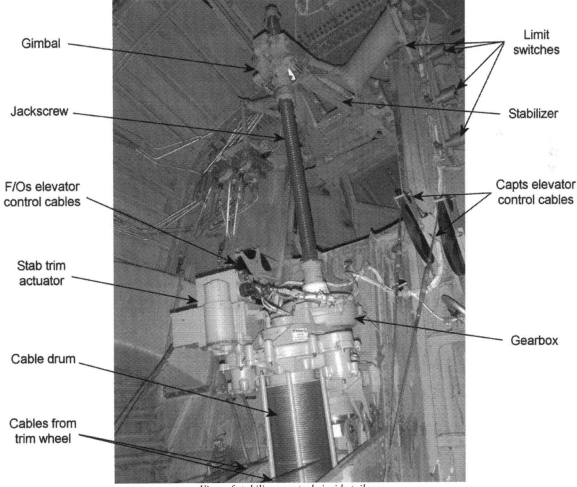

View of stabilizer controls inside tailcone

Elevator Tabs

For Originals and Classics, the elevator tabs always function as balance tabs, ie the tabs move in the opposite direction to the elevators.

On the NG series the elevator balance tabs become anti-balance tabs when the flaps are not up, this makes the aircraft less sensitive in pitch. However, during manual reversion, the tabs revert to balance tabs regardless of flap position to assist with pitch control. Notice from the series of photos below that in the unusual case of flaps not up and a single flight control system in STBY RUD or OFF, one tab will be balance and one anti-balance. This configuration does not give any significant unwanted roll as elevator and hence tab deflections are usually moderate.

Balance tab mode
Flaps up, hydraulics in any configuration

Anti-Balance tab mode
Flaps not up, hydraulics normal

LH Balance tab & RH Anti-Balance tab
Flaps not up and Hyd A in Standby Rudder

There is a speed limitation for speedbrake extension of 300kts on NG series aircraft without stiffened elevator tabs. All aircraft should have been modified to stiffened tabs by Sept 2006 (see SB 737-55A1080). You can tell if this has been done externally by counting the balance tab hinges, you should see six; if you only see four then your aircraft has not been modified and the speedbrake limitation applies.

Speed trim is applied to the stabilizer automatically at low speed, low weight, aft C of G and high thrust – i.e. on most take-offs. Speed trim is a dual channel system. Sometimes you may notice that the speed trim is trimming in the opposite direction to you, this is because the speed trim is trying to trim the stabilizer in the direction calculated to provide the pilot with positive speed stability characteristics. The speed trim system adjusts stick force so the pilot must provide significant amount of pull force to reduce airspeed or a significant amount of push force to increase airspeed. Whereas, pilots are typically trying to trim the stick force to zero. Occasionally these may be in opposition.

As the Mach increases, so the centre of pressure moves aft and the nose of the aircraft will tend to drop (Mach tuck). **Mach trim** is automatically applied above M0.615 (Classics & NGs), M0.715 (-1/200) to the elevators to counteract this and to provide speed stability.

Feel

The flying controls are hydraulically powered which tends to mask any control forces. This is resolved by adding artificial feedback called "feel". For the ailerons and rudder this is reproduced by spring pressure. The aileron feel springs can be seen in the wheel-well at the base of the aileron input shaft located on the forward wall of the wheel-well.

Also on the aileron input shaft, adjacent to the feel springs is the centering unit. This rotates the input shaft, adjusting the neutral position, in response to aileron trim inputs. You can see both of these functioning if you stand in the wheel-well whilst the aileron & trim are moved.

Aileron feel and centering unit in the wheel-well

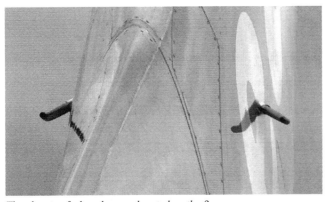

The elevator feel probes are located on the fin

Elevator feel is hydraulic, controlled by a computer. Since it must increase with airspeed there are two dedicated elevator pitot probes, located on the fin, to sense local airflow as near to the elevators as possible. The FEEL DIFF PRESS light is armed when the trailing edge flaps are up and will illuminate if either hydraulic system or an elevator pitot fails. The elevator feel system will continue to function on the remaining hydraulic system.

Leading Edge Devices

Are comprised of 4 Krueger flaps inboard of the engines and 6 slats (8 NG) outboard of the engines. The LE flaps are extended whenever the TE flaps are not up. The slats will be at EXT when the TE flaps are between 1 and 5, and will be at FULL EXT when the TE flaps are beyond 5 (25 for SFP). Slat numbers 1 & 6 (the outboard slats) move a few degrees less than slats 2 to 5 when at full extend, causing the leading edge to look slightly disjointed in this configuration; this is a throwback to the design of the original series and is normal.

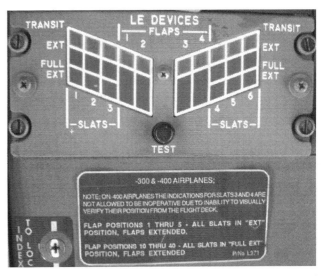

737-Classic LED Panel

737-1/200 basic LED Panel

737-NG LED Panel

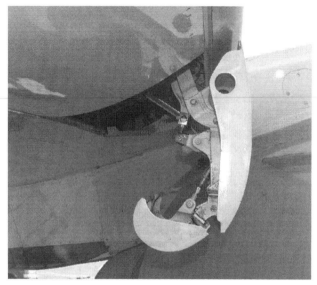

Side view of partially extended Krueger flap

The normal power for the LEDs is hydraulic system B (System A on the -1/200). If this fails the LEDs may be extended but not retracted with the standby hydraulic system using the ALTERNATE FLAPS switch on the flight controls panel. Note that when alternate flap is used, all LEDs will continue to travel to the FULL EXTEND position regardless of switch or flap position and cannot be retracted (unless hydraulic B is restored).

There are four LE Krueger flaps, two inboard of each engine. They are either fully extended or retracted (the photo shows a flap in transit). When the flap is retracted, the folding nose section rotates and is stored under the wing. The LE flaps do not have anti-ice.

The autoslat system will fully extend the slats for stall protection whenever flap is selected and the slats are not already at the full extend position (ie flap 1 to 5). If system B pressure is lost, system A can pressurise system B fluid for autoslat via the PTU. The LE FLAP TRANS light is inhibited during autoslat operation.

Underneath each slat are six connections:

1) A single hydraulic ram actuator, which is normally powered by hyd system B. The actuator is powered (to the full extend position only) by the standby hydraulic system when the ALTERNATE FLAPS control is selected to DOWN.

2) Two main tracks (outer) and two auxiliary tracks (inner) to help hold the slats in the three positions.

3) A bleed air duct for wing anti-ice and exhaust holes in the underside of the slat. Note NGs do not have wing anti-ice on their outboard slats.

737 NG slat components

Vortilons

The NGs have several vortilons on the slats and flaps. Also known as "underwing fences", they act as both aerodynamic fences to restrict the spanwise flow of air and vortex generators to create vortices on the underside of the wing at low angles of attack and on the upper surface at high angles of attack.

There are three on the underside of the outer two slats. These improve the handling characteristics near the stall, by reducing roll off / wing drop. There are also seven on the leading edge of the flaps which energise the airflow over the flaps.

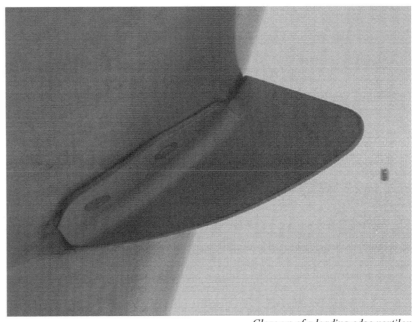

Close up of a leading edge vortilon

The discontinuity between the outer slat and the wingtip is effectively a "dogtooth" which will create a single vortex over the aileron.

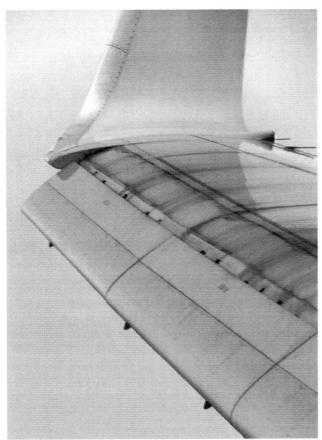

Leading edge vortilons with slats at full extend

Flap vortilons with landing flaps set

Trailing Edge Flaps

Triple slotted TE flaps – 737 Classic   *Double slotted TE flaps – 737 NG*

The trailing edge (TE) flaps are triple slotted (-1/2/3/4/500) / double slotted (-NG). The Originals and Classics have two exhaust gates either side of the aft engine fairing. They are connected by pushrods to the adjacent midflaps & foreflaps which cause them to droop when the flaps travel beyond 25.

An asymmetry condition on the classic is deemed to be a split of 22+/-5 degrees on the flap position indicator. On the NGs the flaps/slats electronics unit (FSEU) will detect an asymmetry at the flap position transmitters at a level which is almost undetectable by eye. In any series of 737, if an asymmetry is detected, hydraulic power is removed. The NG FSEU will also detect a flap skew, if detected the flap position indicators will display a 15 degree split.

The normal power for the flaps is hydraulic system B. If this is not available, the flaps & slats can be moved electrically with ALTERNATE FLAPS. There is no asymmetry protection with alt flaps and the LE flaps and slats can be extended but not retracted. The duty cycle limitations are for a complete extension and retraction e.g. if flaps moved from 0 to 15 and back to 0, then must wait 5mins before using alternate flap again.

A flap load limiter (-3/4/500) / FSEU (-NG) will automatically retract the flaps from 40 to 30 (-3/4/500) / 30 to 25 (NG) / 10 or greater (SFP) if the limit speed is exceeded. The flaps will extend again when speed is reduced. This feature is on all aircraft although the FLAP LOAD RELIEF light is optional.

The basic 737-1/200s also had a green LE FLAPS FULL EXT light.

Flap position indicator. Only early models had the Flap Load Relief light although all had the system

Trivia: Although the flap placard limit speeds are different for each 737NG variant, the structural limit speed for the flaps is equal to the placard speeds (175k – F30, 162k – F40) for the heaviest variant (737-800/900). The Flap Load Relief trigger speeds (176k – F30, 163k – F40) are set to allow all variants to fly to the structural limit speed without system activation. Setting lower flap placard speeds for the –600 and –700 variants allows for greater service life of flap components due to the larger margins to the structural design speed.

Flap track fairings at the flap 40 position on a 737-300.

Aft view of the flaps at the 40 position on a 737-300 showing the exhaust gate for the engine.

Flap Modifications

A trailing edge flap modification from AeroTech has been available for the 737-2/3/4/500 series since 2004. The modification slightly extends and lowers the aft segments of the trailing edge flaps thereby increasing wing area, camber and importantly lift-to-drag ratio. The original modification had one fixed position for the aft flap segments. However it is now adjustable/tunable to fit a wide spectrum of flight operations. The cruise angle of attack is reduced by approximately 0.5 degree which reduces induced drag. There are no changes to operational procedures and every pilot that has flown this mod (FAA, test pilots, and line pilots) says that the aircraft handles better. Fuel savings vary with series and route structure but may average up to 4.3%.

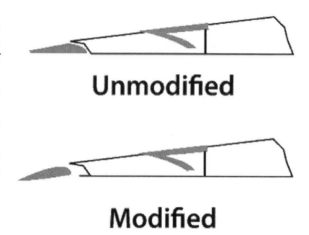

Unmodified

Modified

Limitations

Max flap extend altitude: **20,000ft** *

In flight, do not extend speedbrake lever beyond FLIGHT DETENT.

Holding in icing conditions with flaps extended is prohibited.

Do not deploy the speedbrakes in flight at radio altitudes less than **1,000ft**.

Although not a limitation, the FCTM advises not to use speedbrake with flaps extended. Also "With flaps 15 or greater speedbrakes should be retracted."

Avoid rapid and large alternating control inputs, especially in combination with large changes in pitch, roll, or yaw (e.g. large side slip angles) as they may result in structural failure at any speed, including below VA.

737-6/900 without stiffened elevator tabs: Do not operate the airplane at speeds in excess of 300 KIAS with speedbrakes extended. WARNING: Use of speedbrakes at speeds in excess of 320 KIAS could result in a severe vibration, which, in turn, could cause extreme damage to the horizontal stabilizer.
Note: All aircraft should have been retrofitted with stiffened elevator tabs by Sep 2006.

Alternate flap duty cycle:

Flap Position	Minutes Off
0 - 15	5
15 - 40	25

* This is taken from the Boeing Airliner magazine:

"Several operators have asked Boeing why the Airplane Flight Manual has a limitation restricting the use of flaps above 20,000 feet. The reason for the limitation is simple; Boeing does not demonstrate or test (and therefore does not certify) airplanes for operations with flaps extended above 20,000 feet.

There is no Boeing procedure that requires the use of flaps above 20,000 feet. Since flaps are intended to be used during the takeoff and approach/land phases of flight, and since Boeing is not aware of any airports where operation would require the use of flaps above 20,000 feet, there is no need to certify the airplane in this configuration."

FLIGHT INSTRUMENTS

Original Flight Instruments

Early 737-200 Instrument panel

This photo is of a 1969 built 737-200, you can see that apart from the new TCAS VSI all the instruments are electromechanical. An early customer option was the 5-inch HSI and ADI with Collins FD110 flight director. These later became standard.

HSI source selector switches

The big Gotcha with SP-177 equipped 737-200Advs and non-EFIS Classics were the HSI source selectors, sometimes referred to as "Killer Switches". These were located either side of the MCP and changed the HSI to show deviation either from the LNAV or ILS/VOR track. It is vitally important that these switches are set to VOR/ILS before commencing an approach otherwise you will still be indicating LNAV deviation rather than LLZ deviation.

Classic Flight Instruments

737-300 Non-EFIS F/O flight instruments

The first 737-300s were not fitted with EFIS and the flight instruments were almost identical to the 737-200Adv. The warning lights above the ADI are the instrument comparator warnings. The FMA annunciations were all contained in the panel above the ASI

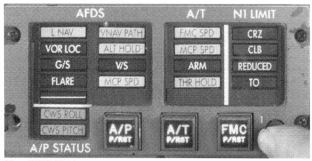
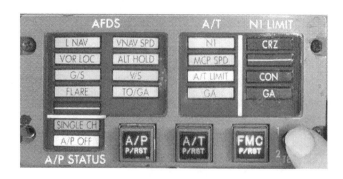

Non-EFIS FMA

The 737-300s were soon available with EFIS, an option which most operators took. The EADI included a speed tape, radio altimeter, groundspeed indicator, and FMA annunciations. The EHSI has a selectable display either to represent the old HSI or a moving map display. See navigation section for details.

737-500 EFIS Captains flight instruments

The flight instruments use information from 2 Air Data Computers -Classics / Inertial Reference Units -NGs, which have separate pitot-static sources. The ADC/ADIRUs are powered whenever the AC buses are powered.

737-3/4/500 Cockpit upgrades

These two cockpit upgrades have been designed to offer 737 Classic operators with the reliability, safety and operational capabilities of a 737NG, at a fraction of the cost. They both feature pilot selectable PFD/NDs similar to the 737NG. The flat panels replace 20 year old, technology which is becoming difficult and expensive to support and also give a ten-to-one improvement in reliability as well as a platform for growth.

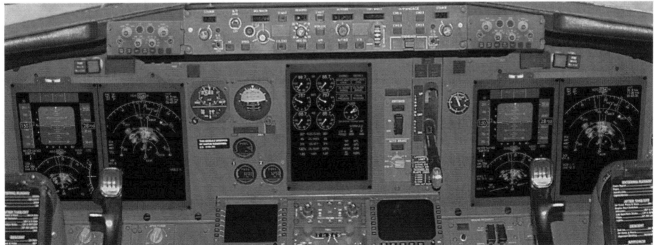

The "Cockpit/IP Flat Panel Display System"   *Photo: IS&S*

The "Cockpit/IP Flat Panel Display System" from Innovative Solutions & Support comprises five DUs (2 PFD/NDs and an EIDS), two CDUs (either side of the MCP) and three data concentrator units.

The "Advanced Cockpit"   *Photo: Universal Avionics*

The "Advanced Cockpit" is a joint venture between ARC Avionics, Commercial Jet, DAC International and Universal Avionics. Although it does not have the EIS, it is a fully integrated package - EFIS, FMS, Synthetic Vision, and Terrain Awareness & Warning System (TAWS) with extensive software to work with existing autopilot, flight director and short range Nav systems.

Each panel has its own CDU located below it. Dual UNS-1F FMS is installed in anticipation of RNP-1, P-RNAV, and WAAS LPV Class III approaches. A single Smiths FMC has been retained for performance and the autothrottle. The Class A TAWS display is pilot selectable between 3D Perspective View, Map View and Vertical Profile View and has the FMC route overlaid on the display. The Synthetic Vision displays Egocentric view (a 3-D view of the terrain ahead, as if the pilot was looking out the flight deck window) on the PFD and Exocentric view (a 3-D view of the aircraft with respect to the flight plan and surrounding terrain) on the ND.

In addition, an Application Server Unit which provides weather graphics, charts, checklists and E-Docs to be displayed on the ND has been certified.

EFIS

If display unit cooling is lost, then after a short time the Electronic Attitude Display Indicator (EADI) colours will appear magenta and the WXR DSPY caption will be shown on the EHSI. This can be rectified by selecting ALTERNATE equip cooling supply and/or exhaust fans. The NGs use Honeywell flat panel displays rather than the CRTs of the Classics and have the advantages of being lighter, more reliable and consume less power, although they are more expensive to produce.

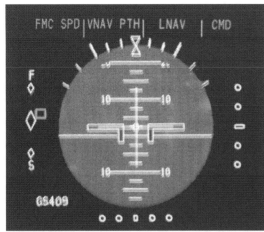

This display is usually seen on older examples of 737s or with operators of mixed fleets of originals or non-EFIS Classics. The main difference is a fast - slow pointer instead of a speed tape to mimic the mechanical ADI.

The 737-3/4/500 EADI display, with fast/slow indicator.

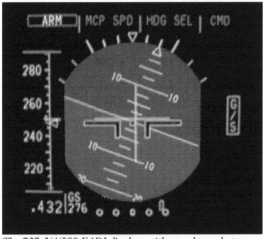

This display was the first to use the speed tape and trend arrow. The tip of the trend arrow shows the predicted speed in 10 seconds. The boxed G/S flag shows that the glideslope display has failed.

The 737-3/4/500 EADI display, with speed tape but no rolling digit curser.

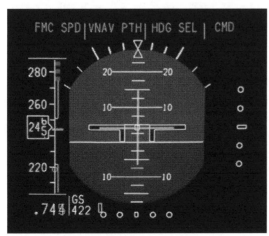

This is the most common EADI in use. Much information is shown on here which greatly helps the instrument scan. It is particularly useful for an ILS approach because no other flight instrument need be referred to, except perhaps N1.

The green circle on the speed tape is the flaps up manoeuvring speed and is only displayed when the flaps are up.

The 737-3/4/500 EADI display, with speed tape.

NG Flight Instruments

The NGs have 6 **Display Units** (DUs). These display the flight instruments; navigation, engine and some system displays. They are controlled by 2 computers - **Display Electronics Units** (DEUs). Normally DEU 1 controls the Captains and the Upper DUs whilst DEU 2 controls the F/O's and the lower DUs. The whole system together is known as the **Common Display System** (CDS).

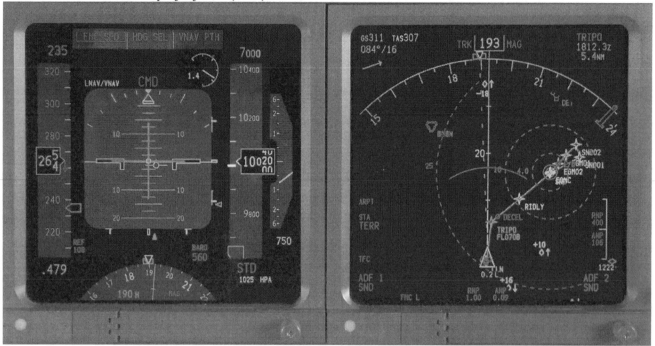

Capts outboard (L) & inboard (R) DUs in the most usual PFD/ND display mode.

The DUs normally display the PFDs outboard, NDs inboard, engine primary display centre (upper) and engine secondary display lower. However some mixed fleet operators have them configured to a round-dial style display to assist standardisation across the different generations.

F/Os outboard (R) & inboard (L) DUs in the optional "Round dial" display mode.

PFD Angle of Attack display

One of the many customer PFD options is an analogue/digital angle of attack display. The red line is the angle for stick shaker activation, the green band is the range of approach AoA.

The speed tape shows minimum and maximum operating speeds. The maximum operating speed provides a 0.3g manoeuvre margin to high speed buffet. The minimum operating speed is computed from the SMYD as follows: The SMYD has two types of Min manoeuvre speed. The first is identified as Vmnvr, the second as Vbl (low speed buffet). The transition from Vmnvr to Vbl is dependent on gross weight, but in general Vmnvr is output below 22,000 feet and Vbl above that altitude. Although not used directly in the calculation of Vmnvr, once the airplane starts flying, gross weight becomes a factor indirectly (in the calculation of Vmnvr) via the load factor calculation. FMC Gross Weight is used by the SMYD in the switching logic from Vmnvr (min man speed) to Vbl.

The CDS FAULT annunciation will only occur on the ground prior to the second engine start; it is probably a DEU failure, but is in any case a no-go item. If a DEU fails in-flight, the remaining DEU will automatically power all 6 DUs and a DSPLY SOURCE annunciation will appear on both PFDs. The displays can be switched around to almost any other position with the DU selector switches in case of a DU failure.

DU Selector switches – NG

The nomenclature requirements for these annunciations were developed by Boeing Flight Deck Crew Operations engineers during the early design phase of the 737NG program. The intent of the design function is as follows:

The CDS FAULT message is intended to be activated on ground to tell the maintenance crew or air crew that the airplane is in a non-dispatchable condition.

The DISPLAY SOURCE message is annunciated in air to tell the crew that all the primary display information is from one source and should be compared with all other data sources (standby instruments, raw data, etc.) to validate its accuracy.

Since the DISPLAY SOURCE message is intended to be activated in air and CDS FAULT is intended to be activated on ground, air/ground logic is used by CDS to determine which message is appropriate. The air/ground logic system uses a number of inputs to determine airplane state. One of the inputs used is "engines running". CDS uses the "engines running" logic as the primary trigger for changing the CDS FAULT message to its in-air counterpart. The "engines running" logic is used in case the air/ground data is incorrect as a result of other air/ground sensing faults.

Instrument transfer switches - NG

The DISPLAYS - SOURCE selector is only used on the ground for maintenance purposes (to power all 6 DUs from either DEU 1 or 2). This may be why the switch is a different shape to the other three; if not, it is still a good way to remember that this is a switch that pilots should not touch!

The DISPLAYS CONTROL PANEL annunciation merely indicates that an EFIS control panel has failed. There is an additional, rather odd, attention getter because the altimeter will blank on the failed side, with an ALT flag, until the DISPLAYS - CONTROL PANEL switch is positioned to the good side. Note that this is not the same as the EFI switch on the -3/4/500s which was used to switch symbol generators.

CDS Block Points

The 737NG Common Display System has had several software updates to incorporate additional features, improvements to existing features and bug fixes. Each new update is known as a Block Point, here is a list of their features to date:

BP 98, Dec 1998

- "IAS DISAGREE" message added.
- "ALT DISAGREE" message added.
- Low Airspeed Alert when the ADIRU CAS is 70% of the displayed minimum maneuver speed
- Flight Path Vector (FPV) function activated.
- EGPWS display enabled.
- Predictive Windshear display enabled.
- "Pop–up" display of TCAS, PWS and EGPWS information when not selected for display.

BP 99, Nov 1999

- Angle of attack (AOA) display on the PFD where the round dial radio altimeter was displayed.
- Increased Auxiliary Fuel Quantity Indication and Auxiliary Fuel Quantity Alerting
- "CDS FAULT" reverse video option
- Dispatch with "CDS MAINT" flag displayed
- In-flight cross bleed start envelope for Double Annular Combustor (DAC) engines – bug fix.
- Display of peaks and obstacles information
- Polar Navigation capability (when a polar navigation capable FMC is installed.)
- Control Surface Position Information display on the systems page
- Airspeed filter display change
- "CDS FAULT" resolution – bug fix
- ILS back course sense anomaly – bug fix

BP 02, Jan 2003

- Flaps 30 amber band
- Ground Speed indication on compacted EFIS/MAP display format
- AOA Vane-to-Body green band and flap reference coefficients revised for the 737–900/-700IGW/-700C airplanes.
- CDS and FMC flap manoeuvre speed differences corrected.
- "ILS" PFD source annunciation

Options:
- CAT IIIB related annunciations
- Quiet Climb annunciation – "Q-CLB".
- Double Derate (Common N1)
- VNAV speed bands – give an indication of the acceptable airspeed range during idle VNAV PTH descents.
- Flap/reference speed

- Horizon line heading scale - a heading scale aligned with the horizon line pitch and roll axis.
- NAV Performance Scales (NPS)
- Selectable compacted engine format
- Vertical Situation Display (VSD)
- GPS landing system
- Integrated Approach Navigation (IAN)
- ATC annunciation
- MDA Upper Limit raised to 20000 feet to support landings at very high altitude airports.
- CWS R and CWS P indications relocated directly beneath the Roll Mode Annunciation (FMA) box to make room for new annunciations.
- Mode change highlight symbol

BP 04, Sep 2005

- Round dial radio altitude option, format toggle improvement
- Vertical Situation Display option, improvements
- Navigation Performance Scale option, improvements
- Integrated Approach Navigation option, improvements
- VREF +15 bug becomes VREF+20 for longer fuselage models
- Adaptive airspeed filter improvement – airspeed indications smoother during gusts.
- Course disagree alert logic improvement
- GPS landing system option improvements

Options
- WXR automatic tilt mode
- LNAV armed roll mode annunciation
- Frequency and Course disagree alerting on EFIS/Map
- Engine oil quantity display in gallons or percent
- Fuel totalizer - a digital readout of individual tank quantity plus a separate total fuel readout.
- Limited COMM messages - ".ACARS" and ".ATC".
- Low fuel threshold – 1000lbs (from 2000lbs).

BP 06, Sep 2006

- Coefficients for flap bug speeds and the AOA for new Short Field Performance (SFP) minor model 737s (OPCs for winglets, 2-postion tail skid)
- Low Airspeed Alert on EFIS/MAP
- AOA Cross Compare Annunciation
- LNAV and VNAV Armed Before Takeoff
- Operational Program Software Data Compression

Options:
- Flight Deck Entry Video Surveillance
- Auxiliary Tank and Fuel Totalizer display
- RDR-4000 Weather Radar Display

New Approach Formats

With increased navigational accuracy available and hardware/software improvements on the 737, many new types of approaches have been developed. Cat IIIb, LNAV/VNAV, RNAV(GPS), RNAV(RNP), IAN, GLS.

Cat IIIb ILS Is very similar to the current ILS display except that rollout guidance will display as "ROLLOUT" (armed) underneath the VOR/LOC annunciation. A new MFD button labelled "C/R" (Clear/Recall) is required to display system messages on the upper display unit. These messages could be either "NO LAND 3" or "NO AUTOLAND". Note Cat IIIa is still possible with a NO LAND 3 advisory. In this case green "LAND 2" annunciations will appear on both PFDs.

LNAV/VNAV most non-precision approaches which are in the FMC database may be flown to MDA in LNAV/VNAV. Look for the coded GP angle in the LEGS pages.

Navigation Performance Scales (NPS) combine the display of ANP/RNP (see page 184) with LNAV/VNAV deviation to give either a Cat I approach of its own or a transition to an approach. Note: NPS provides crew awareness of airplane position with respect to the intended path and RNP. They are not required for VNAV approaches, which may be flown with standard displays.

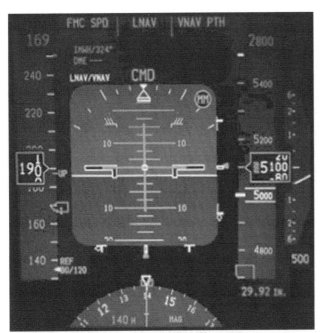
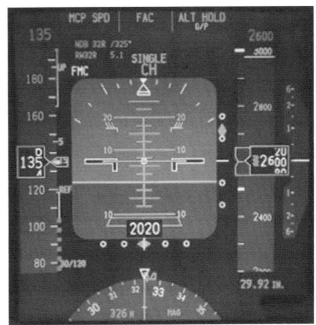

The PFD with NPS during an LNAV/VNAV transition   *The PFD with IAN during an NDB approach*

Integrated Approach Navigation (IAN) gives an ILS look-alike display and allows the pilot to fly the approach like an ILS, ie by selecting APP on the MCP. It is a Cat I only approach system which uses the FMC to transmit IAN deviations to the autopilot and display system. Flight path guidance is from navigation radios, FMC or a combination of both. The type of approach must first be selected in the FMC. The flight mode annunciations will vary depending upon the source of the navigation guidance as follows:

	Approach	FMA	ADI ID
Localiser based approaches:			
	ILS	VOR/LOC & G/S	ILS
	GLS	VOR/LOC & G/S	GLS
	ILS with G/S out, LOC, LDA, SDF	VOR/LOC & G/P	LOC/ G/P
	B/C LOC	BCRS & G/P	LOC/ G/P
FMC course guidance:			
	GNSS, RNAV	FAC & G/P	FMC
	VOR, NDB, TACAN	FAC & G/P	FMC

Where FAC = Final Approach Course and G/P = Glidepath.

GNSS Landing System (GLS) Approaches use GPS and a ground based augmentation system (GBAS) to give signals similar to ILS signals and will probably replace ILS in the future. Certified May 2005, it is initially Cat I but will become Cat IIIB and should have the capability for curved approaches.

Most of the above approaches require FMC U10.5+, CDS BP02+, FCC -709+ and DFDAU & EGPWS.

Instrument Transfer

Overheating of an individual display unit will cause that unit to blank until it cools down when it will return. If 2 display units on one side blank then the problem is with that symbol generator, SG FAIL will annunciate in the centre of both displays. The display can be restored by using the EFIS transfer switch. This will enable the remaining symbol generator to display onto both sides; the output is controlled through the EFIS control panel of the good side. Caution: the autopilot will disengage when the EFI switch is repositioned. If either Nav receiver fails, the VHF NAV transfer switch may be used to display the functioning Nav information onto both EFIS and RDMIs. With Nav transferred, the MCP course selector on the serviceable side becomes the master, but all other EFIS selections remain independent. If an IRS fails, the IRS transfer switch is used to switch all associated systems to the functioning IRS

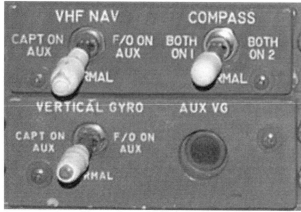

Instrument transfer switches – 1/200

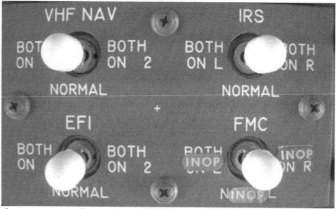

Instrument transfer switches - 3/4/500

Manual Selection of N1 and Speed Bugs

Normally the N1 and V-speed bugs are set by the FMC (the AUTO position). However, since it is permissible to despatch with the FMC inop or if it failed in-flight, facility is provided to set the bugs manually using data from the QRH. NGs use the panel shown here, on Classics this is done by pull to set knobs on the primary EIS and ASI.

Manual N1 & speed bug selection knobs (NG)

Head-Up Guidance System (HGS)

HGS, was certified for the 737 by the US FAA in 1994 to allow Cat IIIA landings down to 200m RVR and take-offs in 90m. The first production 737 HGS was fitted to a 737-300 of Morris Air (later bought by Southwest) delivered September 1995.

The HGS comprises of an overhead unit which contains the CRT and projector; a combiner, also known as the Head Up Display (HUD), which combines the projected display with the outside view; a control panel for data entry and to select mode of operation and an annunciator panel on the F/Os instrument panel.

HGS Control Panel

HGS Annunciator panel

Standby Flight Instruments

The standby airspeed indicator & altimeter use auxiliary pitot & alternate static sources, not the ADC/ADIRUs. The altimeter is fitted with a vibrator, which can be heard when the flightdeck is quiet, to prevent errors from mechanical linkage friction.

Classic/NG standby flight instruments

The Integrated Standby Flight Display

The Integrated Standby Flight Display was introduced in 2003 to replace the mechanical standby artificial horizon and ASI/altimeter. Personally, I find the new ASI & altimeter much easier to read but the ILS more difficult. The + - buttons are brightness controls.

The ISFD also sends inertial data to the FCCs which use the data during CAT IIIB approaches, landings and go-around.

Surprisingly, the ISFD cannot be switched off from the flightdeck - even by pulling the ISFD c/b on the p18 panel. It has its own dedicated battery and the ISFD c/b only removes power from the battery charger. Let us hope that one does not start to smoke in-flight! The battery will give 150 minutes of power.

Electronic Flight Bag (EFB)

A Class 3 EFB

EFB is becoming the latest "must-have" device in the cockpit. They have the ability to do the following tasks:
- Calculate take-off or landing performance.
- Calculate weight & balance.
- Contain the aircraft technical log.
- Store navigation charts & plates.
- Store company manuals, FCOMs, crew notices, etc.
- Retrieve & display weather.
- Display checklists.
- Display on-board video surveillance cameras.

The advantages to crew are the accuracy of the data and ease of use. The advantages to the airlines are the cost benefits of a less paper cockpit and real time data transfer.

There are three classes of EFB:
- Class 1: Fully portable. Eg a laptop.
- Class 2: Portable but connected to the aircraft during normal operations. Eg tablet & docking station.
- Class 3: Installed (non-removable) equipment.

Flight Data Recorder (FDR)

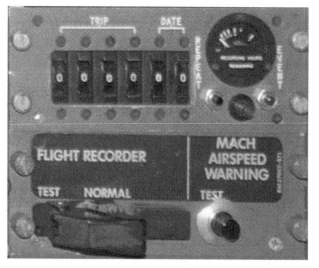

737-200 FDR panel

The FDR is located in the ceiling above the rear galley. There have been several different models of FDR in the life of the 737. They can collect anything from 30 minutes to hundreds of hours of data and 8 to hundreds of parameters. Most FDRs start recording as soon as the first engine oil pressure rises.

Early FDRs, such as that fitted to 737-200s, comprised metal scribes which etched their data into a 150ft long roll of metal foil. These would last about 300hrs but only recorded vertical acceleration, heading, IAS and altitude, plus binary traces such as date, flight number and time of R/T transmissions. A gauge on the panel (see left) shows the recording hours remaining before the foil spool needs replacing.

Later Digital FDRs (late 200s & Classics) record onto a 1/4-inch wide, 450 feet long magnetic tape and the newest Solid State FDRs (later Classics & NGs) record data onto memory chips.

Late model 737-3/4/500 FDRs record 25 hours of data. The protective casing includes an inner aluminium cover, isothermal protection shield, an outer stainless steel casing and an exterior stainless steel dust cover. This enables it to withstand a crush force of 20,000 pounds per axis and provides impact protection of 1000g for 5 msec. It is protected from heat by an isothermal insulation which maintains the inner chamber at a safe temperature. It also has an underwater location device that transmits under water for a minimum of 30 days.

The FDR of the 737-NG is similar to that described above but can withstand 3400g of impact, 20,000ft depth of water and temperatures of 1,100C for 30mins.

Limitations

Altitude Display Limits for RVSM Operations

- Standby altimeters do not meet altimeter accuracy requirements of RVSM airspace.

- The maximum allowable in-flight difference between Captain and First Officer altitude displays for RVSM operations is **200 feet**.

- The maximum allowable on-the-ground altitude display differences for RVSM operations are:

737-1/500

Field Elevation	Max Diff between Captain & F/O	Max Diff between Captain or F/O and Field Elevation
Sea Level	40 ft	75 ft
5,000ft	45 ft	75 ft
10,000ft	50 ft	75 ft

737 NG

Field Elevation	Max Diff between Captain & F/O	Max Diff between Captain or F/O and Field Elevation
Sea Level to 5,000ft	50 ft	75 ft
5,001 to 10,000ft	60 ft	75 ft

HGS System

- Option - With HGS 4000 Phase I: AIII mode approach and landings are not approved for airplanes with Flight Dynamics Model 4000 Phase I HGS installed.

- With HGS 2350 and polar navigation: Do not use HGS System at latitudes greater than 85 degrees latitude or when the Heading Reference Switch is in the TRUE position.

Integrated Approach Navigation (IAN)

Do not use Integrated Approach Navigation Final Approach Course or Glide Path guidance when any altitude constraint specified by the approach procedure for a final approach fix, or for waypoints between a final approach fix and a runway, has been modified by the flight crew.

FLIGHT MANAGEMENT COMPUTER

First introduced on the series 200 in Feb 1979 as the Performance Data Computer System (PDCS), the Flight Management Computer (FMC) was a huge technological step forward. Smiths Industries (formerly Lear Seigler) of Grand Rapids, Michigan has supplied all FMCs installed on the 737.

The PDCS was developed jointly by Boeing and Lear Seigler in the late 1970's. It enabled EPR and ASI bugs to be set by the computer and advise on the optimum flight level, all for best fuel economy. It was trialled on two in-service aircraft, a Continental 727-200 and a Lufthansa 737-200 for nine months in 1978 with regular line crews and a flight data observer. The 737-200 showed average fuel savings of 2.95% with a 2 minute increase in trip time over an average 71 minute flight. The 727 gave a 3.94% fuel saving because of its longer sector lengths. The PDCS quickly became standard fit and many were also retrofitted. By 1982 the autothrottle had been devised and thrust levers could be automatically driven to the values specified by the PDCS.

The true FMC was introduced with the 737-300 in 1984. This kept the performance database and functions but also added a navigation database which interacts with the autopilot & flight director, autothrottle and IRSs. The integrated system is known as the Flight Management System (FMS) of which the FMC is just one component. Most aircraft have just one FMC, but there is an option to have two, this is usually only taken by operators into MNPS airspace eg Oceanic areas. The FMS can be defined as being capable of four dimensional area navigation (latitude, longitude, altitude & time) while optimising performance to achieve the most economical flight possible.

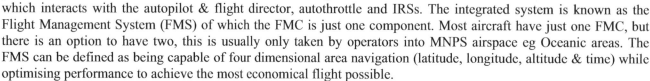

The photograph above is of a Control Display Unit (CDU), which is the pilot interface to the FMC. There are 2 CDUs but normally only one FMC, think of it as having two keyboards connected to the one PC. A CDU is recognisable by having a DIR INTC key and the space key is blank. A Multifunction CDU (MCDU) has a MENU key to give access to the subsystems such as FMC, ACARS, DFDAU, ALT NAV, etc and a space (SP) key for ACARS and ATC datalink message writing. CDUs can be either CRT (shown above), which display green print; or LCD, which display grey print. The LCD CDUs have the facility to display in colour; you can see this briefly during the boot-up sequence (shown left). Update 10.2+ allows the option of a colour CDU display (see page 337).

In its most basic form, the FMC has a 96k word navigation database, where one word is two bytes (ie a 16 bit processor). This was increased to 192k words in 1988, 288k in 1990, 1 Mega word in 1992 and is now at 8M words for the 737-NG with Update 10.8. The navigation database is used to store route information which the autopilot will fly when in LNAV mode. When given data such as ZFW & MACTOW, it takes inputs from the fuel summation unit (or FQIS on NGs) to give a gross weight and best speeds for climb, cruise, descent, holding, approach, driftdown etc. These speeds can all be flown directly by the autopilot & autothrottle in VNAV mode. It will also compute the aircrafts position based upon inputs from the IRSs, GPS and radio position updating.

The latest FMC – Model 2907C1, has a Motorola 68040 processor running at 60MHz (30Mhz bus clock speed), with 4Mb static RAM and 32Mb for program & database.

FMC Inputs

The FMC gets inputs from the aircraft in the form of "sensors" and "discretes" to enable it to function and make any necessary computations. Sensors include: VOR, DME, ADIRS, MMR (ILS & GPS), DFCS, FQIS, Clock and CDS DEU. Discretes are listed at INIT / REF INDEX - MAINT – FMCS – DISCRETES as follows:

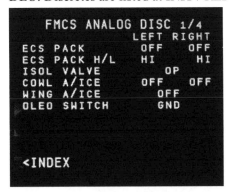

Page 1 is essentially pneumatics.

This shows pack valve positions and flow modes. Cowl and wing anti-ice simply look at switch position on the P5 overhead panel. "Oleo switch" shows if the aircraft is in the ground or air mode. This input comes from the proximity switch electronic unit (PSEU).

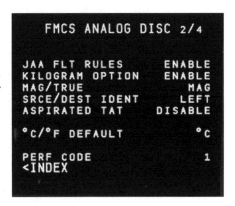

Page 2 shows some customer and hardware options.

FLT RULES - can be FAA or JAA; this sets the default CRZ CG (shown on PERF INIT page) for the applicable regulatory flight rules. See page 285 for more details.
ASPIRATED TAT - is enabled when an aspirated TAT probe is installed. This will display the OAT on the N1 LIMIT page.
PERF CODE - This sets the performance code for the FMC, it is normally 1.

![FMCS ANALOG DISC 3/4 screen showing: MODEL/ENG VALID; NO.1 NO.2; ENGINE BLEED ON ON; SEL CONFIG MODE ENABLE; <INDEX]

Page 3 shows the things set through the programmable switch modules.

MODEL/ENGINE - Shows the current status of the engine / airframe program pins to the FMC.
ENGINE BLEED - Shows the position of the engine bleed air switches on the P5 overhead panel. I don't know why this is not on page 1 with the other pneumatics.
SEL CONFIG MODE - This enables the loadable software configuration.

Page 4 shows FMC and CDS display options:

VOR INHIBIT - Disables VOR inputs for navigation.
FLIGHT NUMBER - Allows entry of the flight number on the RTE page.
TOGA RW POS UPD - Lets the FMC do a position update when the TOGA switches are pushed on the ground.
TAKEOFF PROFILE - Gives FMC control of the altitude at which takeoff thrust is reduced to climb thrust.
TAKEOFF SPEEDS - Enables the FMC to calculate the takeoff speeds based on the QRH and shows them on the TAKEOFF page.
NAVAID SUPPRESS - Suppresses autotuned navaids on the CDS map display.
SEL CRS INHIBIT - Suppresses selected course radials for manually tuned navaids on the CDS map display.
ACARS INSTALLED - This option enables the ACARS/FMC interface.

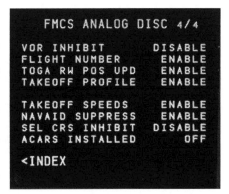

FMC Databases

An FMC has three databases: Software options (OP PROGRAM), Model/Engine Data Base (MEDB) and Navigation Data Base (NDB), all of which are stored on an EEPROM memory card. These databases can all be updated via the data loader (see page 73).

Model/Engine Data Base (MEDB)

The MEDB holds all the performance data for V speeds, min & max speeds in climb, cruise & descent, fuel consumptions, altitude capability etc. This can be refined for individual aircraft by the Performance Factors page at INIT / REF INDEX - MAINT – FMCS – PERF FACTR. To change any data on this page you must enter "ARM" into 6R, this should not normally be done by crew.

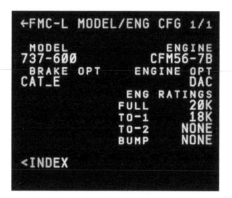 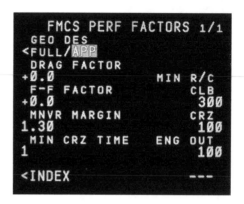

Software Options (OP PROGRAM)

The Software options database includes the operational program and its update (eg FMC Update 10.7), plus any company specified options. The enabled options can be viewed from INIT / REF INDEX - MAINT – FMCS – SW OPTIONS. For a full list of all FMC software updates and their features, see page 327.

FMC Software Options	
ABEAM WAYPOINTS (U10.0+)	GROSS WEIGHT INHIBIT (U10.4+)
ADDITIONAL FIX PAGES (U10.6+)	High idle descent (U7.0+)
ALTERNATE DESTINATION (U10.0+)	INTEGRATED APPR NAV (U10.5+)
ALT/SPD INTERVENTION (U10.0+)	MANUAL RNP ENTRY (U7.0+)
AOC DATALINK (U10.0+)	MANUAL TAKEOFF SPEEDS (U7.0+)
ATC light & chime on FIXED OUTPUTS (U10.4+)	MESSAGE RECALL (U10.0+)
ATC message on map displayed (U10.4+)	MISSED APPROACH COLOR (U10.4+)
CDU COLOR (U10.2+)	Optional quieting gradient (U10.3+)
CERTIFIED TAKEOFF SPEEDS (U7.0+)	Pilot defined company routes (U10.3+)
COMMON VNAV (U10.6+)	PLAN FUEL ENTRY (U10.0+)
DEFAULT DME UPDATE OFF (U10.?+)	Position update – runway offset (U7.0+)
DIRECT GPS INPUT (U10.0+)	QFE ALTITUDE REFERENCE (U10.0+)
ENGINE OUT SID (U10.3+)	Quiet climb system (U10.3+)
FANS-1 ATS DATALINK (U10.4+)	RWY OFFSET IN FEET/METERS (U7.0+)
FMS RNAV ILS look alike approach (U10.5+)	RWY REMAINING IN FEET/METERS (U7.0+)
GEOMETRIC PATH DESCENT (U10.3+)	SATCOM interface (U10.0+)
GPS landing system approach (U10.5+)	TAKEOFF DERATE INHIBIT (U10.4+)
GPS operational mode bits 1 & 2 (U7.0+)	Vertical RNP values loadable by default (U10.5+)
GPS WITH INTEGRITY (U10.0+)	VNAV ALT (U8.0+)

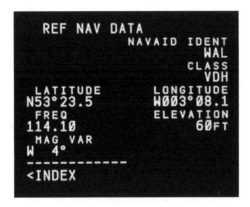

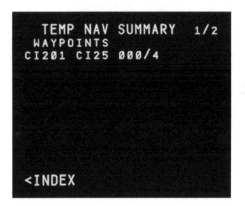

Navigation Data Base (NDB)

The NDB is comprised of Permanent, Supplemental and Temporary. The Permanent database cannot be modified by crew. There are four types of data: Waypoint, Navaid, Airport and Runway. Runway data is only held in the permanent database.

There is capacity in the SUPP and REF databases for up to 40 waypoints, 40 navaids and 6 airports. SUPP data can only be entered on the ground. It is then stored indefinitely but crew may delete individual data or the whole database. Any existing SUPP data should be checked for accuracy before flight using the SUMMARY option (U6+ only) or DELeted and re-entered, cross-checking any Lat & Longs between both crew members. All Temporary (REF) data is automatically deleted after flight completion.

When entering navaids into either the REF NAV DATA or SUPP NAV DATA database, you will be box prompted for a four letter "CLASS" classification code. The following table should be used:

Navaid Classification Codes	Box Numbers			
VHF Navaids	**1**	**2**	**3**	**4**
VOR	V			
TACAN Ch 17-59, 70-117		T		
Military TACAN Ch 1-16, 60-69		M		
DME		D		
ILS/DME		I		
Terminal			T	
Low Altitude			L	
High Altitude			H	
Use unrestricted by range or altitude			U	
Scheduled Weather Broadcast				B
No Voice on Navaid Frequency				W
Automatic Trans Weather Broadcast				A

Eg. To create a new en-route VOR/DME, the Class code would be VDHW.

FMC Pages

Pre-Flight Preparation Pages

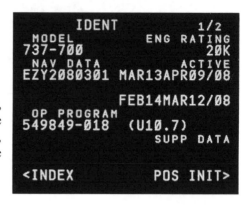

IDENT: This may be considered as the title page.
Check that the aircraft model and engine rating is what you are flying, especially if your airline operates a mixed fleet. Also check the database effectivity and expiry dates. Software update number is given in brackets, here U10.7. U10 was specifically designed for the 737-NG but can be used on Classics.

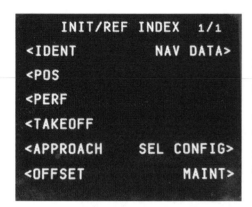

If IDENT can be regarded as the title page, INIT/REF is the contents. The SEL CONFIG & MAINT options are only available on the ground, see page 144 for some of these pages.

The NG, U10+ (shown left) has some extra options available:
The OFFSET function is used in-flight to fly parallel to a portion of the route. This may be used to avoid the turbulent wake of the aircraft ahead or to increase separation with opposite direction traffic.

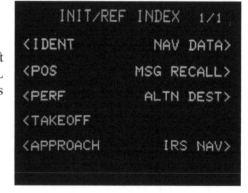

This INIT/REF INDEX page (shown right) is from an U6.2 aircraft equipped with ACARS and IRS navigation. Selecting MSG RECALL allows the recall of deleted CDU scratchpad messages whose set logic is still valid. ALTN DEST allows the entry of selected alternate airports.

POS INIT is used to enter the aircraft position into the IRSs for alignment. The Lat & Longs of REF AIRPORTS & GATES are held in the database and do not need typing in but should be cross-checked against published data. The date & time information comes from the Captains clock or the GPS if installed.

Routes are usually entered by inputting the CO ROUTE. Eg AMSLPLRPL = Amsterdam to Liverpool Repetitive Flight Plan. If the company route is not recognised the route can be entered manually by filling in the VIA & TO columns.

In-flight an OFFSET feature is available on U7 onwards. See page 143 to use this to access the hidden NEAREST AIRPORTS and ALTERNATE DESTS pages.

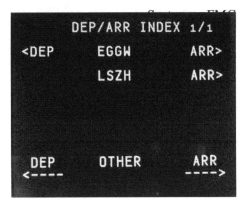

After the route is entered, press the DEP ARR button on the CDU to access the departures and arrivals for the ends of the route.

You can select the SID/STAR, transition, type of approach and runway from the available list. They are denoted by <ACT> when they have been selected.

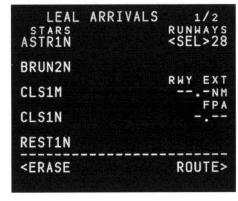

If the approach was not already in the database, selecting the runway will allow the input of a final approach distance (RWY EXT) and, from U10.6, flight path angle (FPA).

After the route has been entered it is worthwhile stepping through the LEGS pages with PLAN mode selected on the EHSI/ND to check the route for errors or route discontinuities.

Notice the DME arc between waypoints 2 & 3. This must be pre-programmed by your database supplier (ie not constructed by you) for RNAV operations.

The PERF INIT page is completed before fight and gives the FMC the data to calculate leg times & fuels and the optimum & max altitudes.

Cost Index is the ratio of the time-related operating costs of the aircraft vs. the cost of fuel. If CI is 0 the FMC gives maximum range airspeed and minimum trip fuel. If CI is max the FMC gives Vmo/Mmo for climb & cruise; descent is restricted to 330kts to give an overspeed margin. The range of CI is 0-200 (Classics) and 0-500 (NGs).

See page 285 for details about CRZ CG entry.

CRZ WIND is actually used for both climb wind and cruise wind. If you enter the forecast top-of-climb wind before departure the FMC will recalculate your climb speed accordingly. When in the cruise if you enter the average cruise wind, the time and fuel calculations will be updated. If you do not enter anything here the FMC will assume still air.

Standard take-off page for PF. Speeds & assumed temperature must be entered manually from performance tables. TO SHIFT should be entered if departing from an intersection as this is used to update the FMC position when TOGA is pressed at the start of the take-off roll (probably not necessary on GPS equipped aircraft).

This screenshot is from an U10.x FMC and shows more data than previous updates, eg optional QRH speeds and a stab trim based upon your entered MACTOW.

```
     TAKEOFF REF    1/2
 FLAPS            QRH    V1
 5°               127   122
 RED 20K N1              VR
 87.1/ 87.1       128   129
 CG      TRIM            V2
 26.1%   5.12     132   133
                    GW / TOW
                 58.0/
 RUNWAY           TO SHIFT
 RW27             RW27 -00M
 --------------------SELECT
 <INDEX           QRH OFF>
```

```
     TAKEOFF REF    2/2
 RW WIND          RW COND
 350°/  3      DRY/WET SK-R>
 RW SLOPE/HDG     SEL/OAT
 U0.1%/257°       +30/ +15°C

 ACCEL HT      EO ACCEL HT
 1500AGL          1000AGL
              THR REDUCTION
              CLB  1500AGL
 -------------------------
 <INDEX
```

Take-off Ref 2/2 (U10.x only) allows you to enter runway conditions to update the QRH V speeds shown on page 1/2.

This photo is from U10.7 in which you can also set the heights for thrust reduction (cutback) altitude, acceleration and engine out acceleration. Quiet Climb parameters are also displayed on this page when installed (not shown here).

This is the N1 Limit page pre-U10. Provides thrust limit and reduced climb thrust selection. This is usually automatic but manual selections can be made here.

The most common use is either to select a reduced climb thrust (1 or 2) after a full power take-off to reduce engine wear or to delete the reduced climb thrust to get a high rate of climb.

```
     N1 LIMIT       1/1
 <AUTO <ACT>

 <GA             98.3/ 98.3

 <CON            98.3/ 98.3

 <CLB            98.3/ 98.3

 <CRZ            94.3/ 94.3
 ------REDUCED CLB--------
 <CLB-1            CLB-2>
```

```
     N1 LIMIT       1/1
 SEL/OAT       RED 20K N1
 +36/ +15°C    87.1/ 87.1
 20K
 <TO  <ACT> <SEL>    CLB>
 18K DERATE
 <TO-1             CLB-1>

                   CLB-2>


 -------------------------
 <PERF INIT      TAKEOFF>
```

From U10 onwards the N1 limit page has this very different appearance when on the ground, displaying similar information to the old Take-off Ref 2/2

In-flight it reverts to the regular N1 Limit display shown above.

One of the most useful pages in the FMC and has just been updated to an optional six pages in update 10.6. The fix can be anything in the database ie airfield, beacon or waypoint. An abeam point can be constructed (as illustrated) or a radial or range circle can be displayed on the EHSI.

```
     FIX INFO       6/6
  FIX   RAD/DIS FR
 ELLX    190/162
 RAD/ DIS ETA    DTG    ALT
 ---

 ---

 ---
 ABEAM
 230/125 1425.1 103 FL400
```

Climb Pages

Standard ECON CLB page (here 290Kts IAS) is highlighted indicating that this is the target speed; this will automatically change over to mach (0.775 in this case) during the climb. Other climb modes are available with keys 5 & 6, L & R.

```
 ACT ECON CLB        1/1
 CRZ ALT
FL370
 TGT SPD           TO FL370
290/.775      1654.4z/ 51NM
 SPD REST
---/-----
-----------         CLB N1
                 98.6/ 98.6%
<MAX RATE           ENG OUT>

<MAX ANGLE              RTA>
```

CLB-1 indicates that the autothrottle is commanding a reduced thrust climb power (reduces N1 by approx 3% = 10% thrust reduction). CLB-2 is a reduction of a further 10% ie 20% total. The reduced climb thrust setting gradually increases to full climb power by 15,000ft.

```
 ACT MAX RATE CLB  1/1
 CRZ ALT
FL370
 TGT SPD           TO FL370
258/.780         .  z/   NM
 SPD REST
---/-----
-----------         CLB N1
<ECON            98.6/ 98.6%
-----------
                    ENG OUT>

<MAX ANGLE              RTA>
```

Selecting MAX RATE from a climb page will show this page with the target speed not highlighted and the ERASE option available until the EXEC button is pressed. If EXEC is pressed you can return to ECON by line selecting it.

Selecting MAX ANGLE from a climb page will show this page with the target speed not highlighted and the ERASE option available until the EXEC button is pressed. If EXEC is pressed you can return to ECON by line selecting it.

```
 ACT MAX ANGLE CLB 1/1
 CRZ ALT
FL370
 TGT SPD           TO FL370
235/.780         .  z/   NM
 SPD REST
---/-----
-----------         CLB N1
<ECON            98.6/ 98.6%
-----------
<MAX RATE           ENG OUT>

                        RTA>
```

```
 ACT MAX ANGLE CLB 1/1
 CRZ ALT
FL370
 TGT SPD           TO FL370
235/.780      1654.8z/ 45NM
 SPD REST
---/-----
-----------         CLB N1
                 98.7/ 98.7%
-----------
<LT ENG OUT   RT ENG OUT>
```

Selecting ENG OUT on a climb page gives the following choice of left or right engines.

When the left (shown here) or right engine has been selected this page shows your MAX ALT (climb or driftdown alt), target speed and max cont N1. This is a useful page to check if flying over terrain above 15,000ft. Remember the altitude penalties for anti-ice.

Before U7.5, the EXEC light was illuminated; if you pressed it you lost all VNAV info. Select ERASE to return to the two engine CLB/CRZ pages.

```
  ENG OUT CLB    1/1
 CRZ ALT         MAX ALT
FL370             FL241
 ENG OUT SPD      CON N1
234KT             98.8%


-----------------------
 LT ENG OUT   RT ENG OUT>
```

The Boeing 737 Technical Guide

Cruise Pages

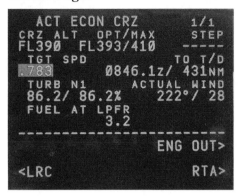

Standard PF in-flight cruise page. The target speed (highlighted) is the ECON speed which is derived from the cost index and winds.

Entering a different altitude in the STEP field will show you the fuel penalty / savings. If the EXEC button is then pressed and the MCP reset, the aircraft will climb / descend at 1000fpm to the new level.

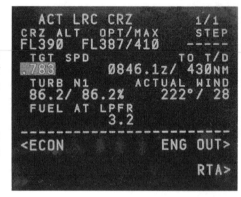

Selecting LRC gives Long Range Cruise speed. This is calculated as 99% of the maximum range speed for a given weight & altitude in still air conditions. It is used in preference to MRC because it is a more stable speed and hence gives less autothrottle movement.

LRC takes no account of winds so it may give a higher fuel burn than ECON. It also takes no account of operating costs; hence it has little practical value.

Selecting ENG OUT on the cruise (ECON or LRC) page will give this page which asks you to select which engine is inoperative.

After selecting the appropriate engine, the FMC will calculate your driftdown target speed (max L/D speed) and stabilisation altitude. It is worthwhile making this check before crossing over areas of high terrain. If doing this with pre- Update 7.5, do not press EXEC at any stage otherwise you will lose VNAV information, these actions can be undone by selecting ERASE.

140

```
   7253 PROGRESS 1/4
  FROM      ALT  ATA  FUEL
MAMUL    FL236 1645z  9.6
  110°      DTG  ETA  FUEL
AMVEL          17 1652z  9.1
  111°
DOLAS          56 1657z  8.8

EPKK          803 1839z  5.6
  TO T/C        FUEL QTY
1655z/    41NM        9.2
  WIND
229°/ 28KT   NAV STATUS>
```

Very useful howgozit page, but don't trust the fuel summation unit (FQIS NGs) - it often over-reads by approx 1-200kgs (2-400lbs). For an accurate arrival fuel, subtract 2-300kgs (4-600lbs) to allow for this and the drag of flying slower than the optimum speed whilst on the approach.

PROGRESS 2/4: Standard landing page for PF. Especially useful for giving crosswind component and SAT in icing conditions.
Confusingly this page used to be at Progress 3/3 before U10, so instead of selecting Progress-Prev Page you now have to select Progress-Next Page.

```
      7253 PROGRESS 2/4
  TAILWIND      CROSSWIND
13KT            R 24KT
  WIND        SAT/ISA DEV
231°/ 28      -37°C/+06°C
  XTK ERROR
R 0.43NM
  GPS-L TRK          TAS
109°T            449KT
```

```
ACT RTA PROGRESS   3/4
  RTA WPT          RTA
DOLAS          1656:41z
  RTA SPD     TIME ERROR
291/.775        ON TIME
  SPD REST          GMT
---/-----       1649:57z
DIST--TO DOLAS --ALT/ETA
  53NM    FL370/1656:41z
FIRST--RTA WINDOW---LAST
1656:31z        1657:55z
------------------------
<LIMITS
```

Very rarely used but can be used to try to reach a waypoint at a specific time. Selecting LIMITS takes you to the PERF LIMITS page (below).

RNP Progress displays RNP information. This page is showing us that the RNP at AMVEL is 2nm and our actual navigation performance (ANP) is 0.06nm. The lowest RNP for the approach is 0.50nm, this figure is from the database because it is written in small font, but can be manually overwritten if a different RNP is required.

```
     RNP PROGRESS    4/4
  111°       9.2NM
AMVEL          .775/FL329
  RNP/ACTUAL
2.00/0.06NM
  XTK ERROR
R 0.59NM



  RNP--APPROACH----------
0.50NM
```

```
  ACT PERF LIMITS   2/2
  TIME ERROR TOLERANCE
  30 SEC AT RTA WPT
  MIN SPD --CLB-- MAX SPD
100/.400        340/.820
        --CRZ--
100/.400        340/.820
        --DES--
100/.400        340/.820


------------------------
<INDEX              RTA>
```

Very rarely used. Sets min & max speed limits for climb, cruise & descent.

Descent Pages

FPA is the actual flight path angle of the aircraft. It is typically 3 to 4 degrees in a descent.

V/B is the required vertical bearing to reach the WPT/ALT ie COL/FL050. This would be an altitude restriction in the LEGS page that you could enter (or delete) manually or with the ALT INTV button on the MCP if fitted.

V/S is the required vertical speed to make good the V/B.

```
ACT 280KT PATH DES 1/1
 E/D ALT            AT COL
   351            228/FL050A
 TGT SPD           TO COL
 .738/280  1328.5z/  20NM
 SPD REST         WPT/ALT
 250/FL100       COL/FL050
 VERT DEV     FPA V/B  V/S
 1888LO       2.8 2.1 1017
------------------------
<ECON               SPEED>

<FORECAST             RTA>
```

```
ACT DES FORECASTS 1/1
 TRANS LVL     TAI ON/OFF
FL066         ----/----
 CABIN RATE   ISA DEV/QNH
313FPM         10°C/ 1018
 ALT-----WIND----DIR/SPD
FL260         330°/ 10KT

FL241         330°/  8KT

FL080         290°/  7KT
```

As you approach your ToD the FPA will remain at zero (because the aircraft is in level flight), but the V/B and V/S will both increase until the V/B is at about 3 to 4 degrees, or whatever the FMC has calculated is the optimum value. If VNAV is engaged the aircraft will descend with a FPA equal to the V/B. The actual V/S will, however, be slightly different to the computed V/S because V/S changes during the descent.

Use the forecasts pages to enter met data to compute a more accurate ToD.

FMC position can be forced to any of the other positions at this page.

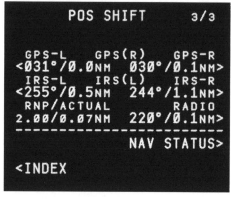

```
POS SHIFT      3/3



 GPS-L    GPS(R)     GPS-R
<031°/0.0NM  030°/0.1NM>
 IRS-L    IRS(L)     IRS-R
<255°/0.5NM  244°/1.1NM>
 RNP/ACTUAL        RADIO
2.00/0.07NM  220°/0.1NM>
------------------------
              NAV STATUS>

<INDEX
```

```
RTE   HOLD   1/1
 FIX         SPD/TGT ALT
NARBO         214/FL390
 QUAD/RADIAL    FIX ETA
N/020°          0817.2z
 INBD CRS/DIR   EFC TIME
200°/R TURN       ----z
 LEG TIME    HOLD AVAIL
1.5MIN            0+16
 LEG DIST    BEST SPEED
--.-NM           214KT
------------------------
<NEXT HOLD
```

You can either select a waypoint from the legs page or use PPOS (Present POSition) on which to base the hold. The database will give the turn direction & inbound course. Leg time will be that appropriate to your altitude.

SPD will default to the current (small font) speed & altitude or speed at the fix (large font)

HOLD AVAIL is calculated from the present fuel, fuel flow and reserves figure entered into the PERF INIT page.

BEST SPEED is the highest of max endurance, min manoeuvre speed or initial buffet.

```
APPROACH REF   1/1
 GROSS WT   FLAPS    VREF
 53.9        15°    132KT

             30°    127KT
EDDK32R
12500FT3810M 40°    124KT
 ILS 32R/CRS     FLAP/SPD
109.70IKEN/317°    30/127
                 WIND CORR
                    +05KT
------------------------
<INDEX
```

APPROACH REF: Standard landing page for PNF. Vref is calculated from the current gross weight, pedants may wish to overwrite the GROSS WT with the predicted GW for landing.

Vref is "hardened" by line-selecting it over itself which will cause it to be displayed on the speed tape.

Optional Features

Each version of FMC software comes with several customer selectable options, many of which carry a surcharge to activate. There are some ingenious ways of getting to these pages by careful use of simultaneous CDU entries.

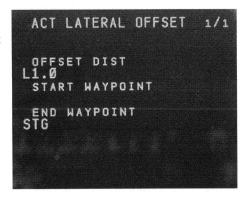

An alternative way to view the ALTERNATE DESTS pages is to start with the OFFSET page either from RTE or INIT/REF INDEX.

1. Have this page displayed on both CDUs.
2. Enter any offset but do not execute.
3. Simultaneously press ERASE on both CDUs.

Unfortunately this entry method was bug-fixed in U10.7.

You can enter up to 5 alternates here. Selecting 1R to 5R against any entered alternate will show the info below...

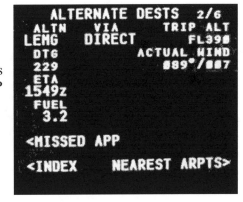

All the diversion data is now shown based on you flying direct to this alternate from present position (VIA DIRECT). Selecting MISSED APP will show the same data but calculated from the missed approach point. Selecting nearest airports will give...

This extremely useful page takes a couple of minutes to calculate but will list the nearest airports in the database in order of DTG.

Once again, line selection of 1R to 5R will give more useful diversion information as shown below.

All the diversion data is now shown based on you flying direct to this airport from present position (VIA DIRECT). Selecting MISSED APP will show the same data but calculated from the missed approach point. Selecting INDEX to return to the normal pages.

```
     NEAREST AIRPORTS  2/6
  ALTN      VIA       TRIP ALT
  LEAS    DIRECT        FL390
  DTG               ACTUAL WIND
  147                  229°/031
  ETA
 0810z
 FUEL
   5.4

 <MISSED APP

 <INDEX              PREVIOUS>
```

Maintenance Pages

The MAINT pages are used for running BITE (Built In Test Equipment) checks, monitoring sensors and setting up various other parameters, eg performance. Needless to say the uninitiated can do a lot of damage here, so these pages should not be accessed without engineering supervision. Examples of the MAINT pages in use are shown below and at pages 51, 67 and 293.

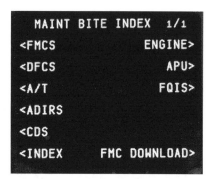

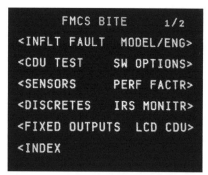

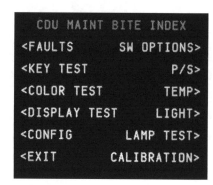

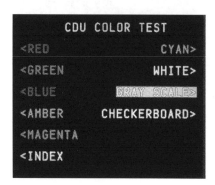

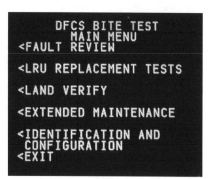

```
     FMCS  H/W CONFIG   1/1
  HARDWARE P/N
  171497-05-01
     SERIAL NO
       2538
  ETI              PROC SPEED
    .                  20 MHz


  RS422-----ISB---ETHERNET

 <INDEX
```

```
    ENGINE 2 BITE TEST
    RECENT FAULTS  2/2
       LONG TIME

 MSG NBR: 73-21352
 THE APL VOLTAGE
 INPUT TO THE EEC
 IS OUT OF RANGE
 FLIGHT LEG (X=FAULT SET)
    0 1 2 3
    X X

 <INDEX          HISTORY>
```

```
      APU BITE TEST
  CURRENT STATUS     1/1
  MAINTENANCE LIGHT
  MAINT MSG  49-41244
  START CONVERTER SHOWS
  FAILED GENERATOR DIODE

  RUN APU AND LOAD APU
  GENERATOR TO CONFIRM
  REPAIR
                    OTHER
 <INDEX       OCCURRENCES>
```

```
    FMCS SENSOR STATUS 1/2
   LRU       LEFT      RIGHT
  VOR        OK          OK
  DME        OK          OK


  ADIRS      OK          OK
  MMR        OK          OK
  DFCS       OK         ----

  FQIS       OK         ----
  CLOCK      OK         ----
 <INDEX
```

The following example shows how maintenance pages are used. The aircraft was in level flight at FL310 with autopilot A engaged when it dropped out without pilot input. After landing the crew called an engineer who stepped through the following pages to ascertain the fault:

INIT/REF INDEX > MAINT (page only available on the ground) > DFCS > FAULT REVIEW >

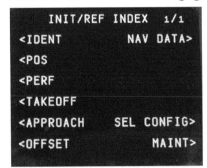
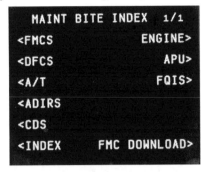
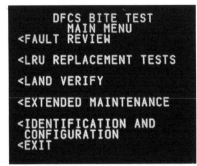

FAULT HISTORY SORTED BY FLIGHT LEG > CHANNEL A (because it was A/P A that disconnected) > LEG 02 (normally leg 1, but I only took these photographs on the next sector) >

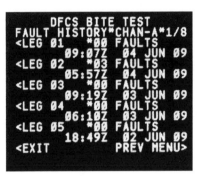

This page verifies that autopilot A disconnected on 4 Jun 09 at 07:29z, 31001ft, 279kts with LNAV engaged. This was due to invalid MCP data and points to the MCP and interface pins listed.

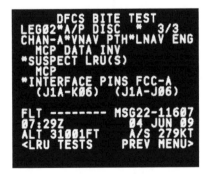

FUEL

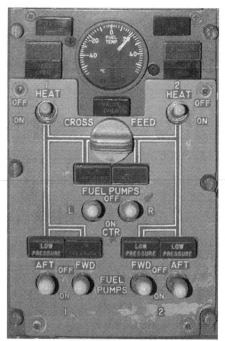
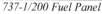

737-1/200 Fuel Panel	*737-Classic 4-Tank Fuel Panel*	*737-NG Fuel Panel*

The maximum declarable fuel capacity for tech log, nav log, etc is 16,200kgs (35,583lbs) for 3-Tank Classics, 20,800kgs (46,063lbs) for NGs and up to 37,712kgs (83,000lbs) for BBJs depending upon how many tanks the customer has specified (max 12). The AFM limits are higher, but not normally achievable with standard specific gravities. The classic fuel gauges will compensate for changes in specific gravity and have an accuracy of ±2½%. The NG fuel quantity indicating system uses a microprocessor to analyse a capacitance signal from units in each tank. The signal contains data on fuel quality and temperature, which is then used to calculate density. The processor then sends the fuel-weight signal to the flightdeck displays and FMC with an accuracy of 1%. That said, it is my experience that the FMC usually over-reads by 100kg from the fuel gauge totals. Furthermore, the centre tank fuel quantity can reduce by almost 200kgs in the climb and increase by the same in the descent. As long as you are aware of these idiosyncrasies they should not cause any problems.

With all pumps on, centre tank fuel is automatically used before main tank fuel. To achieve this, the NG centre tank pumps deliver at a higher pressure (23psi) than the main tank pumps (10psi). On the 1-500's the centre tank pump check valves operate at a much lower pressure (1.3psi cf 12psi) allowing fuel to flow from this tank first.

The fuel panels for the various series have not changed much over the years. The NGs have separate ENG VALVE CLOSED & SPAR VALVE CLOSED lights in place of FUEL VALVE CLOSED. The -1/200 panel also has blue VALVE OPEN lights similar to that on the crossfeed valve. The FILTER BYPASS lights were FILTER ICING on the 1/200.

The 1/200s had heater switches; these used bleed air to heat the fuel and de-ice the fuel filter. They were solenoid held and automatically moved back to OFF after one minute.

NG: The engine spar valves and APU are normally powered by the hot battery bus but have a dedicated battery to ensure that there is always power to shut off the fuel in an emergency.

Fuel Gauges

Digital Sunburst Fuel Gauges - Simmonds 4 Tank
-3/4/500s

Analogue Fuel Gauges with total fuel weight calculator
-1/200s without a PDC

Digital Sunburst Fuel Gauges – Smiths
- 3/4/500s

DU Fuel Gauges
– NGs

On the Digital Sunburst fuel gauges, pressing the "Qty test" button will start a self test of the display and the fuel quantity indicating system. After the test, each gauge will display any error codes that they may have.

Note: The gauges are still considered to be operating normally with error codes 1, 3, 5 or 7 on the Simmonds gauges or error codes 1, 3 and 6 on the Smiths gauges. ie If the gauge is indicating (rather than zero) the gauge may be used.

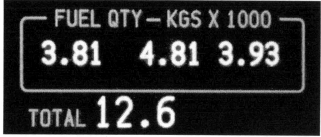

NG fuel totalizer (optional)

This new fuel totalizer display is an optional feature on NGs available with CDS Blockpoint 2004. It replaces the current "round dial" fuel gauges on the DU.

Vref & Total Fuel Quantity indicator

This useful little gauge was the predecessor (by 30 years) of the new NG fuel totalizer. It also combined the APPROACH REF page of the FMC long before the FMC was available. To use it, you set the ZFW into the top window with the bottom right knob (here 76,100lbs) and the fuel summation unit would constantly display the total fuel quantity in the bottom window (here 6,500lbs). You then selected either flap 25, 30 or 40 on the selector bottom left and the needle would display your Vref.

Digital Fuel Quantity Indicator Error Codes

Digital Fuel Quantity Indicator Error Codes - Simmonds			
Error Code	Fuel Quantity Indicator Reading	Probable Cause	Gauges Serviceable?
0	Zero	Missing or disconnected tank unit	
1	Normal	Tank contamination	Yes
2	Zero	Bad HI-Z lead	
3	Normal	Bad compensator unit wiring	Yes
4	Zero	Bad tank unit wiring	
5	Normal	Bad compensator unit	Yes
6	Zero	Bad tank unit	
7	Normal	Contamination/water in compensator	Yes
8	Zero	Bad fuel quantity indicator	
9	Normal or zero	Improperly calibrated indicator	
	Blank	Bad fuel quantity indicator	

Digital Fuel Quantity Indicator Error Codes - Smiths			
Error Code	Fuel Quantity Indicator Reading	Probable Cause	Gauges Serviceable?
1	Normal	Open or short in compensator LO-Z wiring	Yes
2	Zero	Short circuit in compensator unit	
3	Normal	Too much leakage in compensator unit	Yes
4	Zero	Open or short circuit in a LO-Z to a tank unit	
5	Zero	Short circuit in a tank unit	
6	Normal	Too much leakage in tank unit	Yes
7	Zero (or ERR in flight)	Calibration unit does not operate correctly	
8	Blank	An error in the DCTU data	
9	Zero (or ERR in flight)	A problem with the indicator memory	
10	Zero	Open or short circuit in the HI-Z line	

Calibration

Fuel quantity is measured by using a series of capacitors in the tanks with fuel acting as the dielectric. Calibration of the fuel gauges is done by capacitance trimmers, these are adjusted to standardise the total tank capacitance and allows for the replacement of gauges. On older aircraft the trimmers were accessible from the flightdeck (below the F/O's FMC) but they have since been removed to a safer place!

Capacitance trimmers

Pumps

There are two AC powered fuel pumps in each tank; there are also EDPs at each engine. Both fuel pump low pressure lights in any tank are required to illuminate the master caution to avoid spurious warnings at high AoAs or accelerations. Centre tank LP lights are armed only when their pumps are ON.

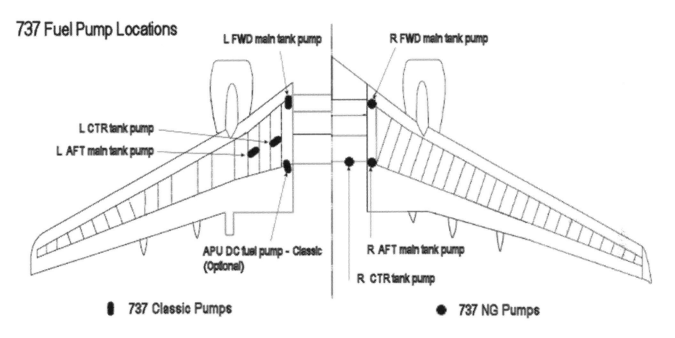

737 Fuel Pump Locations

L FWD main tank pump R FWD main tank pump

L CTR tank pump
L AFT main tank pump

APU DC fuel pump - Classic (Optional) R AFT main tank pump

R CTR tank pump

■ 737 Classic Pumps ● 737 NG Pumps

Leaving a fuel pump on with a low pressure light illuminated is an explosion risk. Also if a pump is left running dry for over approx 10 minutes it will lose all the fuel required for priming, this will render it inoperative even when the tank is refuelled. If you switch on the centre tank pumps and the LP lights remain illuminated for more than 19 seconds then this is probably what has happened. The pumps should be switched off and considered inop until they can be re-primed.

On the 1-500s, the centre tank pumps are located in a dry area of the wing root but on the NGs the pumps are actually inside the fuel tank (see photo). This is why only the NGs are affected by AD 2002-19-52 which requires the crew to maintain certain minimum fuel levels in the center fuel tanks.

Right centre tank fuel pump on the forward wall of the wheel well - NGs only

All aircraft delivered after May 2004 were fitted with a modified system whereby the centre tank fuel pumps automatically shut off when they detect a low output pressure. The same modification should be available to series 1-500 aircraft from 2009.

You can see the location of the centre tank pumps on the forward wall of the wheel well on the NGs, since the forward wall is actually the back of the centre fuel tank.

Centre Tank Scavange Pumps

These transfer fuel from the centre tank into tank 1 at a minimum rate of 100kg/hr, although usually nearer 200kg/hr. The trigger for the scavenge pump is different for the series as follows:

Originals: Only fitted after l/n 990 (Dec 1983). Operates the same as the Classics.
Classics: Switching both centre tank pumps OFF will cause the centre tank scavenge pump to transfer centre tank fuel into tank 1 for 20 minutes.
NGs: The centre tank scavenge pump starts automatically when main tank 1 is half full. Once started, it will continue for the remainder of the flight.

On the Classics, when departing with less than 1,000kg (2,200lbs) of fuel in the centre tank, an imbalance may occur during the climb. This is because the RH centre tank pump will stop feeding due to the body angle so number 2 engine fuel is drawn from main tank 2, while engine 1 is still drawing fuel from the centre tank. When this "runs dry" the scavenge pump will also transfer any remaining centre tank fuel into main tank 1, thereby exacerbating the imbalance.

APU Fuel

The APU uses fuel from the number 1 tank. If AC power is available, select the No 1 tank pumps ON for APU operation to assist the fuel control unit, especially during start. Newer Classics have an extra, DC operated APU fuel pump in the No 1 tank which operates automatically during the start sequence. The APU burns about 160kgs (350lbs)/hr with electrics and an air-conditioning pack on and this should be considered in the fuel calculations if expecting a long turnaround, waiting with pax on board for a late slot or anticipating a bleeds off take-off. Oddly, the NG APU uses less fuel if running two packs rather than one. This can be seen in the maint pages of the FMC.

Fuel Temperature

737NG fuel temperature gauge

Max fuel temp +49°C, Min fuel temp -45°C (-43°C NGs) or freezing point +3°C, whichever is higher. Typical freezing point of Jet A1 is minus 47°C. If the fuel temp is approaching the lower limits (only likely on a long flight) you could descend into warmer air or accelerate to increase the kinetic heating. Note that the NG main tanks have a tendency to cool much faster than earlier series due to their shape and size (See page 168).

Fuel temp is taken at main tank 1 because this will be the coldest due to less heat transfer from the smaller hydraulic system A.

Cold Soaked Fuel Frost

See page 168 under "non-environmental icing".

Auxiliary Fuel System

The standard number of fuel tanks is three. The 737-200Adv and Classics could be fitted with an auxiliary fourth tank in the forward end of the aft hold, which was controlled from the main panel. The capacity of the 737-200Adv auxiliary tank was either 1,475 or 3,065 litres (810 USgal). The capacity of the Classics auxiliary tank was 1,475, 1,893, 3,065 or 3,785 litres (1,000 USgal). Of the NGs, only the BBJ and 737-7/900ER have been fitted with aux fuel tanks as standard.

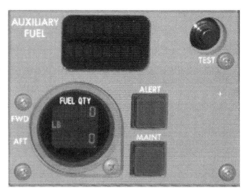

BBJ Aux fuel panel

The BBJ can have up to 9 aux fuel tanks giving it a maximum fuel quantity of 40,485 litres (10,700USG), although in practice this would probably take you over MTOW if any payload was carried. This fuel would give a theoretical range in excess of 6,200nm. The aux tanks are located at the rear of the fwd hold and the front of the aft hold, this reduces the C of G movement as fuel is loaded and used.

Refuelling of the aux tanks is done by moving the guarded switch in the refuelling panel to AUX TANKS. The controls for main tanks 1 and 2 change to Aft Aux (AA) and Fwd Aux (FA) respectively.

Refuelling panel with aux tank capability and fuel quantity preset knobs on the gauges.

The aux fuel system is essentially automatic. It works by transferring fuel from the aux tanks into the centre tank where it is then fed to the engines in the normal way. Flight crew can select fwd or aft tanks, but normal practice is to use both to maintain C of G balance. The fwd and aft tanks are switched off when the ALERT light illuminates on the main panel.

Aux fuel control panel (Overhead panel)

There are no pumps in the aux fuel system. Cabin differential pressure (and bleed air as a backup) is used to maintain a head of pressure in the aux tanks to push the aux fuel into the centre tank.

Aux fuel control panel (Aft overhead panel)

Centre Fuel Tank Inerting

To date, two 737s, 737-400 HS-TDC of Thai Airways on 3 Mar 2001 and 737-300 EI-BZG operated by Philippine Airlines on 5 Nov 1990, have been destroyed on the ground due to explosions in the empty centre fuel tank. The common factor in both accidents was that the centre tank fuel pumps were running in high ambient temperatures with empty or almost empty centre fuel tanks.

Even an empty tank has some unusable fuel which in hot conditions will evaporate and create an explosive mixture with the oxygen in the air. These incidents, and 15 more on other types since 1959, caused the FAA to issue SFAR88 in June 2001 which mandates improvements to the design and maintenance of fuel tanks to reduce the chances of such explosions in the future. These improvements include the redesign of fuel pumps, FQIS, any wiring in tanks, proximity to hot air-conditioning or pneumatic systems, etc.

737s delivered since May 2004 have had centre tank fuel pumps which automatically shut off when they detect a low output pressure and there have been many other improvements to wiring and FQIS. But the biggest improvement will be centre fuel tank inerting. This is universally considered to be the safest way forward, but is very expensive and possibly impractical. The NTSB recommended many years ago to the FAA that a fuel tank inerting system be made mandatory, but the FAA have repeatedly rejected it on cost grounds.

Boeing has developed a Nitrogen Generating System (NGS) which decreases the flammability exposure of the center wing tank to a level equivalent to or less than the main wing tanks. The NGS is an onboard inert gas system that uses an air separation module (ASM) to separate oxygen and nitrogen from the air. After the two components of the air are separated, the nitrogen-enriched air (NEA) is supplied to the center wing tank and the oxygen-enriched air (OEA) is vented overboard. NEA is produced in sufficient quantities, during most conditions, to decrease the oxygen content to a level where the air volume (ullage) will not support combustion. The FAA Technical Center has determined that an oxygen level of 12% is sufficient to prevent ignition, this is achievable with one module on the 737 but will require up to six on the 747.

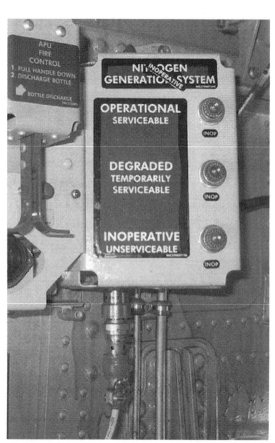

NGS provision in the wheel-well
(Photo: Lonnie Ganz)

On 21 Feb 2006 the Honeywell NGS was certified by the FAA after over 1000hrs flight testing on two 737-NGs. Aircraft have been NGS provisioned since l/n 1935 with full production cutover scheduled for l/n 2620 onwards. The NGS requires no flight or ground crew action for normal system operation and is not dispatch critical.

Limitations

Max tank fuel temp:	**+49°C**

Min tank fuel temp (prior to take-off and in-flight): 737-1/500: **-45°C** or freeze pt +3°C
737-6/900: **-43°C** or freeze pt +3°C

Max quantity:
1/200: 4,300 + 4,100 + 4,300 = **12,700kg**
1/200: 9,580 + 9,000 + 9,580 = **28,160lbs**
(2 bag ctr bays)

200Adv: 4,300 + 5,400 + 4,300 = **14,000kg**
200Adv: 9,580 + 11,900 + 9,580 = **31,060lbs**
(3 bag ctr bays)

200Adv: 4,300 + 7,000 + 4,300 = **15,600kg**
200Adv: 9,580 + 15,430 + 9,580 = **34,590lbs**
(3 bag integral)

3/4/500: 4,600 + 7,000 + 4,600 = **16,200kg**
3/4/500:10,043 + 15497 + 10,043 = **35,583lbs**

NGs: 3,900 + 13,000 + 3,900 = **20,800kg**
NGs: 8,630 + 28,803 + 8,630 = **46,063lbs**

Max lateral imbalance:
1/200: 680kg; All other series: **453kg (1,000lbs)**
NB: Greater allowance given for unsymmetrical loads in cargo version.

Main tanks to be full if centre contains over 453kg (1,000lbs)

Fuel Quantity Indication System Accuracy Tolerances

737-100/-200: +/- 3.0%

737-300/-400/-500: with digital indicators: +/- 2.5 %, with analogue indicators: +/- 3.0%

737-NG with densitometer: +/- 1.0% overall
Main tanks > 50%, -1 to 5 deg pitch, +/- 1 deg roll: +/- 1.5%
Main tanks < 50%, -1 to 5 deg pitch, +/- 1 deg roll: +/- 1.0%

737-NG without densitometer: +/- 2.0% overall
Main tanks > 50%, -1 to 5 deg pitch, +/- 1 deg roll: +/- 2.5%
Main tanks < 50%, -1 to 5 deg pitch, +/- 1 deg roll: +/- 2.0

The total tolerance is based on a full tank. For example, if the fuel tank maximum capacity is 10,000Kg, then the tolerance of the gauging is 0.02 (NG without a densitometer) * 10000 = 200Kg. The system tolerance is then +/- 200Kg at any fuel level within the tank.

The accuracy tolerance of the fuel flow transmitter is a function of the fuel flow. At engine idle, the system tolerance can be 12%. During cruise, the tolerance is less than 0.5% (1.5% Classics). The fuel flow indication is integrated over time to calculate the fuel used for each engine.

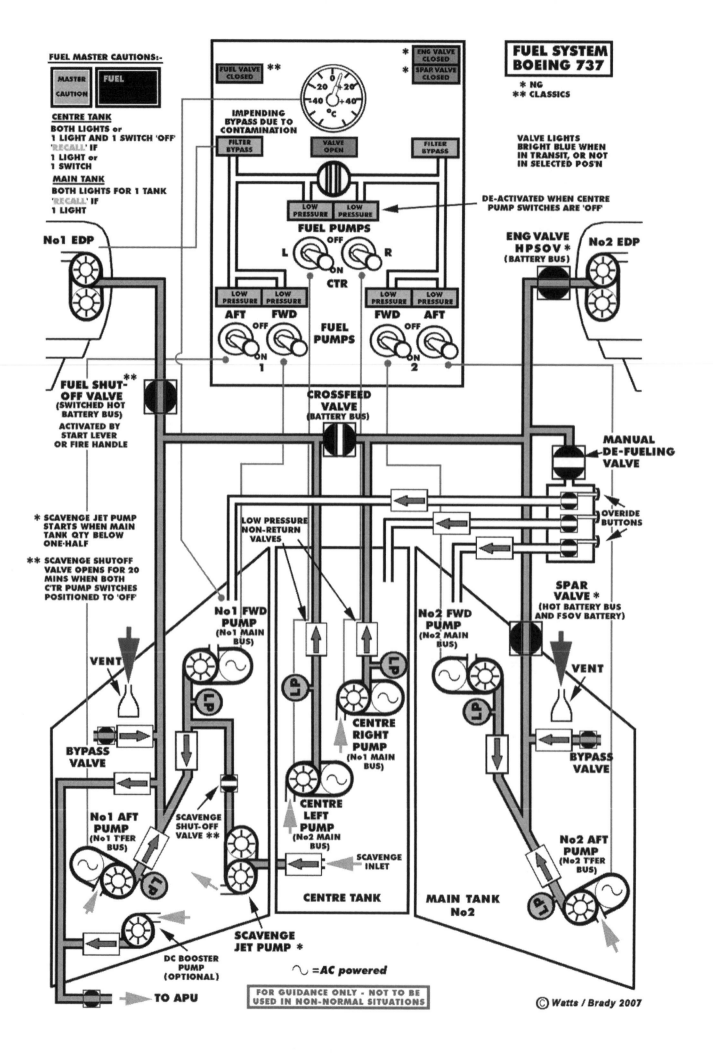

FUEL MASTER CAUTIONS:-

MASTER CAUTION   FUEL

CENTRE TANK
BOTH LIGHTS or
1 LIGHT AND 1 SWITCH 'OFF'
'RECALL' IF
1 LIGHT or
1 SWITCH
MAIN TANK
BOTH LIGHTS FOR 1 TANK
'RECALL' IF
1 LIGHT

FUEL SYSTEM
BOEING 737

* NG
** CLASSICS

* ENG VALVE CLOSED
** SPAR VALVE CLOSED

FUEL VALVE CLOSED **

VALVE LIGHTS
BRIGHT BLUE WHEN
IN TRANSIT, OR NOT
IN SELECTED POS'N

IMPENDING
BYPASS DUE TO
CONTAMINATION

FILTER BYPASS

VALVE OPEN

FILTER BYPASS

DE-ACTIVATED WHEN CENTRE
PUMP SWITCHES ARE 'OFF'

LOW PRESSURE   LOW PRESSURE

FUEL PUMPS

No1 EDP

ENG VALVE
HPSOV *
(BATTERY BUS)

No2 EDP

OFF
L   R
ON
CTR

LOW PRESSURE   LOW PRESSURE   LOW PRESSURE   LOW PRESSURE

AFT   FWD   FWD   AFT

OFF   OFF
ON   FUEL   ON
1   PUMPS   2

FUEL SHUT-
OFF VALVE **
(SWITCHED HOT
BATTERY BUS)

ACTIVATED BY
START LEVER
OR FIRE HANDLE

CROSSFEED
VALVE
(BATTERY BUS)

MANUAL
DE-FUELING
VALVE

OVERIDE
BUTTONS

* SCAVENGE JET PUMP
STARTS WHEN MAIN
TANK QTY BELOW
ONE-HALF

** SCAVENGE SHUTOFF
VALVE OPENS FOR 20
MINS WHEN BOTH
C'TR PUMP SWITCHES
POSITIONED TO 'OFF'

LOW PRESSURE
NON-RETURN
VALVES

No1 FWD
PUMP
(No1 MAIN
BUS)

No2 FWD
PUMP
(No2 MAIN
BUS)

SPAR
VALVE *
(HOT BATTERY BUS
AND FSOV BATTERY)

VENT

BYPASS
VALVE

LP

LP

LP

CENTRE
RIGHT
PUMP
(No1 MAIN
BUS)

LP

VENT

BYPASS
VALVE

No1 AFT
PUMP
(No1 T'FER
BUS)

SCAVENGE
SHUT-OFF
VALVE **

LP

CENTRE
LEFT
PUMP
(No2 MAIN
BUS)

SCAVENGE
INLET

No2 AFT
PUMP
(No2 T'FER
BUS)

LP

CENTRE TANK

MAIN TANK
No2

DC BOOSTER
PUMP
(OPTIONAL)

SCAVENGE
JET PUMP *

TO APU

∿ = AC powered

FOR GUIDANCE ONLY - NOT TO BE
USED IN NON-NORMAL SITUATIONS

© Watts / Brady 2007

FUSELAGE

The fuselage is a semi-monocoque structure formed from circumferential frames, longitudinal stringers and skin with "waffle" doublers, which act as tear stoppers. There are pressure bulkheads behind the radome and passenger cabin which form a pressure vessel. It is made from various aluminium alloys except for the following parts.

- **Fiberglass**: radome, tailcone, centre & outboard flap track fairings.
- **Kevlar**: Engine fan cowls, inboard track fairing (behind engine), nose gear doors.
- **Graphite/Epoxy**: rudder, elevators, ailerons, spoilers, thrust reverser cowls, dorsal of vertical stab.

Aft Body Vortex Generators

Vortex generators became an optional retrofit in November 1968 to reduce "vertical bounce" during cruise flight which was caused by airflow turbulence and separation from the body beneath the horizontal stabilizer.

The 1968 modification was to install the three lower aft body vortex generators and another on the underside of the horizontal stabiliser near the body. The stabiliser vortex generator was removed in 1971 due to higher than expected in-flight loads and the four upper aft body vortex generators were added instead.

These modifications were all optional because not all aircraft exhibited the vertical bounce. The full set of vortex generators were installed at production from 1971.

737-200 aft body vortex generators

Early 737-300 without aft body vortex generators

Classics were initially produced without any aft body vortex generators (see photo). However the upper vortex generators were reinstated after line number 2277 (May 1992 onwards). This was to reduce elevator and elevator tab vibration during flight to increase their hinge bearing service life.

The CDL says that if any of these vortex generators are not fitted or missing *"occasional vertical motions may be felt which appear to be light turbulence These motions are characteristic of this airplane and should not be construed to be associated with Mach buffet."*

The 737MAX has a redesigned tailcone and will not require any vortex generators.

Radome

The radome (RAdar DOME) is an aerodynamic fairing that houses the weather radar and ILS localiser (lower) and glideslope (upper) antennas. Unlike the rest of the fuselage it is made of fibreglass to allow the RF signals through.

Fibreglass is non-conductive which would allow the build up of P-static (static due to the motion of the aircraft through precipitation). This would in turn cause static interference on the antenna within so the radome is fitted with six conductive diverter strips on the outside to dissipate P-static into the airframe.

Weather radar and localizer antennae in the radome

Lap Joints

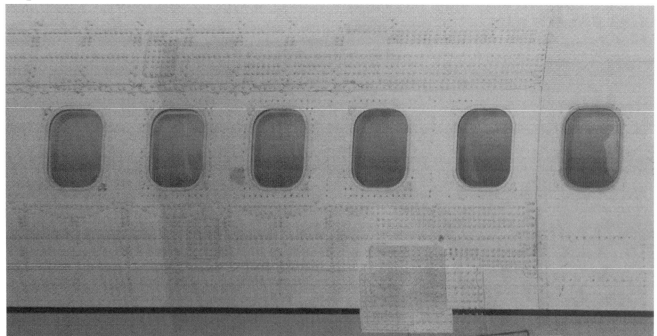

After the Aloha 737-200 accident, in which a 12ft x 8ft section of the upper fuselage tore away in flight, all 737s with over 50,000 cycles must have their lap joints reinforced with external doublers. This tired old aircraft is a 737-200 and the patching is clearly visible. This modification takes about 15,000 man hours and unfortunately has sometimes been the source of another problem - scoring. This is when metal instruments instead of wooden ones have been used to scrape away excess sealant or old paint from the lap joints which create deep scratches which may themselves develop into cracks.

Windows

The flightdeck windows are made of layers of glass and vinyl. The outer pane is a rigid hard scratch resistant surface. The inner pane is structural and carries the aircraft pressure load. The Vinyl layer prevents a broken window from shattering. Windows 1, 2, 3 & 4 have a conductive layer between the outer pane and the vinyl layer this both de-ices the windscreen and makes the windows less brittle. Note that windows 1 & 2 are designed to withstand a birdstrike even without window heat.

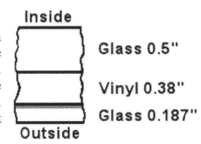

Construction of windows 1 & 2

A 2006 737NG without eyebrow windows and with noise reducing vortex generators

On the 3rd Feb 2005, 737-700, N201LV, L/N 1650, was the first ever 737 to fly without eyebrow windows (window numbers 4 & 5). They have been standard in Boeing aircraft back as far as WW2 bombers to give better crew visibility. Now they have been declared obsolete and removed from production. The design change reduces aircraft weight by 9kg (20lbs) and eliminates approximately 300 hours of periodic inspections per airplane. Retrofit kits to cover eyebrow windows will be available mid-2006 for the in-service 737 fleet.

Notice the 10 small vortex generators above the radome; these reduce the cockpit noise from the windshield by 3dB. See page 207 for more details.

Airframe Drainage

There are many drain holes around the underside of the fuselage, wing and empennage. They prevent fluids such as leaking waste water, condensation, or any other fluid, collecting in the fuselage which may become a corrosion or fire hazard.

Drain holes in pressurised areas have a valve behind them which seals in-flight and re-opens on the ground. Drain holes in other areas are always open.

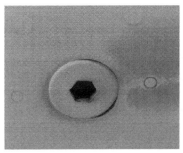

Drain hole

HYDRAULICS

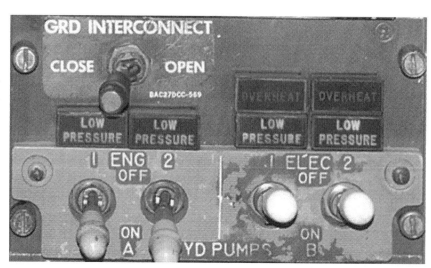

The 737-1/200 had system A powered by the two Engine Driven Pumps (EDPs) and system B powered by the two Electric Motor Driven Pumps (EMDPs).

The ground interconnect switch allows system B (elec pumps) to pressurise system A (eng driven pumps) when on the ground with the parking brake set (ie if engines not running).

The hydraulic pump panel -1/200

737-1/200 Hydraulic Services Supplied		
System A	**System B**	**Standby**
Autopilot "A"	**Autopilot "B"**	
Ailerons	**Ailerons**	
Rudder	**Rudder**	**Rudder**
	Yaw damper	**Standby yaw damper (as installed)**
Elevators & Elevator feel	**Elevators & Elevator feel**	
Inboard flight spoilers	**Outboard flight spoilers**	
Ground spoilers		
T/E & L/E flaps and slats		**L/E flaps and slats (extension only)**
Thrust reversers		**Thrust reversers (slow)**
Nose wheel steering		
Inboard brakes	**Outboard brakes**	
	Autobrakes	
Landing gear		
	Cargo Door (As installed)	

From the 737-300 onwards each hydraulic system had both an EDP and an EMDP for greater redundancy in the event of an engine or generator failure.

The EDP's are much more powerful, having a hydraulic flow rate of 22gpm (Classics) / 37gpm (NG). The EMDP's only produce 6gpm. The standby system output is even less at 3gpm.

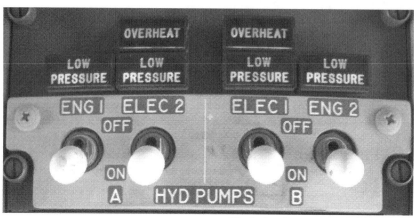

The hydraulic pump panel -300 onwards

Hydraulic Services Supplied, 737-300 Onwards		
System A	**System B**	**Standby**
Autopilot "A"	**Autopilot "B"**	
Ailerons	**Ailerons**	
Rudder	**Rudder**	**Rudder**
	Yaw damper	**Standby yaw damper (as installed)**
Elevators & Elevator feel	**Elevators & Elevator feel**	
Inboard flight spoilers	**Outboard flight spoilers**	
Ground spoilers	**Two position tailskid (as installed)**	
	T/E & L/E flaps and slats	**L/E flaps and slats (extension only)**
PTU for autoslats	**Autoslats**	
No1 thrust reverser	**No2 thrust reverser**	**Nos 1 & 2 thrust reversers (slow)**
Nose wheel steering	**Alt nose wheel steering**	
Alternate brakes (man only)	**Normal (auto & man) brakes**	
Landing gear	**Landing gear transfer unit (retraction only)**	

The hydraulic reservoirs are pressurised from the pneumatic manifold to ensure a positive flow of fluid reaches the pumps, A from the left manifold and B from the right. 737s from mid 2003 onwards have had their hydraulic reservoir pressurisation system extensively modified to fix two in-service problems:

1) Hydraulic vapours in the flight deck caused by hydraulic fluid leaking up the reservoir pressurisation line back to the pneumatic manifold giving hydraulic fumes in the air-conditioning and
2) Pump low pressure during a very long flight in a cold soaked aircraft. The latter is due to water trapped in the reservoir pressurisation system freezing blocking reservoir bleed air supply.

Aircraft which have been modified (SB 737-29-1106) are recognised by only having one reservoir pressure gauge in the wheel well.

Also in the wheel well can be seen hydraulic fuses. These are essentially spring-loaded shuttle valves which close the hydraulic line if they detect a sudden increase in flow such as a burst downstream, thereby preserving hydraulic fluid for the rest of the services. Hydraulic fuses are fitted to the brake system, L/E flap/slat extend/retract lines, nose gear extend/retract lines and the thrust reverser pressure and return lines.

Hydraulic Fuses in the wheel-well

Hydraulic Indications

In the 737-1/200 there is only a quantity gauge for system A, this is because the B system is filled from system A reservoir. System B quantity is monitored by the amber "B LOW QUANTITY" light above. The hydraulic brake pressure gauge has two needles because system A operates the inboard brakes and system B the outboard brakes, each has its own accumulator.

On pre-EIS classic aircraft (before 1988) the hydraulic gauges were similar to the 737-200. There are now separate quantity gauges since the reservoirs are not interconnected and the markings have been simplified. There is now just a single brake pressure gauge showing the normal brake pressure from system B.

With EIS the quantity became a simple digital number and on the NG both pressure and quantity are digital numbers.

737-1/200 Hydraulic Gauges

737-3/400 Hydraulic Gauges (Non EIS A/C)

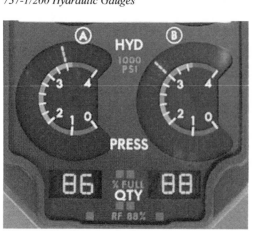

737-3/4/500 EIS Hydraulic Gauges

737-NG DU Hydraulic Gauges

160

Hydraulic Quantity

This table shows the nominal quantities at different levels in the reservoirs

System		Originals	Classics	NGs
A	Full level	3.6 USG	100%	100% (5.7Gal / 21.6Ltrs)
	Refill	2.35 USG	88%	76%
	EDP Standpipe		22%	20%
	EMDP Standpipe	N/A	0%	0%
B	Full level	Full (1.3 USG)	100%	100% (8.2Gal / 31.1Ltrs)
	Refill	3/4	88%	76%
	Fill & balance line (to standby reservoir)		64%	72%
	EDP Standpipe	N/A	40%	0%
	EMDP Standpipe		11%	0%

For instance, if you are in a 737-300 and you notice the System B hydraulic quantity drop to 64%; then from the table above, you may suspect a leak in the balance line or standby reservoir.

The indicated hydraulic quantity will vary with the aircraft configuration as the services use hydraulic fluid. Hence the refill figure is valid only when airplane is on ground, clean with the engines shutdown. The following table shows how indicated hydraulic quantity varies with configuration in the 737NG.

Configuration change	Hyd A	Hyd B
Extending flaps before take-off (slats to extend)	-	-10%
Retracting gear	-20%	-
Thermal contraction in cruise	-5 to -20%	-
Use of flying controls	-	-
Use of speedbrake (flight spoilers)	-6%	-6%
Extending flaps before landing (slats to full extend)	-	-20%
Extending gear	+20%	-
Use of speedbrake on landing (flight & ground spoilers)	-10%	-6%
Retracting flaps after landing	-	+20%

Note that only the LE devices cause a change in the reservoir level. The TE flaps have no effect because they are driven by a hydraulic motor, not an actuator.

The hydraulic reservoirs can be filled from the ground service connection point on the forward wall of the starboard wheel well.

Normal hydraulic pressure is 3000 psi

Minimum hydraulic pressure is 2800 psi

Maximum hydraulic pressure is 3500 psi

Brake accumulator precharge is 1000 psi

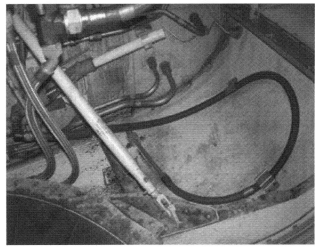

Hydraulic ground service connection

The alternate flap system will extend (but not retract) LE devices with standby hydraulic power. It will also extend or retract TE flaps with an electric drive motor but there is no asymmetry protection for this.

Landing Gear Transfer Valve

Formerly known as the Landing Gear Transfer Unit (LGTU), this makes system B hydraulic pressure available for gear retraction when Engine No1 falls below 50% N2 (NG) or 56% N2 (Classics). Without the LGTV/LGTU, gear retraction would be very much slower as Engine No1 engine driven pump would not be available. So, gear retraction would only be done by the slower electrical driven pump which would compromise engine out performance. It is for this reason that despatch is not permissible on Classics with the Engine No1 tachometer u/s. Other series do not have this restriction because Originals do not have the LGTU and NGs send engine running data digitally to the PSEU.

Methods for Transfer of Hydraulic Fluid

It should go without saying that if a hydraulic system is low on quantity then you should top up that system with fresh fluid (and find out why it was low!) to avoid cross contamination. However, if you really want to move fluid from one system to another here is how to do it.

A to B (1% transfer per cycle)
- Chock the aircraft & ensure area around stabiliser is clear.
- Switch both electric hydraulic pumps OFF.
- Release parking brakes and deplete accumulator to below 1800psi by pumping toe brakes.
- Switch Sys A pump ON and apply parking brakes.
- Switch Sys A pump OFF and depressurise through control column. (Use stabiliser rather than ailerons to prevent damage to equipment or personnel)
- Switch Sys B pump ON and release parking brakes. (Sends the fluid back to system B because the shuttle/priority valves send the fluid back to the normal brake system.)

B to A (4% transfer per cycle)
- Ensure area around No1 thrust reverser is clear.
- Switch both electric hydraulic pumps OFF
- Switch either FLT CONTROL to SBY RUD.
- Select No1 thrust reverser OUT (uses standby hyd sys)
- Switch FLT CONTROL to ON.
- Switch Hyd Sys A pump ON.
- Stow No 1 thrust reverser (using sys A)

Limitations

Originals
Minimum fuel for ground operation of electric pumps is 760kgs (1675lbs) in fuel tank No. 2.

Classics
Minimum fuel for ground operation of electric pumps is 760kgs (1675lbs) in the related main tank.
Minimum hydraulic quantity required for despatch is 88%.

NGs
Minimum fuel for ground operation of electric pumps is 760kgs (1675lbs) in the related main tank.

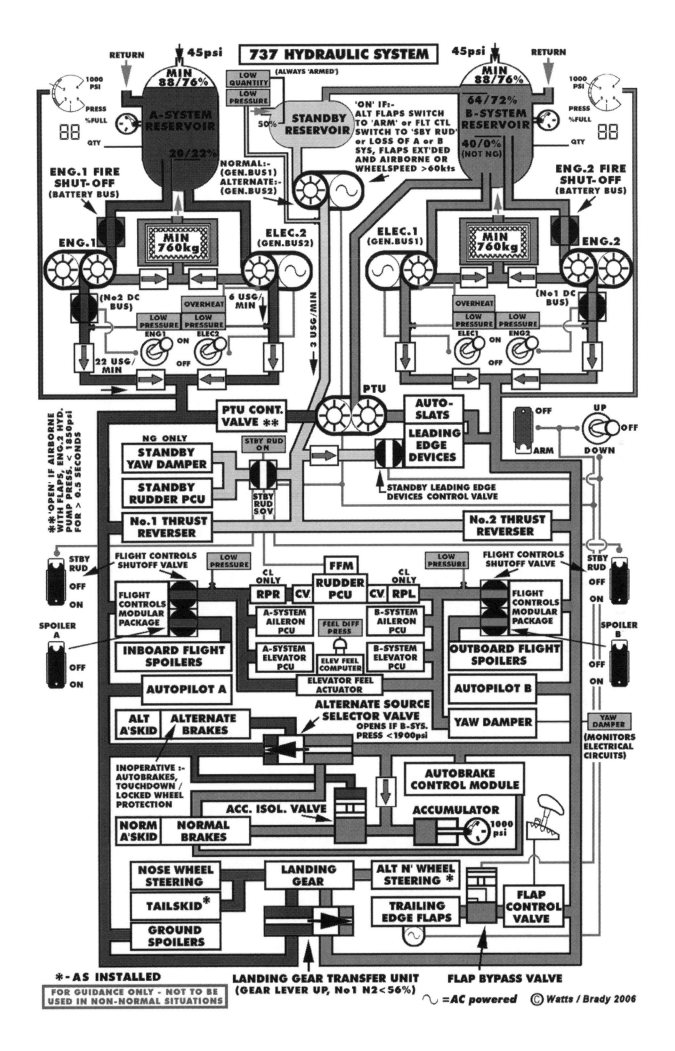

737 HYDRAULIC SYSTEM

ICE AND RAIN PROTECTION

737-1/200 Ice & Rain Panel

Differences from Classics:
1) No alpha vanes
2) WAI has ground test position
3) Engine anti-ice captions are:
   - COWL VALVE OPEN
   - R VALVE OPEN
   - L VALVE OPEN.

The R & L VALVE OPEN captions indicate that 8th stage bleed air is being used for the compressor area and EPR probe anti-icing.

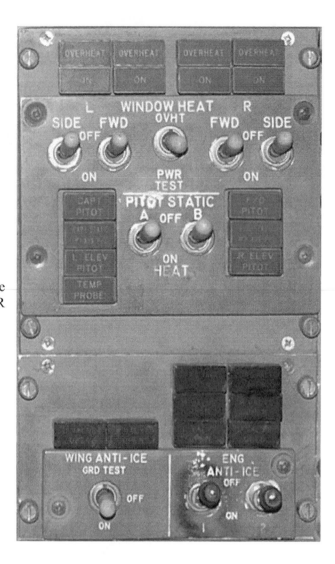

737-3/4/500 Ice & Rain Panel

Differences from originals:
1) Alpha vanes added
2) Now only one temp probe (many 1/200s had two)
3) TAT TEST button (ie aspirated probe). If there is no TAT TEST button you have an unaspirated TAT probe.

737-NG Ice & Rain Panel

Differences:
1) Now called probe heat (was previously pitot static heat).
2) Static ports not heated (hence the name change).
3) Aux pitot added.
4) Optional green ON lights.
5) Probe heat AUTO/ON (instead of OFF/ON) from 2010

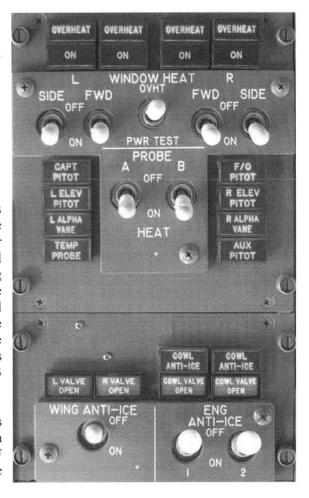

Window Heat

If window heat is switched ON but the ON light is extinguished, this means that heat is not being applied to the associated window. This could be because the heat controller has detected that the window is becoming overheated (normal on hot days in direct sunlight) and can be verified by touching the window. The heat will automatically be restored when the window has cooled down. To verify that window heat is still available a PWR TEST should illuminate all ON lights if the window heat switches are ON. The PWR TEST forces the temperature controller to full power but overheat protection is still available. There is an option for window heat on the No3 windows.

If an OVERHEAT light illuminates, either a window has overheated or electrical power to the window has been interrupted. The affected window heat must be switched OFF and allowed 2-5mins to cool before switching ON again. The OVHT TEST simulates an overheat condition.

Probe Heat

In 2010 the probe heat switches changed from a simple OFF/ON to AUTO/ON. The AUTO position automatically applies probe heat whenever the respective engine is running. This modification was made in response to reports that these switches were occasionally forgotten by crew.

See Instrument Probes, page 258, for further explanation.

Engine Anti Ice

Engine anti-ice (EAI) uses 5th stage bleed air, augmented by 9th stage as required, from the associated engine to heat the engine cowl to prevent ice build-up. The fan blades and spinner are not heated. EAI is an anti-ice system and should be used continuously in icing conditions.

Amber COWL ANTI-ICE lights will illuminate if an over-temperature 440C (not NGs) or over-pressure (65psig) condition exists in either duct. In this situation thrust on the associated engine should be reduced until the light extinguishes.

Wing Anti Ice

Wing anti-ice (WAI) is very effective and is normally used as a de-icing system in-flight, in applications of 1 minute. On the ground it should be used continuously in icing conditions, except after de-icing with fluid on the wings.

The WAI switch logic is interesting. On the ground, the WAI valve (not switch) will cut-off if either thrust lever is above the take-off warning position (32deg Classics, 60deg NGs) or if an over-temperature (125C) is detected, but will be restored after the thrust is reduced or temperature drops. This allows you to perform engine run-ups etc without having to check that the WAI is still on afterwards. The switch is solenoid held and will trip off at lift-off; this is for performance considerations as the bleed air penalty is considerable.

Note that on early systems, ie those with a GND TEST position, with the WAI switch ON on the ground, the WAI is inhibited until lift-off ie "armed". This is opposite to the present system.

WAI, unlike engine anti-ice, uses bleed air from the main pneumatic manifold, this is to ensure a source of bleed air during engine out operations. Only the leading edge slats have WAI (ie not leading edge flaps). The NG series outboard slat has no wing anti-ice facility believed to be due to excessive bleed requirements. However in June 2005 it was announced that the 737-MMA will have raked wingtips with anti-ice along the full span. This is because the MMA will be spending long periods of time on patrol at low level where it will be exposed to icing conditions.

When the QRH ENGINE FAILURE/SHUTDOWN drills ask "If wing anti-ice is required:", if icing conditions are anticipated, these actions should be completed in preparation for WAI use to prevent asymmetric application. There is no bleed penalty for this reconfiguration until WAI is actually used.

Wing ice on the outboard slat of a 737-700

On the NG, if WAI is used for more than 5 secs in-flight, the SMYD will increase the stick shaker speeds and manoeuvre speed bars by enough to allow for 3 inches of rough ice on the leading edge.

Wing and engine VALVE OPEN lights use the bright blue/dim blue - valve position in disagreement / agreement logic. The wing L and R VALVE OPEN lights in particular may remain bright blue after start and during taxi. This is because they are pneumatically operated; they can be made to open with a modest amount of engine thrust.

Use of wing anti-ice above FL350 may cause bleed trip off and possible loss of cabin pressure. (SP.16.8)

Airframe Visual Icing Cues

There is no ice detection system on the 737 so it is up to the crew to spot ice formation and take the necessary action. The following photos show some of the places where ice accretion is visible from the flight deck. Note engine anti-ice should be used whenever the temperature and visible moisture criteria are met and not left until ice is seen, to avoid inlet ice build up which may shed into the engine.

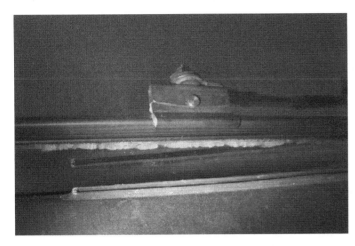

Ice under the windscreen wiper blades.
This is one of the first places that ice will form. Precipitation falls on the bottom of the windscreen and runs up to the wipers.

This is not an accurate indication of the amount of icing on the airframe because of the stagnation point where the blade and windscreen meet and also because the windscreen is heated.

I would describe conditions where ice forms here as LIGHT ICING.

Ice on the wiper nut
This is my preferred indication of airframe ice accretion. If ice is seen here, it is surely also on other parts of the airframe.

The weight and aerodynamic effect of all this ice on the the airframe and control surfaces is why there is the "residual ice" penalty of several tons on the landing performance graphs "If operating in icing conditions during any part of the flight when the forecast landing temperature is below 8C, reduce the normal climb limited landing weight by xxxxkg." (FPPM 1.3.3).

I would describe conditions where ice forms here as MODERATE ICING.

Ice on the windscreen
For ice to form on a flat heated windscreen, conditions must be bad. You can see how the shape of the formation follows the airflow lines. You can imagine how much ice is on the rest of the aircraft, especially when you consider that most of it is unheated, particularly on the fin and stabiliser.

Vol 1 SP.16.8 states "Avoid prolonged operation in moderate to severe icing conditions." This photo was taken at about 20,000ft climbing through the tops of rain bearing frontal cloud. The ice shown here formed in under a minute.

I would describe conditions where ice forms here as SEVERE ICING.

Wiper Controls

The windscreen wipers on the 737 are one of the aircrafts worst features. They are so noisy that some consider them to be more of a distraction than a benefit. If you attempt to use them on HIGH they seem like they are about to fly off!

737-1/200   *737-3/4/500*

One of the most welcome new features of the 737-NG is the vast improvement to the windscreen wipers. They can now be independently controlled, have an intermittent position and best of all - are almost silent.

737-NG

Non-environmental Icing

The NGs have a problem with frost forming after landing on the wing above the tanks where fuel has been cold soaked. This is officially known as "Wing upper surface non-environmental icing" or "**Cold-soaked fuel frost**". The reason is the increased surface area of the fuel that comes into contact with the upper surface of the wing. This is because the shape of the wing fuel tanks was changed (moved outboard) to accommodate the longer landing gear that was in turn required for the increased fuselage lengths of the NG family to reduce the risk of tailstrikes! The only solution until recently has been to limit your arrival fuel to less than approx 4,000kg (9,000lbs). Now Boeing has issued guidelines on the acceptable location and amount of upper wing frost and aircraft delivered after July 2004 have the wings marked with the permissible area for cold-soaked fuel frost.

The Boeing advice is as follows:
"Flight crews should visually inspect the lower wing surface. If there is frost or ice on the lower surface, outboard of measuring stick 4, there may also be frost or ice on the upper surface. The distance the frost extends outboard of measuring stick 4 can be used as an indication of the extent of frost on the upper surface. It should be noted that if the thickness of the frost on the lower surface of the wing is 1/16 inch (1.5 mm) thick or less, the thickness of the frost on the upper surface will be less than 1/16 inch (1.5 mm) thick. If the thickness of the frost on the lower surface is greater than 1/16 inch (1.5 mm), then a physical inspection of the upper surface frost is required."

Takeoff with light coatings of cold-soaked fuel frost on upper wing surfaces is permissible, provided the following are met:

- Frost on the upper surface is less than 1/16 inch (1.5 mm) in thickness
- The extent of the frost is similar on both wings
- The frost is on or between the black lines defining the permissible cold-soaked fuel frost area with no ice or frost on the leading edges or control surfaces
- The outside air temperature is above freezing
- There is no precipitation or visible moisture

Rain Repellent

The rain repellent has been removed due to worries about the environmental effects of the "RainBoe" fluid it uses which contains Freon 113 (a CFC). Furthermore Freon 113 is poisonous and has been blamed in at least 12 deaths in industrial settings and the rain repellent canisters, which were stored behind the Captain, had been known to leak. In 1991 Boeing added D-limonine which has a strong smell of orange peel into RainBoe so that leakage could be detected. There are no plans to replace the rain repellent with another liquid product even though there are safe alternatives eg "Le Bozec".

On 25 May 1982, a 737-200Adv (PP-SMY) was written off by a heavy landing in a rainstorm. One report stated that *"The pilots' misuse of rain repellent caused an optical illusion"*.

Since early 1994 all Boeing aircraft have been built with Surface Seal coated glass from PPG Industries which has a hydrophobic coating. The coating does deteriorate with time depending upon wiper use and windscreen cleaning methods etc, but can be re-applied.

Rainboe was stored behind the Captains seat

Limitations

- Engine anti-ice must be on during all ground and flight operations when icing conditions exist or are anticipated, except during climb and cruise below -40°C SAT.
- Engine anti-ice must be on prior to and during descent in all icing conditions, including temperatures below -40°C SAT.
- Do not use wing anti-ice on the ground when the OAT is above 10C.

737-1/200:

- Minimum N1 for operating in icing conditions except for landing: 40% when TAT between 0 and 10C; 55% when TAT below 0C; 70% in moderate to severe icing conditions when TAT below -6.5C.
- Window heat inoperative: max speed 250kts below 10,000ft.
- Gravel Protect switch: ANTI-ICE position when using engine inlet anti-ice.

737-6/900 without stiffened elevator tabs:

- After any ground deicing/anti-icing of the horizontal stabilizer using Type II or Type IV fluids, airspeed must be limited to 270 KIAS until the crew has been informed that applicable maintenance procedures have been accomplished that would allow exceedance of 270 KIAS. Once the applicable maintenance procedures have been accomplished, exceeding 270 KIAS is permissible only until the next application of Type II or Type IV deicing/anti-icing fluids.

LANDING GEAR

The landing gear on the NG has been extensively redesigned. The nose gear is 3.5" longer to relieve higher dynamic loads and the nose-wheelwell has been extended 3" forward. The main gear is also longer to cater for the increased fuselage lengths of the -8/900 series and is constructed from a one piece titanium gear beam based on 757/767 designs. There is an externally mounted trunnion bearing on the gear, a re-located gas charging valve, and the uplock link is separate from the reaction link. It is fitted with 43.5" tyres and digital antiskid.

Unfortunately the 737-700 was particularly prone to a dramatic shudder from the main landing gear if you tried to land smoothly. Fortunately Boeing started fitting shimmy dampers to this series from L/N 406 (Nov 1999) and a retrofit was made available.

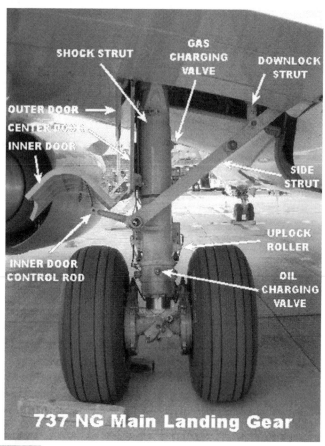

737 NG Main Landing Gear

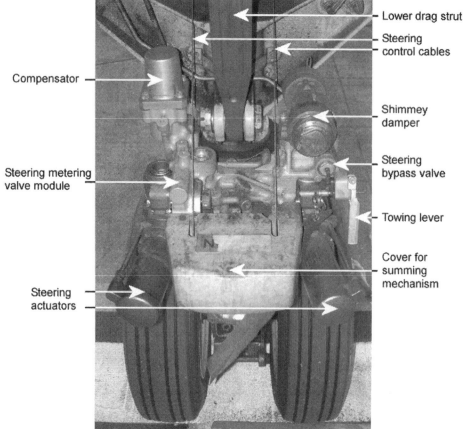

737 NG nose gear

One of the peculiarities of the 737 is that it often appears to crab when taxiing.

Theories for this include: A slightly castoring main gear to increase the crosswind capability; Play in the scissor link pins; Weather-cocking into any crosswind impinging on the fin; Torque reaction from the anti-collision light!

Engineers will tell you that it is due to the main gear having a couple of degrees of play due to the shimmy dampers.

The landing gear panel is located between the engine instruments and F/O's instrument panel.

The Green lights tell you that the gear is down and locked and the red lights warn you if the landing gear is in disagreement with the gear lever position. With the gear UP and locked and the lever UP or OFF, all lights should be extinguished.

On a couple of occasions I have seen 3 reds and 3 greens after the gear has been selected down. This was because the telescopic gear handle had not fully compressed back toward the panel. If this happens to you, give it a tap back in and the red lights should extinguish.

737s used for cargo operations have an extra set of green "GEAR DOWN" lights on the aft overhead panel. This is because with the cabin filled with freight, the main gear downlock viewer could not be guaranteed to be accessible in-flight.

The NGs also have these lights because they do not have gear downlock viewers installed.

Main gear viewer

Gear Viewers (not NGs)

If any green gear lights do not illuminate after the gear is lowered, you might consider a visual inspection through the gear viewers. The main gear viewer is in the cabin and the nose gear viewer is on the flight deck. The main gear viewers are not installed on NG series aircraft.

This is the main gear viewer and it is located in the aisle, just behind the emergency exit row.

The first time you look through a viewer it will probably take you several minutes to find what you are looking for, hardly ideal if you are in the situation for real so it is worth acquainting yourself with its use.

There are two prisms, one for each main gear leg. Don't forget to switch on the wheel-well light at night.

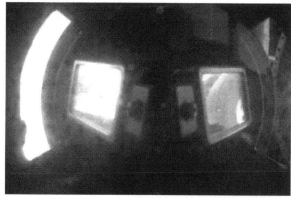

Main gear viewer prisms (from cabin)

Main gear viewer prisms (from wheel-well)

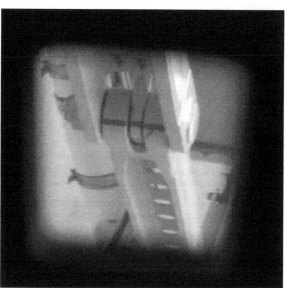

Main gear locked marks as seen through the viewer

Eventually, you should be able to see three red marks on the undercarriage, if they line up then your gear is certainly down and probably locked.

Nosegear viewer

The nosegear viewer is located under a panel toward the aft of the flightdeck. There is no prism, just a long tube to direct your eye exactly toward the correct place to see the downlock marks. It is usually much dirtier than the main gear viewer due to its proximity to the nose wheel. The example in the photograph is an unusually clean view because the aircraft had just come off a C check.

The nosegear down marks are two red arrows pointing at each other.

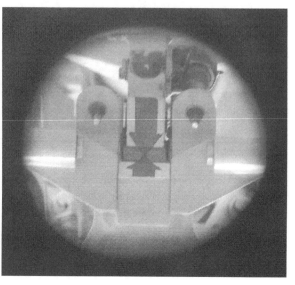

Nose gear downlock markings

Landing Gear Manual Extension

If the gear fails to extend or hydraulic system A is lost, the gear can be manually extended by pulling the manual gear extension handles, located in the flight deck. This should be done in accordance with the QRH procedure to ensure that the gear lever is in the OFF position on Classics. The handles will function with the landing gear lever in any position on the NG.

The hatch on the NG series has a micro switch which prevents landing gear retraction if the hatch is not fully closed.

Manual gear extension access hatch

Tyre damage fitting - NG only

Tyre Damage Pin

This pin is designed to detect any loose tyre tread during gear retraction. If any object impacts on it during retraction, then the gear will automatically extend. The affected gear cannot be retracted until this fitting is replaced. There is one pin at the aft outside of each main wheel well.

Tyres

737 tyres have always been bias-ply but from 2013 an option for radial tyres will be offered for aircraft with carbon brakes. Radial tyres have a longer life, better FOD resistance and are lighter. All tyres are tubeless and inflated with nitrogen. Pressures vary with series, maximum taxi weight, temperature and size of tyres. Unfortunately, this large variation in tyre pressures makes it difficult to know your aquaplaning speed. The table below should prove helpful, notice how the aquaplaning speeds are all just below the typical landing speeds. Note: Once aquaplaning has started, it will continue to a much lower speed. The maximum speed rating of all tyres is 225mph (195kts).

Series	Main Gear Pressure	Aquaplaning Speed	Nose Gear Pressure	Aquaplaning Speed
Originals	96 - 183psi	84 - 116Kts	125 - 145psi	96 - 104Kts
Classics	185 - 217psi	118 - 128Kts	163 - 194psi	111 - 121Kts
NGs	117 - 205psi	93 - 123Kts	123 - 208psi	95 - 124Kts

Another oddity of the 737 is the resonant vibration during taxiing that occurs at approx 17kts in Classics and 24kts in NGs. This is due to tyre "cold set". This is a temporary flat spot that occurs in tyres with nylon chord (ie all Boeing tyres) when hot tyres are parked and they cool to ambient temperature. Hence the reason why the flat spot is most pronounced in cold weather and tends to disappear during taxiing as the tyres warm up again.

The wheel well of a 737-500 in-flight

Gear Seals

The 737 series has never had full main gear doors. Instead the outer walls of the tyres meet with aerodynamic seals in the wheel well to make a smooth surface along the underside of the aircraft. The first few 737s had inflatable seals which were inflated by bleed air when the gear was either up or down and deflated during transit. The landing gear panel had a NOT SEALED caption which would illuminate during transit (normal), if it illuminated at any other time you could have a puncture and the seal could be depressurised with the GEAR SEAL SHUTOFF switch to save bleed requirements. These were soon dropped as being too complicated and a similar drag and noise advantage was achieved with the present fixed rubber seals.

The 737 has never had full main gear doors

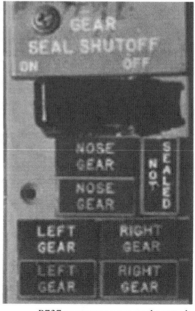

B737 prototype gear seal controls

Brakes

The brakes are multi-disc and are comprised of 4 rotors keyed to the wheel and 5 stators keyed to the axle. Brake pressure is applied by 6 pistons and there are 5 adjustor pins to keep the running clearance equal around the disc. Brake categories on the NG are from A to G depending upon the max taxi weight.

The standard 737 brakes are a steel alloy called Cerametalix(R) with versions made by either Goodrich or Honeywell. Since 2008 the 737NG has had a carbon brake option from either Goodrich with Duracarb(R) or Messier-Bugatti with SepCarb® III-OR. They are both about 300kgs lighter than steel and last twice as long.

Each wheel has 4 fuse pins which will melt at 177C (351F) or 7.5 units on the CDS BTMS. These will deflate the tyre to prevent it from exploding.

737 brake units showing the rotors, stators and pistons

Accumulator

The accumulator operates through the normal brake system and has three functions:

1. It uses its stored gas (nitrogen) pressure to give emergency brake pressure in the event of loss of hydraulic system A and B pressure. It can provide six full brake applications - even if all hydraulic power is lost. If you should get into that situation just apply the brakes and hold them on, don't cycle or pump the brakes because you only get six applications.
2. It dampens pressure surges and assures instantaneous flow of fluid to the brakes regardless of other hydraulic demands (for normal braking only).
3. It powers the parking brake through the normal system. A fully charged accumulator will keep the brakes pressurised for at least 8 hours

Accumulator servicing point – 737 Classic

There are two brake accumulators on the -1/200, one for inboard brakes & one for outboard brakes and just one on all later series.

The brake pressure gauge merely shows the pressure of the nitrogen side of the accumulator and should normally indicate 3000psi. The normal brake system and autobrakes are powered by hydraulic system B. If brake pressure drops below 1500psi, hydraulic system A automatically provides alternate brakes which are manual only (ie no autobrake) and the brake pressure returns to 3000psi. Antiskid is available with alternate brakes, but not touchdown or locked wheel protection on series before the NGs.

If both system A and B lose pressure, you are just left with residual hydraulic pressure and the accumulator pre-charge. The gauge will initially indicate approx 3000psi (residual pressure) and should provide a minimum of 6 full applications of brake power through the normal brake lines (so full antiskid is available). As the brakes are applied the residual pressure reduces until it reaches 1000psi at which point you will have no more braking available.

If the brake pressure gauge ever shows zero, this merely indicates that the pre-charge has leaked out; normal and alternate braking are unaffected if you still have the hydraulic systems (see QRH). The accumulator also provides pressure for the parking brake.

Brake Pressure Indication (psi)	Condition	Implication
3000	Hydraulics Normal	Braking normal (from hyd system B with accumulator damping)
3000	No hyd system B	Alternate braking available (from hyd system A)
3000 reducing with each brake application	No hydraulics	Minimum 6 full applications of brakes available (from accumulator)
1000	No hydraulics	Accumulator used up, no further braking available.
Zero	Hydraulics Normal	No pre-charge, normal braking available. (from hyd sys A or B)

Note that on the 737-1/200, hydraulic system A operates the inboard brakes and system B operates the outboard brakes. Both brake pressures are indicated on the single hydraulic brake pressure gauge.

Autobrake & Antiskid

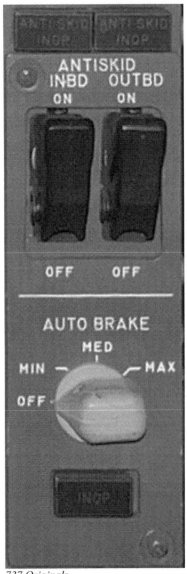

737 Originals

737-Classics

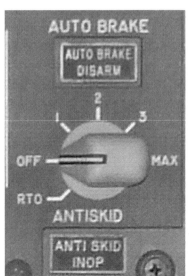

737-NGs

In all series, each main wheel has antiskid protection. Antiskid is available in both normal and alternate brake systems and is still available with loss of both hydraulic systems. Note that antiskid must be operative for autobrake operation (but not vice versa). Pre-NG aircraft had antiskid on/off switches, this is a throwback to early Boeing aircraft where antiskid could be selected off during taxi. Antiskid is automatically deactivated below 8kts.

After landing, there is an "on ramp" period where autobrake pressure is applied over a period of time. Approximately 750psi is applied in 1.75 sec, and then the pressures in the table are reached in another 1.25sec for autobrakes 1, 2, or 3 and approx. 1.0 sec for autobrake MAX.

Notice from the table below that autobrake Max does not give full brake pressure. For absolute maximum braking on landing, select autobrake Max to assure immediate application after touch down then override with full toe brake pressure. Using high autobrake settings with idle reverse is particularly hard on the brakes as they will be working for the given deceleration rate without the assistance of full reverse thrust.

Autobrake selector	Max pressure applied at brakes (PSI)	Deceleration rate (ft/sec/sec)
1	1250	4
2	1500	5
3	2000	7.2
Max	3000	12 (below 80kts)
Max	3000	14 (above 80kts)
RTO	Full	Not Controlled

To cancel the autobrake on the landing roll with toe brakes you must apply a brake pressure in excess of 800psi (ie less than that required for autobrake 1). This is more difficult on the NGs because the feedback springs on the brake pedals are stiffer. Autobrake can also be cancelled by putting the speedbrake lever down or by switching the autobrake off. I would advise against the latter in case you accidentally select RTO and get the full 3000psi of braking!

Occasionally you may see the brakes (rather than the cabin crew!) smoking during a turnaround. This may be due to hard braking at high landing weights. But the most common reason is that too much grease is put on the axle at wheel change so that when the wheel is pushed on, the grease is deposited inside the torque tube; when this gets hot, it smokes. It could also be contamination from hydraulic fluid either from bleeding operation or a leak either from the brakes or another source.

System	Power source
Normal brakes	Hydraulic system B
Alternate brakes	Hydraulic system A (if hyd B pressure falls below 1500psi)
Accumulator	Pre charge / Hydraulic system B
Autobrake	Hydraulic system B (requires normal brakes operative)
Antiskid	Electro-hydraulic (28V DC bus 1 & 28V DC Bat bus)
Parking brake	Hydraulic system B or accumulator

Limitations

- Do not apply brakes until after touchdown.
- Operation with assumed temperature reduced takeoff thrust is not permitted with anti-skid inoperative.
- Max speed for gear extension: 270kt / M0.82
- Max speed for gear retraction: 235kt
- Max speed with gear extended: 320kt / M0.82

NAVIGATION

The aircraft has several nav positions, many of which are in use simultaneously! They can all be seen on the POS REF page of the FMC.

IRS L & IRS R Position: Each IRS computes its own position independently; consequently they will diverge slightly during the course of the flight. After the alignment process is complete, there is no updating of either IRS positions from any external sources. Therefore it is important to set the IRS position accurately in POS INIT.

GPS L & GPS R Position: (NG only) The FMC uses GPS position as first priority for FMC position updates. Note this allows the FMC to position update accurately on the ground, eg if no stand position is entered in POS INIT. This practically eliminates the need to enter a take-off shift in the TAKE-OFF REF page.

Radio Position: This is computed automatically by the FMC. Best results are achieved with both Nav boxes selected to AUTO, thus allowing the FMC to select the optimum DME or VOR stations required for the position fix. Series 500 aircraft have an extra dedicated DME interrogator (hidden) for this purpose and NGs have two. Radio position is found from either a pair of DME stations that have the best range and geometry or from DME/VOR or even DME/LOC.

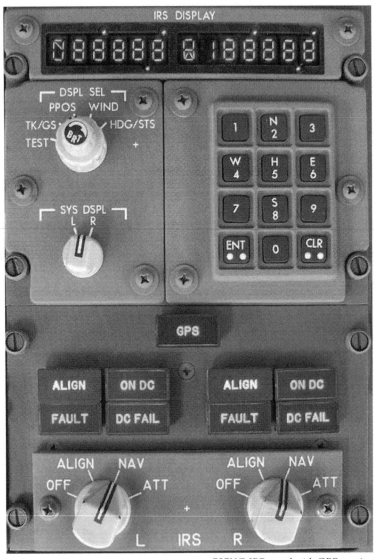

737NG IRS panel with GPS caption

FMC Position: FMC navigational computations & LNAV are based upon this. The FMC uses GPS position (NGs only) as first priority for FMC position updates, it will even position update on the ground. If GPS is not available, FMC position is biased approximately 80:20 toward radio position and IRS L. When radio updating is not available, an IRS NAV ONLY message appears. The FMC will then use a "most probable" position based on the IRS position error as found during previous monitoring when a radio position was available. The FMC position should be closely monitored if IRS NAV ONLY is in use for long periods.

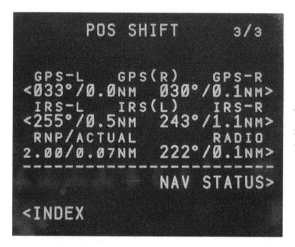

The NAV STATUS page shows the current status of the navaids being tuned. Navaids being used for navigation (ie radio position) are highlighted (here NTS & QPR).

If a navaid or GPS system is unreliable or giving invalid data then they can be inhibited using the NAV OPTIONS page. Note that LOC inhibit is only available from U10.7.

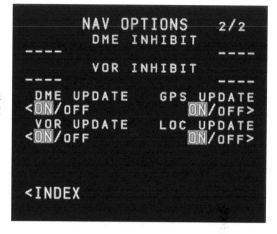

The POS SHIFT page shows the bearing & distance of other systems positions away from the FMC position. Use this page to force the FMC position to any of those offered.

Actual Navigation Performance (ANP) is the FMC's estimate of the quality of its position determination. The FMC is 95% certain that the aircraft's actual position lies within a circle of radius ANP centred on the FMC position. Therefore the lower the ANP, the more confident the FMC is of its position estimate. The 737 was the first aircraft to have the capability to calculate ANP and display it along with RNP at all times. The combination of the advanced navigation logic of the Smith's FMC, combined with dual GPS and dual scanning DMEs, can provide ANPs as low as 0.05nm (300ft).

Required Navigation Performance (RNP) is the desired limit of navigational accuracy and is specified by the kind of airspace you are in or type of approach that you are making. For instance in Oceanic airspace the RNP is 4nm, enroute is 2nm, terminal area 1nm and for an approach is 0.3 nm. The RNP may be overwritten by crew if a different value is required. ACTUAL should always be less than RNP.

Nav Tuning

The first thing to understand about the Nav panels (and the ADF, radio and transponder panels) is that they are only an interface with the receivers which are located in the E&E bay.

Original 737-200 Nav Control Panel

The Nav panels have changed in appearance with an increase in functionality over the years. In the original -1/200s they simply tuned either a VOR or ILS which was then displayed on the RMI and HSI.

Most Classics have an AUTO / MANUAL button to select between these modes. The Auto mode is used during the cruise and allows the FMC to auto-tune suitable nearby navaids to triangulate the radio position. The auto window will either show the frequency the FMC has tuned or dashes during "agility tuning". Agility tuning is when only one radio is in AUTO and is switching between two suitable beacons for triangulation. If the AUTO window shows "108.00" it means that radio is unreliable. The priority is for DME, then VOR, then LLZ.

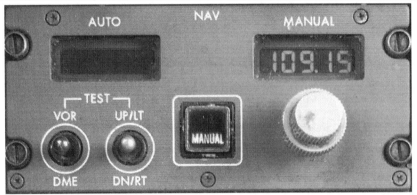

Typical 737 Classic Nav Control Panel

When in Manual mode the panel forces the receiver to the frequency that you select, usually in preparation for an SID or an approach. On some aircraft if Manual is left selected, FMC radio position updating cannot occur which can lead to positional inaccuracies, especially if you are also out of range of the selected frequency and your IRS's have drifted.

The latest NGs have Multi-mode receiver panels. These can tune VOR, ILS or GLS selectable with the MODE arrows.

Latest 737NG Multimode Nav Control Panel

EHSI & Navigation Display

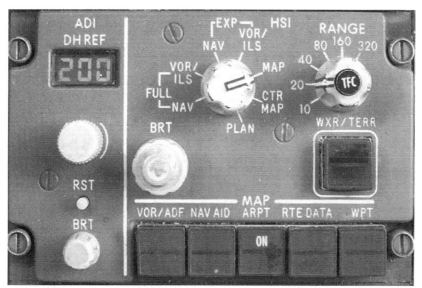

737-3/4/500 EFIS Control Panel

The 737 classic EFIS Control Panel allows you to select the type of display and overlays for the EHSI and DH on the EADI.

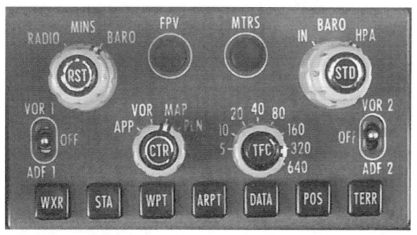

737-NG EFIS Control Panel

The NG EFIS control panel has the additional features of a flight path vector, feet/meters display, Inches/mb and EGPWS terrain overlay.

Full VOR / ILS Mode

This mode displays slightly differently when tuned to a VOR than an ILS. It appears very similar to the RMI/HSI displays found in round dial aircraft and general aviation. Most pilots prefer to use the expanded rather than full mode.

The Classic (shown here) and NG display are very similar, perhaps the main difference is that the NG will also show the beacon ident.

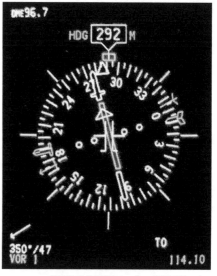

EHSI full VOR mode – 737 Classics

Expanded VOR / ILS Mode

This mode, when mastered, is far more accurate and easier to use than full mode. In particular the track line makes tracking beacons very easy, simply fly the heading that places the track line on the desired track/radial/localiser and away you go. It also unfortunately magnifies any inaccuracies in your tracking.

Full / Expanded Nav Mode

These modes are very rarely used. They give a raw data style display for FMC waypoints. They will show NAV instead of VOR 1 and display the waypoint name in the top left corner in magenta.

EHSI Expanded VOR mode

Map Mode

This is the most used mode; almost the entire flight is spent in map mode. The versions shown below are known as "Heading up" ie the map is oriented to the aircrafts heading. Some operators use "Track up" it is only a pin change to switch between the two but it can be disorientating for a pilot who is used to the other type of display.

The -3/4/500 EHSI Map mode

737-NG Navigation Display in Map mode

Centre Map Mode

The point of this mode is to "see" behind you, which can be useful for orientation. This is not a mode used by everybody, but I will often use it if a turn of more than 90° is expected eg when downwind or during a SID or when in the hold.

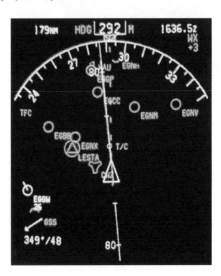

EHSI Centre Map mode

Plan Mode

As the name suggests, this mode is only usually used at the planning stage, to check the route when it is entered into the FMC. A visual gross error check of the route can easily be performed and any discontinuities, duplications or dog-legs removed. It may be used in-flight to inspect a later part of the route or for reference during an approach briefing.

Note that Plan mode is always oriented North up. The actual aircraft heading will still be displayed at the top.

This photograph shows the option with the arrival track in magenta and the missed approach track in cyan. The missed approach track turns magenta if TOGA is pressed on the approach.

737-NG Navigation Display in Plan mode

*** WARNING ***

NG ND & FMC showing DME disagreement

The ND DME readout below the VOR may not necessarily be that of the VOR which is displayed.

This photograph shows that Nav 1 has been manually tuned to 110.20 as shown in 1L of the FMC. DVL VOR identifier has been decoded by the auto-ident facility so "DVL" is displayed in large characters both on the FMC and the bottom left of the ND. Below this is displayed "DME 128" implying that this is the DME from DVL VOR.

However it can be seen on the ND that the DVL VOR is only about 70nm ahead. In fact DVL is only a VOR station and it has no DME facility, the DME was from another station on 110.20. The second station could be identified aurally by the higher pitched tone as "LRH" but was not displaying as such in line 2L of the FMC.

I only discovered this by chance as I happened to be following the aircraft progress by tuning beacons en-route (the way we used to do!). This illustrates the need to aurally identify any beacons, particularly DME, you may have to use, even if they are displayed as decoded.

Vertical Situation Display

The VSD, now certified on NGs, gives a graphical picture of the aircraft's vertical flight path. The aim is to reduce the number of CFIT accidents and profile related incidents, particularly on non-precision approaches and earlier recognition of unstabilised approaches.

The VSD works with the Terrain Awareness and Warning System (TAWS) to display a vertical profile of the aircrafts predicted flight path on the lower section of the ND. It is selected on with the DATA button on the EFIS control panel.

VSD can be retrofitted into any NG but it requires software changes to the displays and FMC and also some additional hardware displays.

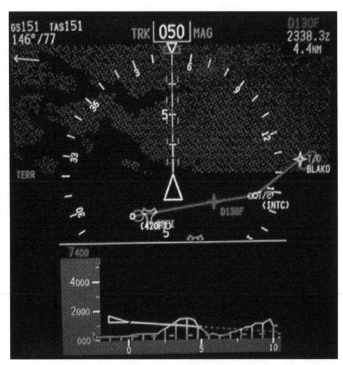

VSD on the Navigation Display

Magnetic Variation (MagVar)

The FMC and IRS's work with reference to True North, Magnetic is only used for pilot convenience. The conversion from True to Magnetic (Variation) is done by reference MagVar tables which are produced by NOAA. The current table is dated 2005 (although it was issued in 2001), the previous tables were dated 1995 and 1980 and the next is scheduled for 2015.

Most functions, such as the display of magnetic heading & track line, and the FCC control of HDG SEL & LNAV, are done with reference to the MagVar table built into each IRS. If the MagVar data is different to the actual variation, this will show as a small difference between track line and course, this can often be seen on expanded ILS mode during an ILS approach. The danger of using an out of date MagVar table is particularly apparent on an NDB approach where the crew fly a magnetic track to/from a beacon. If this track is misaligned due to variation error it could take the aircraft close to terrain.

The variation applied to the FMC legs comes from a variety of sources. The active leg uses IRS MagVar but the downstream legs use the FMC's own MagVar table. Furthermore, any legs originating at points with assigned MagVar or station declination, such as a VOR, use the published variation of that station which is held in the navigation database of the FMC; this data is usually kept up date monthly (see IDENT page).

Polar Navigation

This option enables NGs to operate in latitudes above 82°N/S (70°N/60°S in some areas). Aircraft with this facility have a HDG REF - NORM / TRUE switch on the aft overhead panel. When in NORM the IRS heading reference will automatically change over to True when the aircraft enters a polar region. Ideally TRUE should be pre-selected "to avoid loss of flight control computer functions."

Future Air Navigation System – FANS

On May 24, 2004, a BBJ became the first business jet to cross the Atlantic with FANS technology. It enables aircraft to communicate with ATC via Controller Pilot Data Link Communications (CPDLC) thereby eliminating the need to monitor HF. To upgrade a regular 737 to FANS-1 requires FMC update 10.5+, ATC datalink box and a simple wire run which provides the required aural chime and ATC alert on the forward display.

IRS Malfunction Codes (Classics)

Align Annunciator	Code	Significance of Annunciator or Malfunction Code	Recommended Action
Flashing (after 10 mins)	None	Failed align requirement	Verify and re-enter present position
----	01	ISDU failed power up RAM test	Replace ISDU
Steady	02	Entered latitude disagrees with latitude calculated by IRU	Verify and re-enter present position. If fault persists do full align or replace IRU
----	02	IRU failure	Replace IRU
Flashing	03	Excessive motion during align	Restart a full align
Flashing (During full align)	04	Lat or Long entered is not within 1 degree of stored value	Re-enter the identical position to the last position entered.
Flashing (During fast realign)	04	Lat is not within 1/2 degree or Long not within 1 deg of stored value	Enter known accurate present position. If align light continues to flash, do full align.
----	05	Left DAA is transmitting a fault	Replace left DAA.
----	06	Right DAA is transmitting a fault	Replace right DAA.
----	07	Selected IRU has detected an invalid air data input.	Replace DADC.
Flashing (after 10 mins)	08	Present position has not been entered	Enter present position
Steady	09	Attitude mode has been selected	Restart a full align. NB if ATT mode is desired, enter magnetic heading in POS INIT 1/2.
----	10	ISDU is not receiving power from both IRUs.	Ensure that both IRUs are ON and receiving power.

ADIRS Malfunction Codes (NGs)

ADIRS Maintenance Code Table			
01	ISDU fail	24	No AoA ref signal
02	IR failure	26	No baro 3 ref signal
03	Excessive motion	27	No pitot ADM data
04	Align fault	28	No static ADM data
07	ADR data invalid	29	No baro 1 data
08	Enter present position	30	No baro 2 data
09	Enter heading	31	No IR data
10	ISDU power loss	32	Pitot ADM data invalid
18	No ADR data	33	Static ADM data invalid
19	IR prog pin invalid	34	Baro 1 data invalid
20	ADR fail	35	Baro 2 data invalid
21	ADR prog pin invalid	36	Baro 3 signal fail
22	TAT probe signal fail	37	IR data invalid
23	AoA signal fail	38	Air/gnd logic invalid

Further IRS / ADIRS maintenance information can be found through the MAINT BITE INDEX pages of the FMC.

Alternate Navigation System - ANS

This is an option for the -3/4/500 series. ANS is an IRS based system which provides lateral navigation capability independent of the FMC. The ANS with the Control Display Units (AN/CDU) can be operated in parallel with the FMC for an independent cross-check of FMC/CDU operation or can be used alone in the event of an FMC failure.

The ANS is two separate systems, ANS-L & ANS-R. Each consists of its own AN/CDU and "on-side" IRS. Each pilot has his own navigation mode selector, next to the MCP, to specify the source of navigation information to his EFIS symbol generator and flight director.

Navigation source selector

The following FMC pages are only available on ANS equipped aircraft.

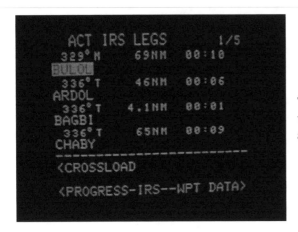

The ANS also performs computations related to lateral navigation which can provide LNAV commands to the AFDS in the event of an FMC failure.

The IRS PROGRESS page is similar to the normal PROGRESS page except that all data is from the "on-side" IRS (L in this example).

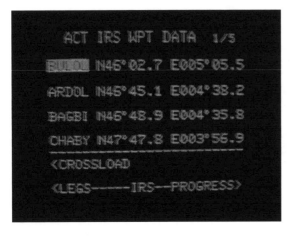

AN/CDU has no performance or navigation database. All waypoints must therefore be defined in terms of lat & long. The AN/CDU memory can only store 20 waypoints; these can be entered on the ground or in-flight and may be taken from FMC data using the CROSSLOAD function.

ETOPS

In 1953, the United States developed regulations that prohibited two-engine airplanes from routes more than 60 min single-engine flying time from an adequate airport (FAR 121.161). These regulations were introduced based upon experience with the airliners of the time, ie piston engined aircraft, which were much less reliable than modern jet aircraft. Nevertheless, the rule still stands.

ETOPS allows operators to deviate from this rule under certain conditions. By incorporating specific hardware improvements and establishing specific maintenance and operational procedures, operators can fly extended distances up to 180 min from the alternate airport. These hardware improvements were designed into Boeing 737-600/700/800/900.

The following table gives some FAA ETOPS approval times & dates:

Aircraft Series	Engine	ETOPS-120 approval date	ETOPS-180 approval date
737-200	JT8D -9/9A	Dec 1985	
	JT8D -15/15A	Dec 1986	
	JT8D -17/17A	Dec 1986	
737-300/400/500	CFM56-3	Sept 1990	
737-600/700/800/900	CFM56-7		Sept 1999
737-BBJ1/BBJ2	CFM56-7		Sept 1999

Limitations

Maximum flight operating latitude – 82° North and 82° South, except for the region between 80° West and 130 ° West longitude, the maximum flight operating latitude is 70° North, and the region between 120° East and 160° East longitude, the maximum flight operating latitude is 60° South.
(The above limitation does not apply to aircraft with polar navigation option)

Do not operate weather radar near fuel spills or within 15 feet of people.

Air Data Inertial Reference Unit: ADIRU alignment must not be attempted at latitudes greater than 78° 15 minutes.

The use of LNAV or VNAV with QFE selected is prohibited.

Use of the vertical situation display during QFE operation is prohibited.

(The QFE limitations are because several ARINC 424 leg types used in FMC nav databases terminate at MSL altitudes. If baro set is referenced to QFE, these legs will sequence at the wrong time and can lead to navigational errors.)

FMC U7.2 or earlier: During VOR approaches, one pilot must have raw data from the VOR associated with the approach displayed on the EHSI VOR/ILS mode no later than the final approach fix.

Enhanced GPWS:
- Do not use the terrain display for navigation.
- Do not use the look-ahead terrain alerting and terrain display functions within 15 nm of takeoff, approach or landing at an airport not contained in the GPWS terrain database

PNEUMATICS

The pneumatic system can be supplied by engines, APU or a ground source. The manifold is normally split by the isolation valve. With the isolation valve switch in AUTO, the isolation valve will only open when an engine bleed air or pack switch is selected OFF.

If engine bleed air temperature or pressure exceeds limits, the BLEED TRIP OFF light will illuminate and the bleed valve will close. You may use the TRIP RESET switch after a short cooling period. If the BLEED TRIP OFF light does not extinguish, it may be due to an overpressure condition.

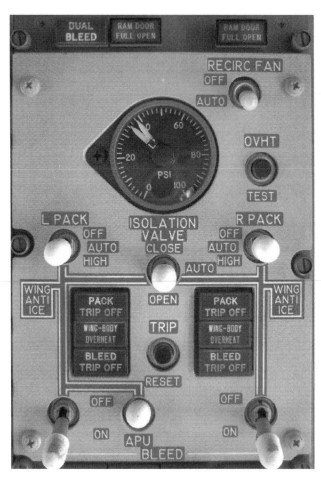

737-3/5/6/700 Pneumatics Panel

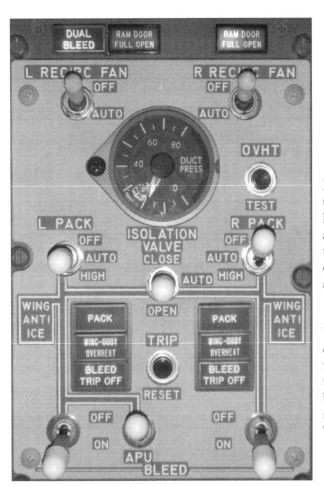

737-4/8/900 Pneumatics Panel

Bleed trip offs are most common on full thrust, bleeds off, take-offs. The reason is excessive leakage past the closed high stage valve butterfly which leads to a pressure build up at the downstream port on the overpressure switch within the high stage regulator. The simple in-flight fix is to reduce duct pressure by selecting CLB-2 and/or using engine and/or wing anti-ice.

WING-BODY OVERHEAT indicates a leak in the corresponding bleed air duct. This is particularly serious if the leak is in the left hand side, as this includes the ducting to the APU. The wing-body overheat circuits may be tested by pressing the OVHT TEST switch for a minimum of 5 seconds; both wing-body overheat lights should illuminate. This test is part of the daily inspection.

Air for engine starting, air conditioning packs, wing anti-ice and the hydraulic reservoirs comes from their respective ducts. Air for pressurisation of the water tank, auxiliary fuel tanks and the aspirated TAT probe comes from the left pneumatic duct. The minimum pneumatic duct pressure to ensure normal operation of these systems is 18psig (with the cowl and wing anti-ice OFF).

External air for engine starting feeds into the right pneumatic duct. Ground conditioned air feeds directly into the mix manifold.

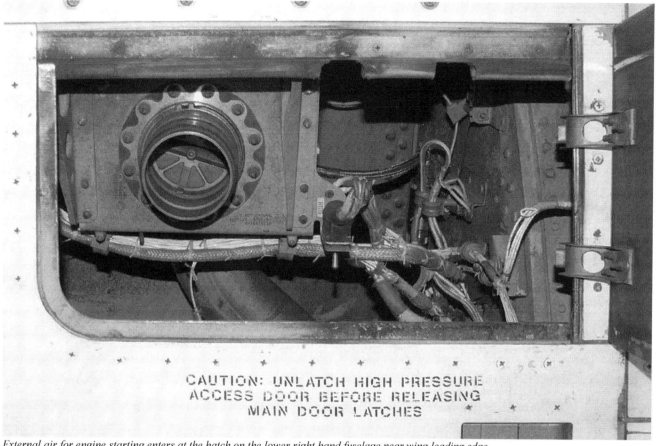

External air for engine starting enters at the hatch on the lower right hand fuselage near wing leading edge

External air for engine starting enters at the hatch on the lower right hand fuselage near wing leading edge. This is very close to the inlet for engine #2 which is why if you are using external air you must start engine #1 first to allow the ground crew to be able to safely disconnect the air.

Differences

1/200s - The PACK switches are simply OFF/ON, rather than OFF/AUTO/HIGH on all other series.

4/8/900s - Have two recirculation fans for passenger comfort and PACK warning lights instead of PACK TRIP OFF. See Air conditioning for an explanation. There are also two sidewall risers either side instead of one on all other series, this is why there appear to be two missing windows forward of the engine inlet.

Limitations

Max external air pressure: 60 psig
Max external air temperature: 450°F / 232°C

POWER PLANT

The original choice of powerplant was the Pratt & Whitney JT8D-1, but before the first order had been finalised the JT8D-7 was used for commonality with the current 727. The -7 was flat rated to develop the same thrust (14,000lb.st) at higher ambient temperatures than the -1 and became the standard powerplant for the -100. By the end of the -200 production the JT8D-17R was up to 17,400lb.st.thrust.

Auxiliary inlet doors were fitted to early JT8Ds around the nose cowl. These were spring loaded and hence opened automatically whenever the pressure differential between inlet and external static pressures was high, ie slow speed, high thrust conditions (takeoff) to give additional engine air. They closed again as airspeed increased causing inlet static pressure to rise.

JT8D with auxiliary inlet doors

The sole powerplant for all 737s after the -200 is the CFM-56. The core is produced by General Electric and is virtually identical to the F101 as used in the Rockwell B-1. SNECMA produce the fan, IP compressor, LP turbine, thrust reversers and all external accessories. The name "CFM" comes from GE's commercial engine designation "CF" and SNECMA's "M" for Moteurs.

The apparently flat-bottomed CFM56

One problem with such a high bypass engine was its physical size and ground clearance; this was overcome by mounting the accessories on the lower sides to flatten the nacelle bottom and intake lip to give the "hamster pouch" look. The engines were moved forward and raised, level with the upper surface of the wing and tilted 5 degrees up which not only helped the ground clearance but also directed the exhaust downwards which reduced the effects of pylon overheating and gave some vectored thrust to assist take-off performance. The CFM56-3 proved to be almost 20% more efficient than the JT8D.

The NGs use the CFM56-7B which has a 61 inch diameter solid titanium wide-chord fan, new LP turbine turbo-machinery, FADEC, and new single crystal material in the HP turbine. All of which give an 8% fuel reduction, 15% maintenance cost reduction and greater EGT margin compared to the CFM56-3.

The core engine (N2) is governed by metering fuel, whereas the fan (N1) is a free turbine. The advantages of this include: minimised inter-stage bleeding, fewer stalls or surges and an increased compression ratio without decreasing efficiency.

These statistics from CFMI issued in July 2006 illustrate the reliability of the CFM56:
- There are almost 16,000 CFM56 engines in service on 6,335 aircraft flying 100,000hrs a day.
- The CFM56 has amassed 325 million flight hours and 190 million cycles.
- The CFM56-3 averaged 18,000 hours before the first shop visit. In June 2012 a CFM56-7B engine delivered in 1999, became the first engine in the world to achieve 50,000 hours without a shop visit.
- The in-flight shutdown rate is 0.003% or better, ie one every 333,333 engine flight hours.
- If a pilot flew 900 flight hours annually he would have, on average, one shutdown every 185 years.

The 737MAX will be fitted with the all new LEAP-1B (Leading Edge Aviation Propulsion) with a 69 inch diameter fan. The turbine has flexible blades manufactured by a resin transfer molding process, which are designed to untwist as the rotational speed increases. The engine is expected to bring a 16% improvement in fuel efficiency, emissions and noise over the CFM-56-7B.

Flameout Problems

In the late 1980's, two 737-300's with CFM56-3-B1 suffered double-engine flameouts in heavy hail whilst descending at idle thrust (21 Aug 87 Air Europe near Thessaloniki and 24 May 88 TACA near New Orleans). Fortunately both crews managed to restart the engines, although the TACA aircraft subsequently made a successful deadstick landing on a levee.

The direct cause of the flameouts was the ingestion of hail into the engine core. Unlike rain, hail is much more effective at entering the core due to it either going between the fan blades and directly entering the core, the so called "Venetian blind effect" or bouncing off the spinner and fan blades and entering the core. At the time of CFM56-3 certification, hail was only considered a FOD hazard and certification focused upon impact damage rather than water ingestion.

> MAINTAIN AT LEAST 45% N1 WHEN OPERATING IN OR NEAR MODERATE TO HEAVY RAIN, HAIL OR SLEET

The 45% N1 placard was in effect from 1988 to 1991

Following the TACA event, AD 88-13-51 was issued to set a minimum 45% N1 during heavy rain, hail or sleet. One month after it was issued a Continental aircraft suffered an engine flameout in similar conditions when this procedure was not followed.

This limitation, which made descent planning very difficult for pilots, lasted for three years whilst CFM produced engine modifications. These included a cutback fan/booster splitter fairing that allowed more ingested rain/hail to be centrifuged out by the fan rotor away from the core and into the fan bypass flow; an elliptical spinner (see below) and variable bleed valve improvements which included increasing the number of variable bleed valve doors to 12 to allow additional rain/hail to be extracted from the core flowpath. In Sept 91, SB-737-77-1031 was issued lifting the minimum 45% N1 restriction in precipitation if the engines had these modifications.

Ice on the new CFM56-3 conical spinner

The original CFM56 spinner was conical (sharp pointed) to minimise the amount of ice that could form on the spinner, unfortunately this shape allowed any precipitation, particularly hail, to pass straight into the core of the engine which caused the flameouts. These were replaced from 1991 with elliptical (round nosed) spinners which succeeded in deflecting any ice or hail away from the core, but because of their larger frontal area, were more prone to picking up ice in the first place. The CFM56-7 spinner has a unique "conelliptical" profile which combines the conical front and elliptical base thereby solving both problems.

On 16 Jan 2002 a Garuda 737-300 with CFM56-3-B1 engines ditched following a double engine flameout 90secs after descending into a thunderstorm with idle thrust, despite having the above modifications. It is believed that the intensity of the hail encountered far exceeded any design standards and has therefore been treated as a "one off".

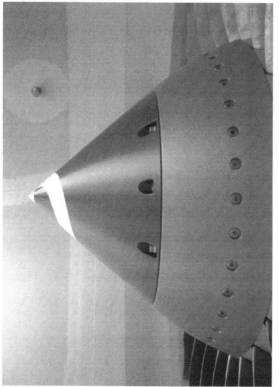

CFM56-7 "conelliptical" spinner

Engine Upgrades

There have been several major upgrades to the CFM 56; each has reduced the EGT margin to reduce wear and increase reliability. This allows a greater time on wing thereby reducing maintenance costs and down time.

The CFM56-3 "**Core**" upgrade became available in 2002. The HP turbine was redesigned and included a new nozzle, blade and shroud materials. EGT margin was increased by 15C and fuel burn by 1%.

From June 2007 the CFM56-7B was delivered with the "**Tech Insertion**" upgrade as standard. The package includes improvements to the HP compressor, combustor and HP & LP turbines. It reduces the EGT margin by up to 20C, fuel burn by 1%, maintenance costs by 5% and oxides of nitrogen (NOx) emissions by 15-20%. TI engines have a "/3" suffix appended to the designation, eg CFM56-7B24/3. TI engines with an optional "/F" rating have a 20C higher redline EGT, although DU & FCOM limits are unchanged.

The CFM56-7BE "**Evolution**" package was delivered from July 2011 with the following improvements:
- HPC outlet guide vane diffuser area ratio improved and pressure losses reduced.
- HPT blades numbers reduced, axial chord increased, tip geometry improved. Rotor redesigned.
- LPT blades & vanes reduced and profiles based on optimized loading distribution. LEAP56 incorporated.
- Primary nozzle, plug & strut faring all redesigned.

The -7BE can be intermixed all other CFM56-7 versions (SAC/DAC/TI) subject to updated FMC, MEDB and EEC.

Fuel

Thrust (fuel flow) is controlled primarily by a hydro-mechanical MEC in response to thrust lever movement, as fitted to the original 737-1/200s. In the –3/4/500 series, fuel flow is further refined electronically by the PMC, which acts without thrust lever movement. The 737-NG models go one stage further with FADEC (EEC).

The 3/4/500s may be flown with PMCs inoperative, but an RTOW penalty (ie N1 reduction) is imposed because the N1 section will increase by approximately 4% during take-off due to windmilling effects (FOTB 737-1, Jan 1985). This reduction should save reaching any engine limits. The thrust levers should not be re-adjusted during the take-off after thrust is set unless a red-line limit is likely to be exceeded, ie you should allow the N1s to windmill up.

Oil

Oil pressure is measured before the bearings, where you need it; oil temperature on return, at its hottest; and oil quantity at the tank, which drops after engine start. Oil pressure is unregulated; therefore the yellow band (13-26psi) is only valid at take-off thrust, whereas the lower red line (13psi) is valid at all times. If the oil pressure is ever at or below the red line, the LOW OIL PRESSURE light will illuminate and that engine should be shut down. NB on the 737-1/200 when the oil quantity gauge reads zero, there could still be up to 5 quarts present.

Ignition

There are two independent AC ignition systems, L & R. Starting with R selected on the first flight of the day provides a check of the AC standby bus, which would be your only electrical source with the loss of thrust on both engines and no APU. Normally, in-flight, no igniters are in use as the combustion is self-sustaining. During engine start or take-off & landing, GND & CONT use the selected igniters. In conditions of moderate or severe precipitation, turbulence or icing, or for an in-flight relight, FLT should be selected to use both igniters. NG aircraft: for in-flight engine starts, GRD arms both igniters.

CFM56 Ignition panel

The **Automatic Ignition** option on the NG has an AUTO position instead of OFF which switches the ignition on whenever flap or engine anti-ice is used.

The 737-NG allows the EEC to switch the ignition ON or OFF under certain conditions:
- The EEC will automatically switch ON both ignition systems if a flameout is detected.
- The EEC will automatically switch OFF both ignition systems if a hot or wet start is detected.

Note that older 737-200s have ignition switch positions named GRD, OFF, LOW IGN and FLT while newer 737s use GRD, OFF, CONT and FLT. This is why QRH uses "ON" (eg in the One Engine Inop Landing checklist) to cover both LOW IGN & CONT for operators with mixed fleets consisting of old and new versions of the 737.

737-200 Ignition panel

Engine Starting

Min duct pressure for start (Classics only): 30psi at sea level, -½psi per 1000ft pressure altitude. Max: 48psi.

Min 25% N2 (or 20% N2 at max motoring) to introduce fuel; any sooner could result in a hot start. Max motoring is when N2 does not increase by more than 1% in 5 seconds.

Aborted engine start criteria:
- No N1 (before start lever is raised to idle).
- No oil pressure (by the time the engine is stable).
- No EGT (within 10 secs of start lever being raised to idle).
- No increase, or very slow increase, in N1 or N2 (after EGT indication).
- EGT rapidly approaching or exceeding 725°C. (Classics only)

An abnormal start advisory does not by itself mean that you have to abort the engine start.

Starter cut-out is approx 46% N2 -3/4/500; 56% N2 -NGs.

The starter duty cycle limit for the 737 NG is maximum 2 minutes on, followed by a minimum of 10 seconds off. The starter duty cycle limit for the Classics and Originals was more restrictive.

Do not re-engage engine start switch until N2 is below 20%.

Engine covers, usually only seen on BBJs, are used to prevent FOD ingestion when parked for long periods.

During cold weather starts, oil pressure may temporarily exceed the green band or may not show any increase until oil temperature rises. No indication of oil pressure by the time idle RPM is achieved requires an immediate engine shutdown. At low ambient temperatures, a temporary high oil pressure above the green band may be tolerated.

When starting the engines in tailwind conditions, Boeing recommends making a normal start. Expect a longer cranking time to ensure N1 is rotating in the correct direction before moving the start lever. A higher than normal EGT should be expected, yet the same limits and procedures should apply.

Engine Indication Systems

"Round Dial" Engine Instruments

The most obvious difference between engine instruments on original series (1/200) and Classics (3/4/500) are the LEDs on the newer gauges. These have digital indication and warning lights for exceedances.

The blanking plate (bottom right) is probably where vibration gauges had been fitted.

-200 Electro-mechanical instruments

The later 737-200Advs also had the gauges with LEDs. So the only way to tell a -200 from a round dial classic was to look for EPR gauges

Round Dial -3/4/500

EIS

The introduction of Engine Instrument System (EIS) in late 1988 gave many advantages over the electromechanical instruments present since 1967. Such as a 10lb weight reduction, improved reliability, reduction in power consumption, detection of impending abnormal starts, storage of exceedances and a Built In Test Equipment (BITE) check facility.

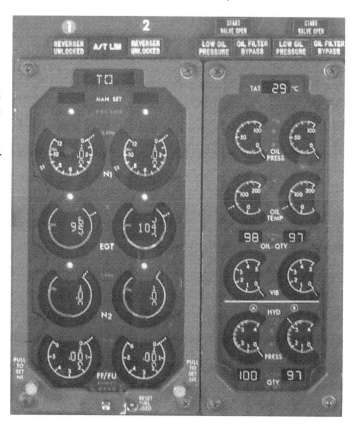

737-3/4/500 EIS

EIS BITE check

The BITE check is accessed by pressing a small recessed button at the bottom of each eis panel; this is only possible when both engines N1 are below 10%. Pressing these buttons will show an LED check during which the various checks are conducted. If any of the checks fail, the appropriate code will be shown in place of the affected parameters readout.

Any exceedance of either N1, N2 or EGT is recorded at 1 sec intervals in a non-volatile memory along with the fuel flow at the time, this data can be downloaded by connecting an ARINC 429 bus reader. Up to 10 minutes of data can be stored. The last exceedance is also put into volatile memory and can be read straight from the EIS before aircraft electrical power is removed. This is done by pressing the primary EIS BITE button twice within 2 seconds; this will then alternately display the highest reading and the duration of the exceedance in seconds.

EIS BITE Codes

Primary EIS BITE Codes

Code	Fault
ROM	Read Only Memory check
RAM	Random Access Memory check
FDC	Frequency to Digital Converter check
ENG	Engine Identity Inputs (not fuel flow)
PWR	Power Monitor
MMF	Maintenance Module Fault (fuel flow only)
RTC	Real Time Clock (fuel flow only)
ERF	Exceedance RAM Full (fuel flow only)
A/D	Analogue to Digital Converter (fuel flow only)
ARF	ARINC Receiver Fault (fuel flow only)
uP	Microprocessor

Secondary EIS BITE Codes

Code	Fault
0-	Microprocessor
1-	Program Memory
2-	Random Access Memory check
3-	Analogue to Digital Converter
4-	Power Monitor
5-	400Hz Reference Voltage
6-	ARINC Receiver Fault

NG Displays

The Compact Display mode can only be shown when the MFD ENG button is pressed for the first time after the aircraft has been completely shut down.

This photo shows the display after one engine has been started. The blank parameters on the number 1 engine are those controlled by the EEC which is not powered until the associated start switch is selected to GND.

During start-up the EECs receive electrical power from the AC transfer buses, but their normal source of power are their own alternators which cut-in when N2 is above 15%.

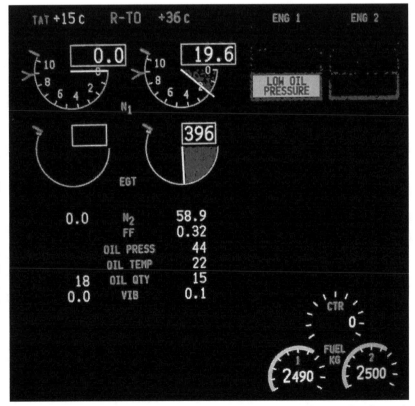

Upper DU in Compact Display mode

The normal display of the upper DU is simply the N1, EGT & fuel quantity.

This de-cluttered (or sparse depending upon your point of view) display can take some getting used to after the earlier series.

If a system is displayed on the lower DU after it has popped up with an anomaly, the upper DU will change to compact mode.

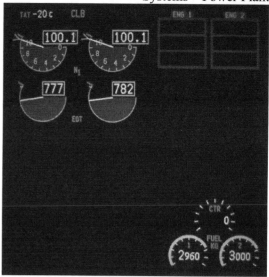

737 NG Upper DU

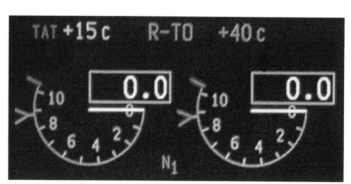

Lower DU

The Lower DU is normally only displayed on the ground during engine start, although it can be shown at any time. When an abnormality is detected, the lower DU will "pop up" automatically.

There is a bewildering array of possible thrust mode annunciations depending upon the type of take-off, climb, phase of flight and customer selectable options. Most of the annunciations refer to assumed temperature, derate or bump take-offs (See page 282). A list of the common options is shown below.

R-TO Thrust mode annunciation does not indicate type of reduced takeoff

TO	Takeoff		CLB	Climb
R-TO	Reduced takeoff		R-CLB	Reduced climb
TO 1	Derated takeoff one		CLB 1	Derated climb one
TO 2	Derated takeoff two		CLB 2	Derated climb two
TO B	Takeoff bump thrust		Q-CLB	Quiet climb
D-TO	Assumed temperature reduced thrust takeoff		CON	Continuous
D-TO 1	Derate one and assumed temperature takeoff		CRZ	Cruise
D-TO 2	Derate two and assumed temperature takeoff		G/A	Go-around
MAN	Indicates the N1 Set Switch is not in AUTO		- - - -	FMC not computing thrust limit

Airborne Vibration Monitors

All series of 737 have the facility for AVM although not all 737-200s have got them fitted. The early 737-1/200s had two vibration pickup points; One at the turbine section and one at the engine inlet there was a selector switch so that the crew could choose which to monitor. Some even had a high and low frequency filter selection switch.

Vibration pickup selector, 737-1/200 only

From Boeing Flt Ops Review, Feb 2003: "On airplanes with AVM procedures, flight crews should also be made aware that AVM indications are not valid while at takeoff power settings, during power changes, or until after engine thermal stabilization. High AVM indications can also be observed during operations in icing conditions."

Dual Annular Combustors (DAC)

The CFM56-7B is available with an optional DAC system, known as the CFM56-7B/2, which considerably reduces NOx emissions. DAC have 20 double tip fuel nozzles instead of the single tip and a dual annular shaped combustion chamber. The number of nozzles in use: 20/0, 20/10 or 20/20, varies depending upon thrust required. The precise N1 ranges of the different modes vary with ambient conditions.

- 20/20 mode - High power (cruise N1 and above)
- 20/10 mode - Medium power
- 20/00 mode - Low power (Idle N1)

This gives a lean fuel/air mixture, which reduces flame temperatures, and also gives higher throughput velocities which reduce the residence time available to form NOx. The net result is up to 40% less NOx emissions than a standard CFM56-7.

The first were installed on the 737-600 fleet of SAS, but unfortunately were subject to resonance in the LPT-1 blades during operation in the 20/10 mode, which occurred in an N1 range usually used during descent and approach. Although there were no in-flight shutdowns, boroscope inspections revealed that the LPT blades were starting to separate. CFM quickly replaced all blades on all DAC engines with reinforced blades and have since replaced them again with a new redesigned blade.

Reverse Thrust

The original 737-1/200 thrust reversers were pneumatically powered **clamshell doors** taken straight from the 727. When reverse was selected, 13th stage bleed air was ported to a pneumatic actuator that rotated the deflector doors and clamshell doors into position. Unfortunately they were relatively ineffective and, because they directed the exhaust vertically down against the runway, they tended to lift the aircraft up when deployed. This reduced the downforce on the main wheels thereby reducing the effectiveness of the wheel brakes.

Clamshell thrust reverser deployment

By 1969 (l/n 135) these had been changed by Boeing and Rohr to the much more successful hydraulically powered **target type thrust reversers**. This required a 48 inch extension to the tailpipe to accommodate the two cylindrical deflector doors which were mounted on a four bar linkage system and associated hydraulics. The doors are set 35 degrees away from the vertical to allow the exhaust to be deflected inboard and over the wings and outboard and under the wings. This ensures that exhaust and debris is not blown into the wheel-well, nor is it blown directly downwards which would lift the weight off the wheels or be re-ingested. The new nacelle was 1.14m longer and improved cruise performance by improving internal airflow within the engine and also reduced cruise drag.

These thrust reversers are locked against inadvertent deployment by both deflector door locks and the four bar linkage being over-centered. To illustrate how poor the original clamshell system was, Boeings own data says target type thrust reversers at 1.5 EPR are twice as effective as clamshells at full thrust!

Target thrust reverser Photo H Pujowinarso

The CFM56 uses **blocker doors** and cascade vanes to direct fan air forwards. Net reverse thrust is defined as: fan reverser air, minus forward thrust from engine core, plus form drag from the blocker door. As this is significantly greater at higher thrust, reverse thrust should be used immediately after landing or RTO and, if conditions allow, should be reduced to idle by 60kts to avoid debris ingestion damage. Caution: It is possible to deploy reverse thrust when either Rad Alt is below 10ft – this is not recommended.

CFM56 Reverse thrust

Engine panel (NGs)

The REVERSER light shows either control valve or sleeve position disagreement or that the auto-restow circuit is activated. This light will illuminate every time the reverser is commanded to stow, but extinguishes after the stow has completed, and will only bring up master caution ENG if a malfunction has occurred. Recycling the reverse thrust will often clear the fault. If this occurs in-flight, reverse thrust will be available after landing.

The REVERSER UNLOCKED light (EIS panel) is potentially much more serious and will illuminate in-flight if a sleeve has mechanically unlocked. Follow the QRH drill, but only multiple failures will allow the engine to go into reverse thrust.

The 737-1/200 thrust reverser panel has a LOW PRESSURE light which signifies that the reverser accumulator pressure is insufficient to deploy the reversers. The blue caption between the switches is ISOLATION VALVE and illuminates when the three conditions for reverse thrust are satisfied: Engine running, Aircraft on ground & Fire switches in normal position. The guarded NORMAL / OVERRIDE switches are to enable reverse thrust to be selected on the ground with the engines stopped (for maintenance purposes).

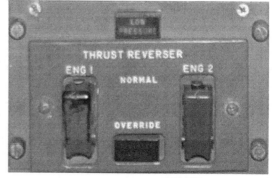

737-200 reverser panel

Hushkits

The first "hushkit" was not visible externally; in 1982 exhaust mixers were made available for the JT8D-15, -17 or -17R. These were fitted behind the LP turbine to mix the hot gas core airflow with the cooler bypassed fan air. This increased mixing reduced noise levels by up to 3.6 EPNdB.

NB The exhaust cone makes a divergent flow which slows down the exhaust and also protects the rear face of the last turbine stage.

JT8D Exhaust Mixer

This is the CFM56-3 turbine exhaust area. Unlike the JT8D, no mixing is required as the bypass air is exhausted coaxially.

CFM56-3 can be fitted with hardwall forward acoustic panels (HWFAP) which absorb noise at the nacelle by 1 EPNdB.

The 737 Classic family received FAA Stage 4 & EASA Chapter 4 noise recertification in 2012.

The view into the CFM56-3 jetpipe

Several different Stage 3 hushkits have been available from manufacturers Nordam and AvAero since 1992. The Nordam comes in HGW and LGW versions. The EU tried to ban all hushkitted aircraft flying into the EU from April 2002 because they use more fuel. This was strongly opposed and the directive has been changed to allow hushkitted aircraft to use airports which will accept them.

JT8D Hushkit from Nordam

Engine Ratings

Maximum Certified Thrust - This is the maximum thrust certified during testing for each series of 737. This is also the thrust that you get when you firewall the thrust levers, regardless of the maximum rated thrust. On the 737NG, the EEC limits the maximum certified thrust gained from data in the engine strut according to the airplane model as follows:

Aircraft Series	Max Certified Thrust
737-600	CFM56-7B22 = 22,700lb.st
737-700	CFM56-7B24 = 24,200lb.st
737-800	CFM56-7B27 = 27,300lb.st
737-900	CFM56-7B27 = 27,300lb.st

Maximum Rated Thrust - This is the maximum thrust for the installed engine that the autothrottle will command. This is specified by the operator from the options in the table below:

Engine	Aircraft Series	Max Thrust (lb.st.)	Bypass Ratio	EGT Margin (C)
JT8D-7/7A/7B	1/200	14,000	1.10	
JT8D-9/9A	1/200	14,500	1.04	
JT8D-15/15A	200Adv	15,500	0.99	
JT8D-17/17A	200Adv	16,000	1.02	
JT8D-17R	200Adv	17,400	1.00	
CFM56-3B4	500	18,500	5.0	90
CFM56-3B1	500/300	20,000	5.0	70
CFM56-3B2	300/400	22,000	5.0	50
CFM56-3C1	400	23,500	4.9	45
CFM56-7B18	600	19,500	5.5	145
CFM56-7B20	6/700	20,600	5.4	135
CFM56-7B22	6/700	22,700	5.3	150
CFM56-7B24	7/8/900	24,200	5.3	104
CFM56-7B26	7/8/900/BBJ	26,400	5.1	85
CFM56-7B27	8/900/BBJ	27,300	5.0	59

There are two fan inlet temperature sensors in the CFM56-3 engine intake. The one at the 2 o'clock position is used by the PMC and the one at the 11 o'clock position is used by the MEC. The MEC uses the signal to establish parameters to control low and high idle power schedules. The CFM56-7 inlet has just one fan inlet temperature probe, which is for the EEC (because there is no PMC on the NGs).

The temperature data is used for thrust management and variable bleed valves, variable stator vanes & high / low pressure turbine clearance control systems.

A subtle difference between the NG & classic temp probes is that the NGs only use inlet temp data on the ground and for 5 minutes after take-off. In-flight after 5 minutes temp data is taken from the ADIRUs.

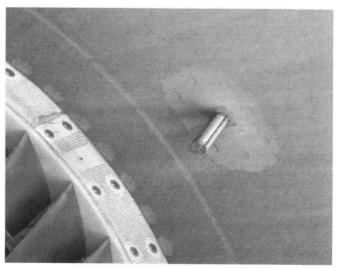

CFM56-3 Inlet Temp Probe

CFM56-3 during overhaul

The photo above shows the CFM56-3 removed from the wing during overhaul. This is becoming a rare sight as continuing improvements to the design increase the time between overhaul. One CFM56-3 on a 737-400 operated by Malev had over 35,000 hours on-wing.

The cascade vanes of the reverse thrust system can be seen opened away from the core. The box at the bottom is the oil tank, the next one up is the PMC and the smaller pair above that are the ignition exciters.

The left hand side of the CFM56-3. The large silver pipe is the start air manifold with the starter located at its base. The unit below that is the CSD.

CFM56-3 with the cowling open

Limitations

- Engine ignition must be on for: Takeoff, Landing, Operation in heavy rain and Anti-ice operation.
- Intentional selection of reverse thrust in flight is prohibited.

Aircraft Series	1/200	1/200	3/4/500	6/7/8/900
Engine	**JT8-9**	**JT8-17A**	**CFM56-3**	**CFM56-7**
Max time limit for take-off or go-around thrust:	5 mins	5 mins	5 mins *	5 mins *
Max N1	100.1%	102.4%	106%	104%
Max N2	100%	100%	105%	105%
Max EGTs:				
Take-off (5 min limit)	580°C	650°C	930°C	950°C
Continuous	540°C	610°C	895°C	925°C
Start	420°C	575°C	725°C	725°C
Oil T's & P's				
Max temperature	157°C	165°C	165°C	155°C
15 minute limit (45 minute limit on NG)	121-157°C	130-165°C	160-165°C	140-155°C
Max continuous	120°C	130°C	160°C	140°C
Min oil press	40psi	40psi	13psi (light) 26psi (gauge)	13psi (light) 26psi (gauge)
Min oil quantity (at dispatch)	2.25 USG	2.25 USG	60% full (12 US Quarts)	**60%** full (12 US Quarts)
Starting pressures prior to starter engagement	30psi -1/2psi per 1000ft amsl			N/A
Starter duty cycle	1st attempt: 2min on, 20sec off 2nd & subsequent attempt: 2min on, 3min off			2mins on, 10secs off.

* May be increased to 10 minutes if certified

WARNING SYSTEMS

Warning Lights

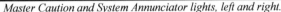

Master Caution and System Annunciator lights, left and right.

The Master Caution system was developed for the 737 to ease pilot workload as it was the first Boeing airliner designed to be flown without a flight engineer. In simple terms it is an attention getter which also directs the pilot toward the problem area. The system annunciators (shown above) are arranged such that the cautions are in the same orientation as the overhead panel e.g. FUEL bottom left, DOORS bottom of third column, etc. The 737-1/200s do not have the IRS or ENG captions.

On the ground, the master caution system will also tell you if the condition is dispatchable or if the QRH needs to be actioned. The FCOM gives the following guidance on master caution illuminations on the ground:

Before engine start, use individual system lights to verify the system status. If an individual system light indicates an improper condition:
- Check the Dispatch Deviations Procedures Guide (DDPG) or the operator equivalent to decide if the condition has a dispatch effect
- Decide if maintenance is needed

If, during or after engine start, a red warning or amber caution light illuminates:
- Do the respective non-normal checklist (NNC)
- On the ground, check the DDPG or the operator equivalent

If, during recall, an amber caution illuminates and then extinguishes after a master caution reset:
- Check the DDPG or the operator equivalent
- The respective non-normal checklist is not needed

Pressing the system annunciator will show any previously cancelled or single channel cautions. If a single channel caution is encountered, the QRH drill should not be actioned.

Master caution lights and the system annunciator are powered from the battery bus and will illuminate when an amber caution light illuminates. Exceptions to this include a single centre fuel tank LOW PRESSURE light (requires both), REVERSER lights (requires 12 seconds) and INSTR SWITCH (inside normal field of view).

When conducting a light test, during which the system will be inhibited, both bulbs of each caution light should be carefully checked. The caution lights are keyed to prevent them from being replaced in the wrong sockets, but may be interchanged with others of the same caption.

Keying of warning lights

- Red lights - Warning - indicate a critical condition and require immediate action.
- Amber lights - Caution - require timely corrective action.
- Blue lights - Advisory - eg valve positions and unless bright blue, ie a valve/switch disagreement, do not require crew action.
- Green lights - Satisfactory - indicate a satisfactory or ON condition.

Aural Warnings

Cockpit aural warnings include the fire bell, take-off configuration warning, cabin altitude, landing gear configuration warning, mach/airspeed overspeed, stall warning, GPWS and TCAS. External aural warnings are: The fire bell in the wheel well and the ground call horn in the nose wheel-well for an E & E bay overheat or IRSs on DC. Only certain warnings can be silenced whilst the condition exists.

Cockpit aural warnings are produced from two sources:

Aural Warning Module.
This unit, located below the F/O's CDU, contains all the electronics and its own speakers to produce aural warnings. It receives discrete inputs from services such as the PSEU, ADIRU, fire detection module, etc to command it to generate the appropriate aural warning. Eg landing config (continuous horn), takeoff config (intermittent horn), cabin pressure (intermittent horn), autopilot disengage (wailer), overspeed (clackers), fire (bells), SELCAL (chimes) & crew call (chimes).

Note that the Partial or Gear Up Landing QRH drill calls for the aural warning c/b to be pulled. This will cause the loss of all horn, chime & wailer warnings, which will include the cabin crew call chime. Worth considering after the gear up landing!

Aural warning module

Remote Electronics Unit

This handles voice warnings. It receives signals from an audio generator from TCAS, GPWS and customer optional calls such as radio altitudes, minimums, bank angle, V1 etc. These are then assessed for priority in accordance with the table below and sent on to the speakers and headsets.

PRIORITY	MODE	DESCRIPTION	ALERT
		AURAL WARNING PRIORITY LOGIC	
1	7	WINDSHEAR WINDSHEAR WINDSHEAR	Wailer
2	1	PULL-UP (SINK RATE)	Wailer
3	2	PULL-UP (TERRAIN CLOSURE)	Wailer
4	2A	PULL-UP (TERRAIN CLOSURE)	Wailer
5	V1	V1 CALLOUT	Intermittent
6	TA	TERRAIN TERRAIN PULL-UP	Wailer
7	WXR	WINDSHEAR AHEAD	Wailer
8	2	TERRAIN TERRAIN	Continuous
9	6	MINIMUMS	Intermittent
10	TA	CAUTION TERRAIN	Continuous
11	4	TOO LOW TERRAIN	Continuous
12	TCF	TOO LOW TERRAIN	Continuous
13	6	ALTITUDE CALLOUTS	Intermittent
14	4	TOO LOW GEAR	Continuous
15	4	TOO LOW FLAPS	Continuous
16	1	SINK RATE	Continuous
17	3	DONT SINK	Continuous
18	5	GLIDESLOPE	Continuous
19	WXR	MONITOR RADAR DISPLAY	Continuous
20	6	APPROACHING MINIMUMS	Intermittent
21	6	BANK ANGLE	Continuous
22	TCAS	RA (CLIMB, DESCEND, ETC.)	Wailer
23	TCAS	TA (TRAFFIC, TRAFFIC)	Continuous
24	TEST	BITE AND MAINTENANCE INFORMATION	Intermittent

Radio Altimeter Callouts

Automatic rad-alt calls are a customer option. Calls can include any of the following:

> 2500 ("Twenty Five Hundred" or "Radio Altimeter"), 1000, 500, 400, 300, 200, 100, 50, 40, 30, 20, 10.
> "Minimums" or "Minimums, Minimums"
> "Plus Hundred" when 100ft above DH
> "Approaching Minimums" when 80ft above DH
> "Approaching Decision Height"
> "Decision Height" or "Decide"

Customers can also request special heights, such as 60ft.

Noise Levels

If is often commented how loud some callouts are. The volume level for these callouts, and any other aural warnings, is set so that they can still be audible at the highest ambient noise levels; this is considered to be when the aircraft is at Vmo (340kts) at 10,000ft.

The design sound pressure level at 35,000ft, M0.74, cruise thrust is 87dB at the Captains seat, compared to 90-93dB in the cabin.

Many pilots consider the 737 flightdeck to be generally loud. This is Boeings response to that charge:
"Using the flight deck noise levels measured by Boeing Noise Engineering during a typical flight profile (entire flight), a daily A-weighted sound exposure was calculated using ISO/DIS 1999 standards. This calculation indicates the time weight noise exposure is below 80dB(A) and should not cause hearing damage. Flight deck noise improvement continues to be a part of current Boeing product quality improvement activities."

That said, in 2004 Boeing introduced the following noise reduction features for the 737NG flightdeck:
- Damping panels added to skin interior between frames.
- 10 small vortex generators added at the base of the windscreen to minimise boundary layer noise (see page 157 for photo).
- Gap around DV windows sealed.
- The noise from the air conditioning packs on the flightdeck has been reduced by rerouting the hoses to reduce bends and designing a new plenum and muffler/diffuser.
- Increased density of insulation blanket.

These were introduced in phases from L/N 1552 and are available for retrofit. The total noise reduction is approximately 3dB(A) in the cruise.

GPWS

To test the GPWS, ensure that the weather radar is on in TEST mode and displayed on the EHSI. Pressing SYS TEST quickly will give a short confidence test, pressing for 10 seconds will give a full vocabulary test.

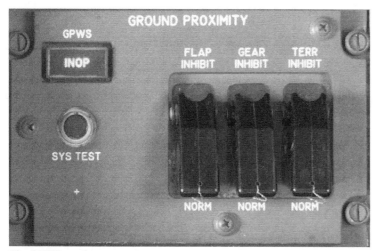

The GPWS panel.

TCAS

Various versions of TCAS have been fitted since its introduction in the 1990's. In the early days of TCAS there were different methods of displaying the visuals. For the Honeywell system (Previously AlliedSignal, previous to that - Bendix/King), their most popular method for non-EFIS airplanes was to install an RA/VSI (see photo right). When activated, this had "eyebrows" on the outer edge directing the pilot to climb (green) or stay away from (red) and use the separate Radar Indicator for the basic traffic display. Even early EFIS aircraft had the RA/VSI due to difficulties (ie expense) integrating them into the EHSI.

Combined TCAS & VSI display

TCAS control is from the transponder panel.

TCAS is now integrated at production into the EFIS displays. The PFD/EADI will display advisories to climb, descend, or stay level since they give the vertical cue to the pilot. The ND/EHSI provides the map view looking down to show targets and their relative altitude and vertical movement relative to your aircraft.

Proximity Switch Electronic Unit

A Proximity Switch Electronic Unit (PSEU) is a system that communicates the position or state of system components eg flaps, gear, doors, etc to other systems. The 737-NGs are fitted with a PSEU which controls the following systems: Take-off and landing configuration warnings, landing gear transfer valve, landing gear position indicating and warning, air/ground relays, airstairs & door warnings and speedbrake deployed warning. The 737-800SFP and -900ER also have a Supplemental PSEU and associated SPSEU light. This monitors the mid-exit locks and tailskid position.

The PSEU light is inhibited in-flight; actually from when the thrust levers are set for take-off power (thrust lever angle beyond 53 degrees) until 30 seconds after landing.

If the PSEU light illuminates, you have a "**non-dispatchable fault**" and the QRH says do not take-off. In this condition the PSEU light can only be extinguished by fixing the fault. However if you only get the PSEU light on recall, you have a "**dispatchable fault**" which it is acceptable to go with. In this condition the PSEU light will extinguish when master caution is reset.

The mysterious PSEU light

Stall Warning

Under some conditions the margin between buffet and stall may be small; therefore artificial stall warners are used. They are simply eccentric weight motors at the base of each control column activated by the SMYD.

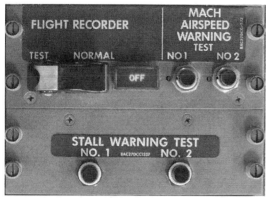

737-3-900 Stall Warning Panel

Stall warning test requires AC power. Also, with no hydraulic pressure, the leading edge flaps may droop enough to cause an asymmetry signal, resulting in a failure of the stall warning system test. If this happens, switch the "B" system electric pump ON to fully retract all flaps and then repeat the test.

Stick shake motor

737-200 Stall Warning Panel

737-1/200: The OFF light may indicate either a failure of the heater of the angle of attack sensor a system signal failure or a power failure. The test disc should rotate, indicating electrical continuity, when the switch is held to the test position.

SMYD

The stall warning system on the NG is far more advanced than in the previous generations due to the Stall Management Yaw Damper (SMYD) computers. The stall warning is triggered by a "speed floor" (Alpha floor?) Angle of Attack (AoA) for the flap setting (between 13.0 – 23.5 AoA) with a bias subtracted for:

- LE flap and slat asymmetry* (between 10.2 - 15.0 AoA)
- High thrust** (between 0 - 13.6 AoA)
- LE uncommanded motion (between 2.3 - 6.7 AoA)
- TAI*** (between 0.8 - 5.3 AoA)

* The leading asymmetry bias is set if one or more LE devices are in a position that disagrees with the TE flaps.
** The high thrust bias is enabled if the offside engine N2 is more than 75% and the onside engine N1 is valid. This prevents a pitch up stall tendency at low speeds and high thrust.
*** The thermal anit-ice (TAI) bias decreases the trip point to account for 3 inches of rough ice on the LE devices. The TAI bias is set when either the WING ANTI-ICE or ENG ANTI-ICE switches are in the ON position. If the WING ANTI-ICE switch is moved to the ON position for greater than 5 seconds, a TAI bias latch is set for the remainder of the flight.

When a stall is reached, the FCC gives a nose-down speed trim input and the elevator feel shift module gives an increased control column back-force gradient. These two effects artificially give the 737NG much better stall recovery characteristics than previous series. These effects were increased on demand from the JAA during 737-700 certification trials but was only mandated to JAA registered aircraft. Therefore not all aircraft (or simulators) will behave the same during a stall.

Weather Radar

The weather radar is an array of radiation slots arranged into a flat plate (see photo on page 156). It produces a beam 3.4 degrees high and 3.6 degrees wide at 9.345GHz. It detects water droplets

Weather radar or terrain can be overlaid onto the EHSI with these switches on the Classics. In the NG the overlay switches are part of the EFIS control panel. The colours may appear similar but their meanings are very different.

There are many different weather radar control panels in use in 737s. You should be familiar with the controls of the panels fitted in your fleet.

Note that there are no radar emissions in the test mode, so it is safe to conduct a test on stand or near people.

WXR/TERR Switching Controls

Typical Weather Radar Controls

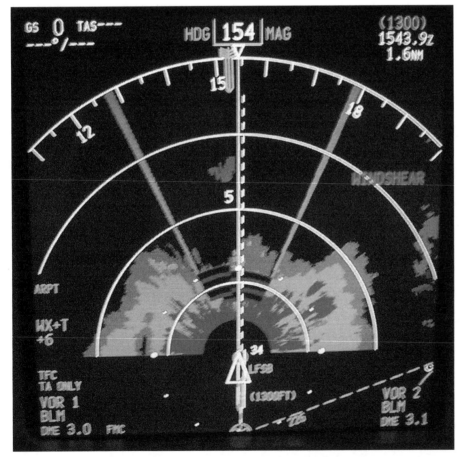

Windshear warning displayed on the ND. Notice the cone and range at which windshear is predicted

737 NGs are fitted with predictive windshear system (PWS). This is available below 2300ft. You do not need weather radar to be switched on for PWS to work, since it switches on automatically when take-off thrust is set. However there is a 12 sec warm up period, so if you want PWS available for the take-off you should switch the weather radar on when you line up.

This photograph shows windshear detected 25 degrees either side of the nose at range 2 to 3 miles (striped area). Fortunately I was still on the ground when I took the photo and I stayed there until it had passed through!

EGPWS - Peaks Display

The Peaks display overlays EGPWS terrain information onto the EHSI or ND. The two overlaid numbers are the highest and lowest terrain elevations, in hundreds of ft amsl, currently being displayed. Here 5900ft and 800ft amsl. One of the main differences between Peaks display and standard GPWS is that it will show terrain more than 2000ft below your level (ie from cruise altitude). This can be very useful for terrain awareness in situations such as descending in the terminal area, off airway deviations around weather or an unplanned descent.

737 Classic EHSI with Peaks terrain display

The colour coding is similar to the weather radar but with several densities of each colour being used. The simplified key is:

Colour	Altitude Diff from Aircraft (ft)
Black	No significant terrain
Cyan	Zero ft MSL (Customer option)
Green	-2000 to -500 (Gear up) -2000 to -250 (Gear down)
Yellow	-500 to +2000 (Gear up) -250 to +2000 (Gear down)
Red	+2000 or above
Magenta	Terrain elevation unknown

EGPWS Limitations
- Do not use the terrain display for navigation.
- Do not use within 15nm of an airfield not in the terrain database.

WINGS & WINGLETS

The wings are an aluminium alloy, dual-path, fail-safe, two-spar structure. Shear loads are carried by the front and rear spars, bending loads are carried by the upper and lower skin panels. These four surfaces form the wing box which also serves as an integral fuel tank. The spars are reinforced by vertical stiffeners and the skin panels are reinforced by spanwise "Z" or "J" section stringers.

The original wing was designed to be short to keep weight to a minimum, but to be large enough to carry fuel for the designed range. It also had to have a short chord to accommodate the engine which was to be flush mounted rather than on a pylon because of ground clearance. It also had much less sweep than its predecessors (25° compared to 35° on the 727) because speed was not considered important for a short range jet. This is why we plod along the airways at M0.74 Classics, M0.78 NGs.

Wing Dimensions	Originals	Classics	NGs
Span (m)	28.35	28.88	34.32
Gross Area (m²)	102.00	105.40	124.58
Aspect Ratio	8.83	8.99	9.45
Taper Ratio	0.339	0.240	0.159
Root Chord (basic)	4.71	4.71	7.877
Mean Aerodynamic Chord (m)	3.80	3.73	3.96
Tip Chord (m)	1.60	1.13	1.251
Dihedral (°)	6	6	6
Incidence at root (°)	1	1	1
¼ Chord Sweep (°)	25.02	25.02	25.02

Having the engines wing-mounted also helped alleviate the bending moment of the wings from the lift, which allowed the spar weights to be reduced. Unfortunately this was initially overdone as a prototype managed to buckle the rear wing spar during high speed tests at 34% above normal operational loads. The spars were strengthened and this was incorporated into all production aircraft.

The basic wing remained the same on all 1/200s until the 3/4/500s which had 0.2m wingtip extensions and a slightly modified aerofoil section (taken from the 757/767) for the LE slats. This gave a 4% improvement in maximum L/D, a 0.02 Mach increase at maximum L/D, and a 3% reduction in block fuel at 1500nm range. The new slats went from engine pylon to wingtip and gave an average chord increase of 4% over the whole wing and gave the Classics similar approach speeds to the originals. The wingtip extension came about when a flutter boom was being designed to eliminate flutter between the wing and the new powerplant & pylons. As it happened the boom was not needed but the wingtip extension was retained and with the new slats gave the Classics an extra 4000ft altitude capability.

The 737-1/2/3/4/500 aerofoil is made up of four specially designed Boeing aerofoil sections splined into one smooth surface as follows:

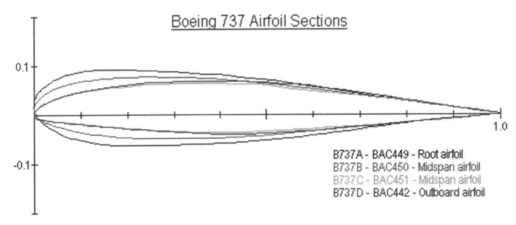

212

Short field take-off performance was achieved by Krueger flaps and leading edge slats. Short field landing performance came from the triple slotted trailing edge flaps (double slotted for NGs). Both of these gave the 737 good low speed handling and allowed it to operate into many fields which had not previously been able to take jets.

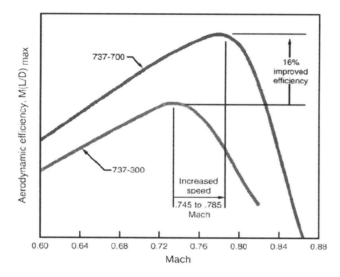

The NG wing has a whole new airfoil section with 25% increase in area, 107 inch semi-span increase, 17 inch chord increase, raked wing-tip and a larger inspar wingbox with machined ribs. Approach speeds have been reduced and the altitude capability increased by 4000ft to FL410.

Blended Winglets

The most noticeable feature to appear on new 737s are the blended winglets. These are wing tip extensions which reduce lift induced drag and provide some extra lift.

They have been credited to Dr Louis Gratzer formerly Chief of Aerodynamics at Boeing and now with Aviation Partners Boeing (APB). They were first flown on a 737-800 in June 1998 as a testbed for use on the BBJ. They became available as a standard production line option in May 2001 on all NGs with the exception of the -600 series, for which Boeing is "continuing to assess the applicability" and the BBJ range which have to be retrofitted.

They are 8ft 2in tall and about 4 feet wide at the base, narrowing to approximately two feet at the tip and add almost 5 feet to the total wingspan. The winglet for the Classic is slightly shorter at 7ft tall.

737NG winglets

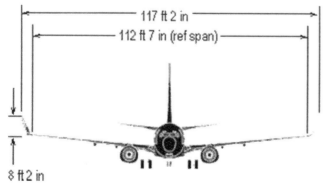

Winglets have the potential to give the following benefits:

- **Improved climb gradient**. This will enable a higher RTOW from climb limited airports (hot, high or noise abatement) or obstacle limited runways.
- **Reduced climb thrust**. A winglet equipped aircraft can typically take a 3% derate over the non-winglet equivalent aircraft. This can extend engine life and reduce maintenance costs.
- **Environmentally friendly**. The derate, if taken, will reduce the noise footprint by 6.5% and NOx emissions by 5%. This could give savings on airport noise quotas or fines.
- **Reduced cruise thrust**. Cruise fuel flow is reduced by up to 6% giving savings in fuel costs and increasing range.
- **Improved cruise performance**. Winglets can allow aircraft to reach higher levels sooner. Air Berlin notes, *"Previously, we'd step-climb from 35,000 to 41,000 feet. With Blended Winglets, we can now climb direct to 41,000 feet where traffic congestion is much less and we can take advantage of direct routings and shortcuts which we could not otherwise consider."*
- **Good looks**. Winglets bring a modern look and feel to aircraft, and improve customers' perceptions of the airline.
- **Higher Aircraft Resale Value**.

Drawbacks

If winglets are so good, you may wonder why all 737s don't have them. In fact 85% of all new 737s are now built with winglets, mostly the 800, 900ER & BBJs and there is a two year wait for Classic winglets. It comes down to cost versus benefits. Winglets cost $850,000USD for NGs and $550,000USD for the Classics and take about one week to install which costs an extra $25-80,000USD. Once fitted, they add 170-235kg (375-518lbs) to the weight of the aircraft, depending upon whether they were installed at production or a retrofit. The fuel cost of carrying this extra weight will take some flying time each sector to recover, although this is offset by the need to carry less fuel because of the increased range. In simple terms, if your average sector length is short (less than one hour) you wont get much the benefit from winglets - unless you need any of the other benefits such as reduced noise or you regularly operate from obstacle limited runways.

Handling Differences

There is a small difference in rotation rate for aircraft with winglets installed and, as a result, crew need to be cautious of pitch rate. There is approximately a ½ unit take-off trim difference so the green band is slightly different and winglet aircraft feel slightly heavier in roll than non-winglet aircraft. Finally, the dry "maximum demonstrated" crosswind limit is slightly reduced with winglets to 34kts. According to APB this is because:

"The FAA will only let us document the max winds experienced during flight test... so if we had been able to find more crosswind, then the 33kts might have been more. There appears to be no weather cocking effect due to winglets."

Blended Winglets are also available for the 737-3/4/500. The first one flew in Nov 2002 and gained its FAA supplemental type certificate (STC) on 30 May 2003. Blended Winglet equipped Classics are known as Special Performance (SP).

737-300SP

737-200 Mini-Winglets

There was a retrofit for 737-200s to be fitted with mini-winglets. This is part of the Quiet Wing Corp flap modification kit which gained its FAA certification in 2005. The package includes drooping the TE flaps by 4 degrees and the ailerons by 1 degree to increase the camber of the wing. Benefits include:

- Payload Increase of up to 5,000 lbs.
- Range Increase / Fuel Savings of up to 3%
- Improved Takeoff / Landing Climb Gradients
- Reduced Takeoff / Landing Field Length
- Improved Hot & High Altitude Takeoff / Landing Capability
- Reduced Stall Speeds by 4-5kts

737-200Adv with mini-winglets

Split Scimitar Winglets

The Split Scimitar Winglet is a retrofit winglet for the 737NG family. It is designed and produced by Aviation Partners Boeing. It was first flown on July 16, 2013 and went operational in early 2014.

Retrofit to an aircraft already equipped with regular blended winglets is done by replacing the existing regular blended winglet aluminium winglet tip cap with a new aerodynamically shaped "Scimitar" winglet tip cap and adding a new Scimitar-tipped ventral strake.

The split scimitar winglet gives up to a 2% fuel burn reduction over the basic blended winglet configuration.

737-800 with split scimitar winglets

737 MAX Advanced Technology (AT) Winglets

Boeing has now developed, built and will be installing their own winglets for the 737 MAX family. The "Advanced Technology" winglet combines rake tip technology with a dual feather winglet concept into one advanced treatment for the wings of the 737 MAX." The AT Winglets measure 8 feet from root to top of winglet and a total of 9 feet 7 inches from bottom of lower tip to top of higher tip. The top portion is 8 feet 3 inches and the bottom portion is 4 feet 5.8 inches. The ground clearance of the bottom tip is 10 feet 2 inches.

Boeing claim they will give 1.5% fuel burn improvement over current technology winglets. They explain this as follows: "The AT winglet further redistributes the spanwise loading, increasing the effective span of the wing. The AT winglet balances the effective span increase uniquely between the upper and lower parts and therefore generates more lift and reduces drag. This makes the system more efficient without adding more weight."

Model of 737MAX AT winglet shown next to a 737-900ER at the Farnborough Air Show, 2012

Clearly there are similarities between APBs Split-Scimitar design and Boeings AT design. According to APB the new split-wingtip designs evolved through a combination of independent engineering and collaboration between Boeing and Aviation Partners, although it's hard to tell how much is which. "APB didn't participate in the Boeing design, and the Boeing designers didn't participate in APB's," said Bill Ashworth, CEO of Aviation Partners Boeing. "The APB design was approved by Boeing engineers, and they participated in evaluation of the test data. They also looked at the design technically, and said it's a good design."

Although Boeing will be using its own AT winglet on all production 737 Max aircraft, Aviation Partners Boeing already has landed 455 firm orders and options for its Split Scimitar Winglets, and expects to get a lot more.

THE RUDDER STORY

The safety record of the 737 has been exemplary, with less than 120 hull losses in almost 40 years. However two mysterious accidents in the early 1990's, which were possibly rudder related, brought the design into sharp focus resulting in a huge redesign & retrofit program which will not end until late 2008.

The Accidents

3 March 1991, UA585: A 737-200Adv crashed on approach to Colorado Springs. The aircraft departed from controlled flight approximately 1,000 feet above the ground and struck an open field. After a 21-month investigation, the Board issued a report on the crash in December 1992. In that report, the NTSB said it, "could not identify conclusive evidence to explain the loss of the aircraft", but indicated that the two most likely explanations were either a malfunction of the airplane's directional control system or an encounter with an unusually severe atmospheric disturbance.

11 Apr 1994, CO: 737-300 at 37,000ft over the Gulf of Honduras, the crew heard a muffled thump and the aircraft rolled violently to the right. For the next 18 minutes the crew struggled to keep the aircraft from rolling to the right and managed to land the aircraft in that condition.
Boeing concluded that hydraulic fluid had leaked from the PCU onto the yaw-damper signalling component, creating an open electrical circuit that inadvertently moved the rudder 2.5 degrees to the left. Such a deflection should have caused the jet to veer only slightly off course, something easily controllable by the pilots. Moreover, Boeing said the problem could not have lasted more than 110 seconds.

8 Sep 1994, US427: A 737-300 was approaching Pittsburgh Runway 28R when ATC reported traffic in the area, which was confirmed in sight by the First Officer. At that moment, the aircraft was levelling off at 6000ft (speed 190kts) and rolling out of a 15deg left turn (roll rate 2deg/sec) with flaps at 1, the gear still retracted and autopilot and auto-throttle systems engaged. The aircraft then suddenly entered the wake vortex of a Delta Airlines Boeing 727 that preceded it by approx. 69 seconds (4.2nm). Over the next 3 seconds, the aircraft rolled left to approx. 18deg of bank. The autopilot attempted to initiate a roll back to the right as the aircraft went in and out of a wake vortex core, resulting in two loud "thumps". The First Officer then manually overrode the autopilot, without disengaging it, by putting in a large right-wheel command at a rate of 150deg/sec. The airplane started rolling back to the right at an acceleration that peaked at 36deg/sec, but the aircraft never reached a wings level attitude. At 19.03:01 the aircraft's heading slewed suddenly and dramatically to the left (full left rudder deflection). Within a second of the yaw onset, the roll attitude suddenly began to increase to the left, reaching 30deg. The aircraft pitched down, continuing to roll through 55deg left bank. At 19.03:07 the pitch attitude approached -20deg, the left bank increased to 70deg and the descent rate reached 3600fpm. At this point, the aircraft stalled. Left roll and yaw continued, and the aircraft rolled through inverted flight as the nose reached 90deg down, approx. 3600ft above the ground. The 737 continued to roll, but the nose began to rise. At 2000ft above the ground the aircraft's attitude passed 40deg nose low and 15deg left bank. The left roll hesitated briefly, but continued and the nose again dropped. The plane descended fast and impacted the ground nose first at 261kts in an 80deg nose down, 60deg left bank attitude and with significant sideslip. All 132 on board were killed.

9 June 1996, the crew of Eastwind Airlines flight 517, a 737-200, experienced two hard, inadvertent rudder deflections as they approached Richmond, Va. Boeing found the yaw damper had been misrigged and did a flight test with the Captain and a Boeing test pilot to replicate the event. The Captain remained unconvinced saying that his incident had been much stronger than that shown on the test flight.

Unfortunately none of the aircraft involved had modern, flight data recorders, so the NTSB staff were forced to make assumptions in developing their hypotheses. The NTSB, FAA, Boeing, US Airways and ALPA all had different opinions about the cause of these accidents:

217

Conflicting Opinions

The US Air View

"The probable cause of this accident was an uncommanded, full rudder deflection or rudder reversal that placed the aircraft in a flight regime from which recovery was not possible using known recovery procedures. A contributing cause of this accident was the manufacturer's failure to advise operators that there was a speed below which the aircraft's lateral control authority was insufficient to counteract a full rudder deflection."

The ALPA View

"Based on the evidence developed during the course of this accident investigation, ALPA believes that the airplane experienced an uncommanded full rudder deflection. This deflection was a result of a main rudder power control unit (PCU) secondary valve jam which resulted in a primary valve overstroke. This secondary valve jam and primary valve overstroke caused USAir 427 to roll uncontrollably and dive into the ground. Once the full rudder hardover occurred, the flight crew was unable to counter the resulting roll with aileron because the B737 does not have sufficient lateral control authority to balance a full rudder input in certain areas of the flight envelope."

The Boeing View

Charlie Higgins, Boeing vice president - Airplane Safety & Performance, said that, "the rudder PCUs from the 737s that crashed near Colorado Springs in 1991 and Pittsburgh in 1994 were both thoroughly examined as a part of the NTSB's accident investigations. No such jam was detected in either unit."

Mike Denton, former chief engineer for the 737 said, "We don't know if there was an airplane system failure; we don't know if the flight crews had their foot on the pedal the full time."

Conclusions from Boeing's report to the NTSB, 30 Sep 1997:
> Several elements leading to this accident are clear:
> 1. The crew was startled by the severity of an unexpected wake vortex encounter.
> 2. A full rudder deflection occurred. However, the events that led to the full rudder deflection are not so clear:
> > • There is no certain proof of airplane-caused full rudder deflection during the accident sequence. The previously unknown failure conditions that have been discovered in the 737 rudder PCU have been shown to not be applicable to Flight 427 or any other conditions experienced in commercial service.
> > • There is no certain proof that the flight crew was responsible for the sustained full left rudder deflection. However, a plausible explanation for a crew-generated left rudder input must be considered, especially given the lack of evidence for an airplane-induced rudder deflection.
> In Boeing's view, under the standards developed by the NTSB, there is insufficient evidence to reach a conclusion as to the probable cause of the rudder deflection.
> 3. The airplane entered a stall and remained stalled for approximately 14 seconds and 4,300 feet of altitude loss.

The NTSB View

NTSB investigator Malcolm Brenner said in explaining the crash of Flight 427, "The pilots were trying to deal with emergencies with reasonable actions but could not understand what was happening in the time available"

Dennis Crider, chairman of the NTSB's Aircraft Performance Group, told the board members "A rudder reversal scenario will match all three events,"

The FAA View

The FAA argues that no one will ever know the cause with any certainty, so it has focused on making the plane safer.

The Airworthiness Directives

3 Mar 1994 - AD 94-01-07: "Within 750 flight hours … Perform a test of the main rudder PCU … to detect internal leakage of hydraulic fluid. Repeat at 750hr intervals unless replaced with new main rudder PCU."

27 Nov 1996 – AD 96-23-51: "Within 10 days … perform a test to verify proper operation of the rudder PCU."

17 Jan 1997 - AD 96-26-07: This introduced, within 30 days, a new QRH recall procedure entitled "UNCOMMANDED YAW OR ROLL" and changed the "JAMMED FLIGHT CONTROLS" procedure to include a section entitled "JAMMED OR RESTRICTED RUDDER".

Note: Following the NTSB report of 16 April 1999, these procedures were replaced by "UNCOMMANDED RUDDER" and "UNCOMMANDED YAW OR ROLL" in AD 2000-22-02.

19 Mar 1997 - AD 97-05-10: "Within 90 days … inspect the internal summing lever assembly of the main rudder PCU."

9 Jun 1997 - AD 97-09-15: "Within 5 years or 15,000 flight hours … Requires a one-time inspection of the engage solenoid valve of the yaw damper to determine the part number of the valve, … and replace if necessary."

1 Aug 1997 - AD 97-14-03: "Within 3 years …. Install a newly designed rudder-limiting device that reduces the rudder authority at flight conditions where full rudder authority is not required (a *rudder pressure reducer (RPR)*)….Install a newly designed yaw damper system that improves the reliability and fault monitoring capability"

4 Aug 1997 - AD 97-14-04: "Within 2 years … Requires replacement of all rudder PCUs … with a newly designed main rudder PCU." And "Replace the vernier control rod bolts … with newly designed vernier control rod bolts."

20 Jan 1998 - AD 97-26-01: "Within 18 months or 4,500 hours … Perform an inspection to detect galling on the input shaft and bearing of the standby rudder PCU… and replace the input bearing of the standby rudder PCU with an improved bearing"

17 Feb 1998 - AD 98-02-01: "Within 3,000 hours … Remove the yaw damper coupler, replace the internal rate gyroscope with a new or overhauled unit, and perform a test to verify the integrity of the yaw damper coupler."

More Incidents

Two further rudder incidents in 1999 caused concern because they involved aircraft retrofitted with a PCU redesigned in accordance with the Accidents to make a valve jam impossible. After the incidents, the dual-servo valves on both airplanes were inspected for cracks but none were found.

19 Feb 1999 - The pilots of the United Airlines Boeing 737-300 reported sluggish rudder control during a ground check, while taxiing at Seattle-Tacoma International Airport. The NTSB said the apparent problem was a mispositioned valve-spring guide in the rudder's power-control unit.

23 Feb 1999 - A Metrojet 737-200, made a precautionary landing in Baltimore after the aircraft rolled slightly and changed direction during cruise flight. The MetroJet rudder actually moved involuntarily at two rates, first slowly and then more rapidly all the way over to the point where the rudder deflects to its maximum extent, known as a "hardover." The pilots could do nothing to make the rudder move, including the initial emergency procedures of turning off the autopilot and the yaw damper. The pilots reported that when they turned off the plane's hydraulic pressure, the rudder "snapped back" into position but continued to "chatter" and vibrate through the emergency landing. No known scenario could cause such an event. The NTSB examined the PCU but did not find evidence of a cause. All that investigators were able to conclude is that a rudder deflection did occur, according to information from the airplane's flight-data recorder.

The Crossover Speed

The Crossover speed is the speed at which full lateral (roll) control is needed to balance the roll due to sideslip caused by full rudder deflection.

It is probably fair to say that before 1999 and the publication of the NTSB recommendations, very few 737 pilots knew about the crossover speed. However, during the investigation of the Colorado Springs accident, Boeing produced data which indicated that crossover speed was an issue at the block speeds of Flaps 1 to Flaps 10. This information was relayed to the NTSB on 20th September 1991 in a letter to Mr. John Clark (NTSB) from Mr. John Purvis (Boeing). It should also be noted that crossover speed increases with g-loading.

At the point of the USAir 427 upset, the aircraft was configured at Flaps 1 and 190kts, which combined with the g of the attempted recovery manoeuvre, would have made recovery almost impossible.

In June 1997, Boeing undertook some additional flight testing in order to further explore the crossover speed. During these tests, it was discovered that operations with flaps up were also impacted by crossover speed. Furthermore, during this flight testing, full rudder hardover malfunctions were conducted in order to quantify B737 handling characteristics and recovery techniques with full rudder deflections. It was determined that for the Flaps 1, 190 knot case, once a full rudder hardover was experienced, the aircraft had to accelerate to well above crossover speed before sufficient lateral control margin was reached and the aircraft could be recovered.

It was not until March 1999 that the increased block speeds for flaps 1 to 10 were incorporated into the Boeing Ops Manual.

On 16 Feb 2000, papers released in a court case for punitive damages by a widow of a US427 passenger, say that US Airways officials asserted that Boeing Co withheld flight test data about the crossover speed.

Boeing denied the US Airways allegations and said the information was made available to the FAA when the airplane was being certified in 1984 for commercial operations. During its flight test program on the 737-300 in 1984, Boeing found that at a speed of 190 knots and the flap-one setting, the plane could not overcome a full rudder deflection by using the ailerons, as would normally be the case. "There is no FAA requirement that says this phenomenon is not acceptable," said Mike Denton, former chief engineer for the 737. In the wake of the Pittsburgh crash and subsequent investigation, airlines have instructed their pilots to fly the 737 at 10 knots faster when first deploying the flaps, so ailerons would be more effective. The higher speed was officially adopted into Boeing's 737 operating manual last March. Denton, who is now an engineering executive with Boeing's product development team, said the information about the 737's crossover speed was not hidden.

James Gibbs, flight manager for USAir at the time of the crash, and Gordon Kemp, 737 flight manager, said in depositions that if the airline had known about certain aerodynamic data concerning the 737, it would have changed its flying procedures before the crash. With the new procedures, "I would have expected the 427 crew to have successfully been able to fly out of the situation,"

Boeing's reply brief effectively said the crew mishandled the situation. Knowledge of how the aircraft reacts at certain speeds "is less important than a pilot's ability to exercise proper recovery techniques during upsets, no matter what the cause."

NTSB Report & Recommendations

Mar 1999 - The NTSB releases a report that says "although there was no hard physical evidence, both the Colorado and Pittsburgh crashes were probably caused by an abrupt rudder movement that surprised the crew and sent the planes spiralling into an uncontrollable dive."

16 April 1999 - The NTSB made the following recommendations to the FAA:

- All existing and future Boeing 737s and future transport category aircraft must have a reliably redundant rudder actuation system.

- Convene an engineering test and evaluation board (ETEB) to conduct a failure analysis to identify potential failure modes. The board's work should be completed by March 31, 2000, and published by the FAA.

- Amend the FARs to require that transport-category airplanes be shown to be capable of continued safe flight and landing after jamming of a flight control at any deflection possible, up to and including its full deflection, unless such a jam is shown to be extremely improbable.

- Revise AD 96-26-07 so that procedures for addressing a jammed or restricted rudder do not rely on the pilots' ability to center the rudder pedals as an indication that the rudder malfunction has been successfully resolved.

- Require all operators of the Boeing 737 to provide their flight crews with initial and recurrent flight simulator training in the "Uncommanded Yaw or Roll" and "Jammed or Restricted Rudder" procedures. The training should demonstrate the inability to control the airplane at some speeds and configurations by using the roll controls (the crossover airspeed phenomenon) and include performance of both procedures in their entirety.

- Require Boeing to update its Boeing 737 simulator package to reflect flight test data on crossover airspeed and then require all operators of the Boeing 737 to incorporate these changes in their simulators.

- Evaluate the Boeing 737's block manoeuvring speed schedule to ensure the adequacy of airspeed margins above crossover airspeed for each flap configuration, and revise block manoeuvring speeds accordingly.

- Require that all Boeing 737s be equipped, by July 31, 2000 with an FDR which records the minimum parameters applicable to that airplane, plus the following parameters: pitch trim; trailing edge and leading edge flaps; thrust reverser position (each engine); yaw damper command; yaw damper on/off discrete; standby rudder on/off discrete; and control wheel, control column, and rudder pedal forces (with yaw damper command; yaw damper on/off discrete; and control wheel, control column, and rudder pedal forces sampled at a minimum rate of twice per second).

The Flight Control Engineering and Test Evaluation Board (ETEB)

The ETEB was formed by the FAA in May 1999 to take a fresh look at the 737 rudder in the most in-depth scientific study ever of any commercial airplane system. FAA officials stressed that the board was formed only to determine what could happen, not to evaluate the probability that it would happen. And they caution that they have found nothing that they believe will be sufficiently probable to warrant grounding the plane or even ordering immediate design changes. Any eventual design change would be required on all existing and newly manufactured 737s.

The FAA used two criteria to assemble the ETEB members, technical expertise and no connection with, or knowledge of, the 737 rudder system. John McGraw, manager of the FAA's Airplane and Flight Crew Interface Branch in Seattle headed the board, which was composed of scientists from the FAA, NASA, the Defence Department, ALPA, the Air Transport Association, the Russian Air Transport Accident Investigation Commission, Ford Motor Co. and Boeing. However, Boeing personnel came from Boeing military and Boeing Long Beach, not from Boeing Seattle where the 737 is made, because Seattle engineers could be too familiar with the rudder.

Ford engineer Davor Hrovat was included on the team because he had developed a way to use ultrasound to determine movements of internal parts, such as the inner slide of the 737's dual concentric servo valve. A high-level, seven-person "challenge team" of outside experts was formed to review every step of the board's work. It included Col. Charles Bergman, the Air Force deputy chief of safety; Vladimir Kofman, chairman of the Russian accident investigation commission; and Tom Haueter, chief of the major investigations division of the NTSB.

The ETEB had full access to everything it needed, including a special test aircraft owned by Purdue University and laboratory space at Boeing. They even constructed a first-of-a-kind test device called a "fin rig", which was a full 737 vertical tail fin and rudder connected to an aircraft engineering simulator. Any rudder control commands by a pilot were mimicked in real time on the rudder, which was placed where the pilot could easily see its movements.

Engineers used a vibration table to give a good shaking to rudder components, a "cold box" that could produce realistic flight temperatures, and a device that could produce sudden spikes in hydraulic fluid temperatures from 65 degrees below zero to 210 degrees above zero. They could also spray water into the rudder mechanism, producing ice. It was during these ice tests that the engineers found a new and unsuspected failure scenario.

During the study, the ETEB brought in 10 flight crews from 4 airlines on 737s to fly the fin rig simulator. They used the existing recovery procedures to deal with about 40 different rudder failure modes. They found, as expected, that any rudder hardover while taking off or landing, moving slowly and at low altitude, would be catastrophic. And they found that these pilots, who fly the 737s routinely for airlines and had normal training, performed poorly in trouble-shooting rudder problems.

Another AD

28 Jun 1999 (Revised 7 Dec 1999) - AD 99-11-05: "Within 16 months … perform repetitive displacement tests of the secondary slide in the dual concentric servo valve of the PCU to detect cracks in a joint in the servo valve that regulates the intake of hydraulic fluid to the PCU."

NTSB Revises UA585 Report

Oct 99 - The NTSB adopted a revised final report on the UA585 and US427 crashes. The Board said that the most likely cause of the accident was the movement of the rudder to its limit in the direction opposite that commanded by the flight crew, "most likely" because of a jam in the device that moves the rudder. The decision tracks information learned from the investigation of UA585, US427 and the Eastwind incident.

ETEB Preliminary Draft Report

12 Apr 2000 - In a preliminary draft report, the ETEB found that the "JAMMED OR RESTRICTED RUDDER" procedures formulated by Boeing and often modified by airlines were "confusing and time consuming." They said the pilots showed a lack of training and "situational awareness" in controlling malfunctions, and as they prepared to land they never checked to be sure the rudder was operating properly. Boeing seemed surprised at this and promised to revise it.

The evaluation board said it had detected 30 failures and jams that could be catastrophic on takeoff and landing. But partly because airplanes travel faster at higher altitudes and other control systems can overcome the force of a rudder, it considered no failures at cruising altitude to be catastrophic. Nonetheless, 16 of those failures and jams at higher altitudes would be "hazardous", meaning they would require prompt pilot action to prevent a crash.

The report said that another 22 "latent failures," such as a cracked part, combined with single failures and jams, could cause catastrophic or hazardous failures. All but three of the failures found by the panel were also present in the new-generation 737-6/7/800. "The large numbers of single failures and jams and latent failure combinations are of concern."

They also found that current maintenance procedures were insufficient to find hidden problems in the rudder system.

The evaluation board's most unexpected finding was that an ice buildup can cause a 737 rudder to malfunction. The pilot's rudder pedals are connected by cables to a linkage in the tail section. A hydraulic servo valve in the linkage powers an actuator that moves the rudder. The linkage includes a summing lever that stops the rudder at the position specified by the pilot. Mechanical stops prevent the summing lever from moving too far, this prevents proper operation of other levers that shut off hydraulic fluid flowing through the dual control valve, allowing fluid to keep pumping until the rudder goes to its maximum deflection. The linkage isn't pressurized or heated, and operates in temperatures as low as -60C. The board found that ice could form in the linkage, jamming the summing lever. Without the equalizing force of the lever, the servo valve could continue providing hydraulic pressure; the rudder then would keep moving as far as it could go in the requested direction, a condition known as a hardover. There is no proof that this malfunction has ever occurred in flight because ice would melt afterwards, leaving no marks. FAA officials stressed that this phenomenon has not yet been tested in flight, but they are nonetheless working on a fix to make certain ice does not form or is cleared away naturally by the movement of the mechanisms.

The evaluation board recommended in the draft that Boeing modify the 737 rudder control system so that "no single failure, single jam, or any latent failure in combination with any single jam or failure will cause Class I [catastrophic] effects."

In the meantime, the draft report recommends alerting flight crews about early signs of rudder malfunctions, most often rudder "kicks" that pilots might attribute to yaw damper problems. It also recommends new maintenance inspection procedures, a new cockpit instrument to tell the pilots exactly how the rudder is moving and a new hydraulic system design to allow hydraulic pressure to be cut off to the rudder without affecting other airplane systems.

Boeing Agree to Redesign the Rudder

13 Sep 2000 - The FAA reaches an agreement with Boeing to redesign the rudder. Once the directive is issued, the company will have about five years to make the changes in planes currently flying. The new design will be required in all newly made 737s. Because the redesign could take years to implement, the FAA said it will also announce new training procedures for pilots to use in the event of rudder problems.

Earlier changes in the design had fixed problems with some control mechanisms and an earlier set of emergency rudder control problem training procedures for pilots was put in place, but these were found by the ETEB to be too complex and pilots had not received enough training to handle them effectively.

So, to summarise so far: Since the first two accident reports were issued there have been no further fatal accidents but there have been some unexplained rudder incidents. Boeing and the FAA maintain that the rudder safety problem was fixed after the first accidents.

The NTSB's chairman, Jim Hall, said that he was pleased Boeing and the FAA have finally agreed that there was a need to redesign the 737 rudder control system. The current design, Hall said, "represents an unacceptable risk to the travelling public." He went on, "I hope this redesign and retrofit can be accomplished expeditiously so that the major recommendation of our accident report last year will be realised, a reliably redundant rudder system for Boeing 737s."

The FAA now says even more must be done to assure the redundancy of the rudder's safety system and to assure it's impossible for it to malfunction. Pilots must have more training to handle a rollover caused by a malfunctioning rudder. Boeing say the new rudder will take three years to develop and be fitted to the first aircraft and the FAA said the last will not be fitted until 2008 because the retrofit work is expected to take as much as 200 hours per plane. Boeing will pay the full cost of the retrofit, estimated at $240 million.

Boeing's Allen Bailey, the engineer in charge of 737 safety certification said, "We are not fixing a safety problem with this enhancement we are making,"

14 Sep 2000 - NTSB Chairman Jim Hall makes a statement on the FAA release of the ETEB rudder study.
"The men and women of the Engineering Test and Evaluation Board can be justifiably proud of the work they have done over the past year and for the final report issued today. The ETEB - made up of representatives from the Federal Aviation Administration and aircraft manufacturers - was the result of a Safety Board recommendation following our investigation of the crash of USAir flight 427. I think I can speak for my NTSB colleagues by saying that we are gratified that the ETEB essentially confirmed our findings in that accident report. The major finding of both reports is that the Boeing 737 rudder control system has numerous potential failure modes that represent an unacceptable risk to the travelling public. The ETEB found dozens of single failures and jams and latent failures in the 737 rudder

system, in addition to the single point of failure we identified in our accident report, that can result in the loss of control of the airplane. Although the failure mechanism that we believed led to the crashes of United Airlines flight 585 in 1991 and USAir flight 427 in 1994, and the near loss-of-control of Eastwind Airlines flight 517 in 1996, appears to have been eliminated through a redesigned rudder power control unit, the results of the ETEB echo our findings that failure modes still exist in the Boeing 737 rudder system. While we are very concerned that some ETEB recommendations will not be adopted - particularly an independent switch to stop the hydraulic flow to the rudder and a rudder position indicator in 737 cockpits - we are pleased that both the FAA and Boeing Aircraft Company agree that there is a need for a redesign to the rudder actuator system. However, before the Board can determine if this will satisfy the goal of our recommendations, we will need to evaluate in detail the proposed design. I hope this redesign and retrofit can be accomplished expeditiously so that the major recommendation of our accident report last year will be realized - a reliably redundant rudder system for Boeing 737s."

QRH Procedures Revised Again (#2)

13 Nov 2000 - FAA issues AD 2000-22-02 this supersedes AD 96-26-07 "To require revising the FAA-approved Airplane Flight Manual (AFM) procedure in the existing AD to simplify the instructions for correcting a jammed or restricted flight control condition. That AD was prompted by an FAA determination that the procedure currently inserted in the AFM by the existing AD is not defined adequately. The actions required by that AD are intended to ensure that the flight crew is advised of the procedures necessary to address a condition involving a jammed or restricted rudder."

In plain English, this AD replaced "JAMMED FLIGHT CONTROLS" which was introduced in 1996 with "UNCOMMANDED RUDDER" and "UNCOMMANDED YAW OR ROLL"

Revised UA585 report

5 Jun 2001 - NTSB adopts revised final report on the 1991 crash of United Airlines Flight 585 in Colorado Springs calls rudder reversal most likely cause.

Extracts follow...
WASHINGTON, D.C. - The NTSB has adopted a revised final report on the 1991 crash of a Boeing 737 near Colorado Springs, Colorado that killed all 25 persons aboard. The Board said that the most likely cause of the accident was the movement of the rudder in the direction opposite that commanded by the flight crew. The decision tracks information learned from the investigation of two fatal 737 accidents - including this one - and a non-fatal incident.
... Break ....
The Board's revised report on the crash of United Airlines flight 585 cites the same probable cause as that of flight 427, that is:
"...a loss of control of the airplane resulting from the movement of the rudder surface to its blowdown limit. The rudder surface most likely deflected in a direction opposite to that commanded by the pilots as a result of a jam of the main rudder power control unit servo valve secondary slide to the servo valve housing offset from its neutral position and overtravel of the primary slide."

In its revised report on flight 585, the Board noted that since the upset occurred less than 1,000 feet above the ground, the pilots had very little time to react to or recover from the event. The Board concluded that the flight crew of United 585 "could not be expected to have assessed the flight control problem and then devised and executed the appropriate recovery procedure for a rudder reversal under the circumstances of the flight." Although training and pilot techniques developed in recent years show that it is possible to counteract an uncommanded deflection of the rudder in most regions of the flight envelope, "such training was not yet developed and available to the flight crews of United flight 585 and USAir flight 427."

The RSEP AD

12 Nov 2002 - FAA issues AD 2002-20-07 R1 (Revision issued a month after original AD):

"Within 6 years … Install a new rudder control system that includes new components such as an aft torque tube, hydraulic actuators, and associated control rods, and additional wiring throughout the airplane to support failure annunciation of the rudder control system in the flight deck. The system also must incorporate two separate inputs, each with an override mechanism, to two separate servo valves on the main rudder PCU; and an input to the standby PCU that also will include an override mechanism."

Note that in the first issue of AD 2002-20-07 it was stated:

"Because of the existing design architecture, we issued AD 2000-22- 02 R1 to include a special non-normal operational "Uncommanded Rudder" procedure, which provides necessary instructions to the flightcrew for control of the airplane during an uncommanded rudder hardover event. The revised rudder procedure included in AD 2000-22-02 R1 is implemented to provide the flightcrew with a means to recover control of the airplane following certain failures of the rudder control system. However, such a procedure, which is unique to Model 737 series airplanes, adds to the workload of the flightcrew at a critical time when the flightcrew is attempting to recover from an uncommanded rudder movement or other system malfunction. While that procedure effectively addresses certain rudder system failures, we find that such a procedure will not be effective in preventing an accident if the rudder control failure occurs during takeoff or landing.

For these reasons, we have determined that the need for a unique operational procedure and the inherent failure modes in the existing rudder control system, when considered together, present an unsafe condition. In light of this, we proposed to eliminate the unsafe condition by mandating incorporation of a newly designed rudder control system. The manufacturer is currently redesigning the rudder system to eliminate these rudder failure modes. The redesigned rudder control system will incorporate design features that will increase system redundancy, and will add an active fault monitoring system to detect and annunciate to the flightcrew single jams in the rudder control system. If a single failure or jam occurs in the linkage aft of the torque tube, the new rudder design will allow the flightcrew to control the airplane, using normal piloting skills, without operational procedures that are unique to this airplane model."

But this statement was withdrawn a month after it was issued saying "Retaining this procedure will ensure that the flightcrew continues to be advised of the procedures necessary to address a condition involving a jammed or restricted rudder until accomplishment of this new AD."

QRH Procedures Revised (#3)

Dec 2003 - The "UNCOMMANDED RUDDER" and "UNCOMMANDED YAW OR ROLL" became "UNCOMMANDED RUDDER/YAW OR ROLL" which made reference to the STBY RUD ON light on the Flight Controls panel. The procedure was after first taking out the A/P & A/T if you had a STBY RUD ON light installed to accomplish the "JAMMED OR RESTRICTED FLIGHT CONTROLS" checklist. If you did not have a STBY RUD ON light then effectively do the old uncommanded rudder procedure.

Dec 2004 - The "UNCOMMANDED RUDDER/YAW OR ROLL" procedure was modified further by adding a stage that called for flaps to be retracted to flap 1 if they were extended, before actioning the STBY RUD ON light options. This stage is not included in the NG checklist because of the additional protection of the FSEU.

The End of the Story?

So by 12 November 2008, 17 years after the Colorado Springs accident, all 737s should have been retrofitted with a new rudder control system. Only then can this whole sorry episode be put to bed.

Technical Description

Yaw control is achieved by a single graphite / composite rudder panel. A single rudder power control unit (PCU) controls rudder panel deflection. A standby rudder PCU provides back up in the event of malfunction of the main rudder PCU. There is no manual reversion for yaw control. The only internal indication of rudder panel deflection is pedal position, which always accurately reflects control surface deflection. Total authority of the control surface is modulated in relation to aircraft IAS using "blowdown", ie a constant pressure is applied to the surface by the actuator, and the movement of the panel reduces accordingly as the dynamic pressure on it increases. For this reason maximum rudder pedal movement is reduced with increasing airspeed. Maximum rudder panel deflection is approximately +/-15 degrees on the ground, reducing to around +/-8 degrees at a typical cruise altitude.

Rudder PCU exposed during maintenance

Power Control Unit (PCU)

The rudder PCU consists of an input shaft / crank mechanism, a **dual concentric servo valve** to control porting of the fluid to the rudder actuator, and a yaw damper actuator. The rudder actuator is a tandem actuator, having two internal piston areas for each hydraulic source (A & B). The actuator is capable of positioning the rudder panel with either one or both main hydraulic sources available, though with one source inoperative a reduced rudder panel deflection would result due to blowdown at higher airspeeds.

Flow of hydraulic fluid to the rudder actuator is controlled by the dual servo-valve. This is a complex dual concentric cylinder with an outer and inner slide. During normal rudder pedal inputs sufficient rudder panel deflection is catered for by the primary (inner) valve alone. However, should a larger panel deflection be required or a higher rate rudder input be commanded, the secondary (outer) sleeve moves in addition to port extra fluid to the actuator. Movement of the outer sleeve is typically no more than 1mm. Position of both sleeves of the servo valve is controlled by a complex mechanism of bell cranks, input rod and summing lever, the geometry of which is such as to provide movement of the sleeves in relation to the body of the valve.

The 737NG also has a standby rudder PCU that moves the rudder during manual reversion operation. The wheel-to-rudder interconnect system (WTRIS) will coordinate (assist) turns by using the standby rudder PCU to apply rudder as necessary based upon the Captains control wheel roll inputs. From experience, I can verify that this makes the NG much easier to handle in manual reversion than previous generations.

Yaw Damper

The yaw damper is incorporated to prevent Dutch roll. It is connected in parallel with the main servo valve and includes its own actuator, powered by hydraulic system B. This actuator applies its own input to the input shaft / crank mechanism to bring about a movement of the servo-valve and hence a rudder panel deflection. No pedal movement results from yaw damper operation. Total authority of the yaw damper is approximately +/-2.5 degrees.

NGs: The 737-NGs also have a standby yaw damper; it uses the standby rudder PCU, with commands from SMYD 2 and powered by the standby hydraulic system. SMYD 1 controls the main yaw damper with hydraulic system B. Note: Only inputs from the main yaw damper are shown on the yaw damper indicator.

Rudder Pressure Reducer (Not NGs)

To limit the effects of various PCU failure modes (pre-RSEP), a rudder pressure reducer (RPR) was fitted to the A-system pressure side of the rudder PCU. This is simply a pressure reducing valve, which operates during the majority of flight phases to reduce the total authority of the rudder panel by approximately one-third. The B-system portion of the rudder PCU is unaffected by the RPR, as is yaw damper operation.

During certain critical phases of flight when full rudder authority may be required, the RPR provides full system pressure (3000psi from 1800psi) to the rudder PCU. These are: -

- a) During take-off below 1000ft RA
- b) During approach below 700ft RA
- c) If a difference of >45% N1 exists between power units.
- d) If B-system hydraulic pressure is lost.

Whilst correct functioning of the RPR is transparent to the flight crew, certain cockpit indications can be helpful in verifying correct operation and faults alike; the A system flight controls low pressure warning light now has two additional functions related purely to the RPR; on initial application of pressure to hydraulic system A, the low-pressure warning light should remain illuminated for 5 seconds. If the light extinguishes immediately then a fault may be present within the RPR. Incorrect mode switching of the RPR is also indicated by illumination of the light on approach below 700ft RA, indicating that full pressure is not available to the A-system portion of the rudder PCU. A further associated failure of hydraulic system B and an asymmetric thrust condition may result in insufficient rudder authority to maintain directional control.

Digital Yaw Damper Coupler

The yaw damper coupler comprises the control electronics and yaw rate gyro. A digital yaw damper coupler helps reduce the possibility of electro-magnetic interference (EMI). Turn co-ordination is provided by reducing the gain of the yaw-rate gyro in proportion to bank angle detected from the IRU. In this way during a turn the yaw damper coupler is "tricked" into believing the aircraft is yawing into the turn and provides an increased rudder input. The coupler is sensitive only to yaw rates that produce Dutch Roll. Note that the yaw damper coupler controls and monitors both the RPR (Sys A) and RPL (Sys B) of the main rudder PCU, a de-activated yaw damper also renders the RPR & RPL inoperative. For this reason higher block manoeuvring speeds are used when the yaw damper is u/s (not NGs).

Rudder System Enhancement Program (RSEP)

The Rudder System Enhancement Program (RSEP) introduced in 2003 (SB 737-27-1252/3/5) must be implemented on all series of 737s by 12 Nov 2008. It replaces the infamous dual concentric servo valve with separate input rods, control valves and actuators; one set for hydraulic system A, and one set for hydraulic system B. The standby PCU is controlled by a separate input rod and control valve powered by the standby hydraulic system. All three input rods have individual jam override mechanisms that allow inputs to be transferred to the remaining free rods if a jam occurs. All 737s must be fitted with the RSEP by Nov 2008. Modified aircraft are identifiable by the STBY RUD ON light on the flight controls panel and new c/bs on the P6-2 panel labelled "Force Fight Monitor" and "Rudder Load Limiter".

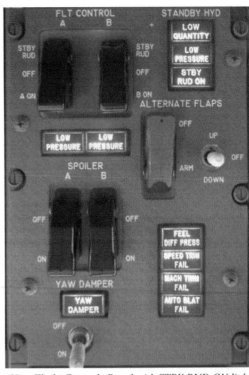

Force Fight Monitor

The main rudder PCU contains a Force Fight Monitor (FFM) that detects opposing pressure (force fight) between A & B actuators, this may happen if either hydraulic system, input rod or control valve has jammed or failed. If this condition is detected for more than 5 seconds, the FFM will automatically turn on the standby hydraulic pump pressurising the standby rudder PCU. This will also illuminate the new STBY RUD ON light on the flight control panel see photo right.

New Flight Controls Panel with STBY RUD ON light

Rudder Pressure Limiter (Not NGs)

This is effectively the B system equivalent of the RPR, except that it is physically part of the main rudder PCU rather than upstream of it. Hydraulic system B pressure is reduced within the main PCU from 3000 to 2200psi it has the same activation criteria as the RPR.

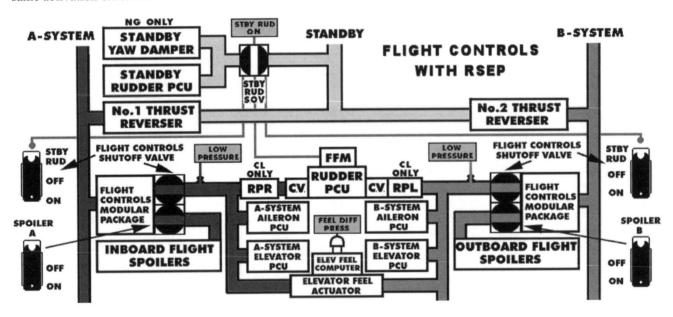

NGs

The NGs do not have an RPR or RPL, but two Load Limiters instead (Shown as "CONTROL VALVE"'s in the FCOM schematics). At speeds above approximately 135kts, hydraulic system A pressure (Pre-RSEP), hydraulic system A and B pressure (Post-RSEP) to the rudder PCU is reduced to 1450psi (Pre-RSEP) / 2200psi (Post-RSEP). They both reduce rudder output force by 25% at blowdown speed. The NGs also gained the FFM and separate input rods, control valves and actuators of the RSEP package.

Analysis of QRH Procedures

The QRH procedures for rudder malfunctions were first introduced in January 1997. Since then they have changed several times, either to simplify the procedures or as a result of hardware changes. They still have many branches depending upon the RSEP status of the aircraft and what condition is detected by the crew.

An uncommanded rudder deflection and/or hardover may be caused by a number of different failure modes within the rudder PCU and/or yaw damper, and the severity of symptoms could differ widely from a nuisance yaw damper deflection to a full-scale rudder hardover resulting in reduced controllability. As identification of the primary cause of such a failure may be impossible in flight, certain procedures have been mandated which aim to minimize the effects of the malfunction. There are two similar QRH drills which cover this situation:

> JAMMED OR RESTRICTED FLIGHT CONTROLS
> Condition: Movement of the elevator, aileron/spoiler or rudder is restricted

> UNCOMMANDED RUDDER/YAW OR ROLL
> Condition: Uncommanded rudder pedal displacement or pedal kicks or uncommanded yaw or roll.

The latter is more appropriate for a rudder hardover, but either procedure will direct you to the correct solution.

AUTOPILOT (if engaged)......................DISENGAGE
AUTOTHROTTLE (if engaged)...............DISENGAGE
Verify thrust is symmetrical.
These are the only recall items. First disengage the automatics, get control of the flight path and verify thrust is symmetrical. If you have a STBY RUD ON light installed (and serviceable), ie an RSEP a/c, then go to the JAMMED OR RESTRICTED FLIGHT CONTROLS checklist. The logic here is that the FFM will either have detected an opposing pressure between A & B actuators and activated the standby rudder PCU or there was no opposing pressure and the problem was a jam rather than a hydraulic problem.

The rest of this section assumes that you do not have a STBY RUD ON light installed (ie Pre-RSEP a/c).

YAW DAMPER................................OFF
The yaw damper is switched OFF. This removes power to the yaw damper actuator, therefore eliminating any input to the main rudder PCU. Whilst this should eliminate any nuisance yaw and secondary roll caused by a failure within the yaw damper or coupler, with its limited authority over main rudder panel deflection it is highly questionable whether this alone could produce a large-scale uncommanded rudder movement.

If the yaw or roll is the result of uncommanded rudder displacement or pedal kicks:
Rudder trim.......................................Center
Rudder pedals....................................Free & center
After verifying zero rudder trim, the checklist calls for a maximum combined effort of both pilots to centre the rudder pedals. The intention of this is to provide a maximum force to shear any metal fragments or "chips" which may be present within the servo-valve. Remember, centred pedals mean a centralised rudder.

If rudder pedal position or movement is not normal:
SYSTEM B FLIGHT CONTROL switch....STBY RUD
During a normal flight phase, the rudder has three separate sources of power; A-system hydraulics, B-system hydraulics and Standby hydraulics. The objective of any such drill would be to reduce the uncommanded deflection of the rudder panel, and to restore some directional controllability. This is achieved within the drill by first reducing the authority of the main rudder PCU by removing the B-system hydraulic source. A-system pressure remains, but at a considerably reduced pressure due to the functioning of the RPR / load limiter, hence blowdown will help to re-align the control surface. Re-positioning the B-system flight controls switch to standby rudder removes B-system hydraulic pressure from the main rudder PCU, and provides a completely independent method of rudder panel control through the standby rudder PCU, further assisting in re-alignment of the control surface.

THE FLIGHT DECK

737-100/200

The 737-200Adv   *Photo: Jorge Albanese*

The main feature of the original 737s was their "round dial" AH, HSI & engine instruments. The early 1/200s had the SP-77 autopilot which was of limited functionality and no FMC.

This photo is of one of the latest 737-200Advs with the SP-177 autopilot. It looks very similar to the early -300s with a full AFDS MCP, Autothrottle, PDCS, 2 FMC CDUs and a digital colour weather radar in between. It also has Collins FD-110 flight director and five-inch ADI & HSI; Dual electric Mach/airspeed, altimeter and VSI with dual digital ADCs; Dual RDMI which combines the RMI and DMI; Digital TAT; Pushbutton audio selector panels; VHF comm and VHF nav panels with preselect controls.

Just about the only certain way to tell that this is a 200 and not a 300 is the EPR gauge and EPR selector on the MCP (JT8Ds only). The early -300s were also non-EFIS and had the narrow (2 box) centre electronics console and round dial engine instruments.

737-3/4/500 (next page)

The most obvious difference of the Classics from the originals is the EFIS displays, although the very early Classics retained the electro-mechanical flight and engine instruments. Each pilot now has an EADI which displays the artificial horizon, speed tape, LLZ, G/S, Rad alt and MCP annunciations. Beneath the EADI is the EHSI which can display either navaid or route data superimposed with beacons, airfields, FMC route (shown on the LH display), wx radar, terrain & TCAS data. The mixture of glass and electro-mechanical instruments makes it an ideal first EFIS aircraft.

This particular -300 dates from 1990 so it has the digital EIS display which replaces 21 individual "round dial" engine instruments. It also has a flap load relief light and had an auxiliary fuel tank which is now removed as can be seen by the blanking plate over the missing fourth fuel gauge and the extra pair of fuel pumps on the overhead panel. Other post 9/11 modifications include a locking flight deck door (aft electronics panel) and CCTV camera displays of the cabin (between the CDUs). The large white knob in the middle of the electrics panel is an optional countdown timer.

The two FMC CDUs are forward of the throttle quadrant. The centre electronics console is one radio box wider than the originals to accommodate the extra EFIS controls; this necessitated the present J-track pilot seats to give us room to get past the console. The original two-box width can be seen by the width of the fire panel.

737-NG (next opposite page)

In the NG, the larger PFD and ND (formerly known as the EADI and EHSI) are now side by side to fit into the space available. The EFIS control panels are now located either side of the MCP, giving rise to a hesitation known to all pilots who fly both generations as they reach the wrong way for the EFIS controls!

The EIS & fuel gauges are both on the central (upper) display unit. There is a sixth screen (lower DU) between the CDUs which is used as a pop-up display if any abnormal parameters are detected. The flat panel displays have the advantage over CRTs of being lighter, more reliable and consume less power, although they are more expensive to produce.

According to Boeing, the requirements from the airlines for the new cockpit were:

- To be easy for current 737 pilots to operate.
- To anticipate future requirements eg transitioning to 777 style flightdecks.
- To accommodate emerging navigation and communication technologies.

Generally speaking I like the NG, its large clear displays and its ability to fly faster, higher, further on less fuel etc. On the downside, the flightdeck is much noisier and personally I prefer the handling of the Classics; also I find the landings less flattering.

737 Flightdeck In-Depth

This section illustrates some of the miscellaneous points of interest around the 737 flightdeck. Most of the photographs were taken on a 737-500 but are applicable to most series.

The hand mike can be used instead of the headset by simply pressing the transmit button on the unit, no extra selections need to be made on the ASP.

Be very careful to stow the sun visor back correctly ie between the sheets behind the door. I have seen at least one aircraft where a control cable (to the nose wheel steering tiller) was frayed almost to the point of failure because a sun visor had been put behind the stowage sheets and had fallen onto the cable where it became bound.

There is also visible a data loader and an ACARS printer. Best of all, a real throwback to the sixties - an ashtray!

The control stand has all the usual controls of any aircraft. Notice how the knobs for the engines, speedbrakes, flap and gear are all shaped like the item they control. There are gates at the flap 1 and 15 positions to mark the go-around flap settings for both engine out and all engines. The Horn cut-out button is to silence the gear warning horn (see warning systems section) which is a particular nuisance on the Classics.

The engine start levers are simply on/off switches for the fuel at the MEC (Classics) / HMU (NGs). It does not vary the amount of fuel like a mixture control on a light aircraft!

The main difference between the Captains side wall and the F/O's for almost all 737s is that the Captains side has a tiller for nose-wheel steering, although some airlines have a tiller on the F/O's side as well. The F/O's sidewall has the spare bulb tray instead of the tiller.

The rudder pedals are adjusted fore & aft with this white knob. Notice the hidden circuit breakers for panel lighting. The unit at the base of the control column is the stick shaker.

Aft Electronics panel

The aft electronics panel, or P8 panel to engineers, was only two boxes wide on the 1/200 and early 300 series. In 1986 the aislestand was made three boxes wide to accommodate the EFIS control panels and so began the long tradition of shin banging and leg stretches to get in and out of your seat, despite new J-shaped seat runners. The fire panel remained two boxes wide as a subtle reminder of how good things used to be on the -200.

737-200

737-Classic

The arrangement & composition of avionics varies with almost every aircraft, but the above photos show typical setups. The boxes in the photograph on the right are as follows from top to bottom:

Left row: VHF1, NAV1, Capt EFIS control panel, Audio Selector (or Control) Panel (ASP / ACP), ADF1, Lighting controls.
Centre row: Cargo fire, Transponder, Countdown timer, Cabin secure reminder, HF, CCTV, Rudder trim indicator, Aileron trim & rudder trim knob.
Right row: VHF2, NAV2, F/O EFIS control panel, Audio Selector (or Control) Panel (ASP / ACP), ADF2, Stab trim override & Flightdeck door lock.

Centre Instrument Panel

Officially the P2 panel, the centre panel contains the engine instruments, landing gear & brakes control & display and the standby flight instruments; ie non-duplicated things that either pilot may need to use.

737-100/200

The 737-1/200 had JT8D engines, placarded here as -9As. EPR was used for power settings instead of N1 and the panel on the right is where EPR was set for the autothrottle along with A/T mode - CRZ / CLIMB / CONT / GA. The sequence of primary engine instruments was: EPR, N1, EGT, N2, Fuel flow. The secondary instruments were oil pressure, oil temperature, oil quantity. Notice that there was no vibration gauge and the hydraulic pressure gauge was to the right of the gear lever. The Reverser Unlocked lights were grouped with the other engine warnings.

The two vertical INOP stickers are for the GRAVEL PROTECT lights which blew bleed air from a probe ahead of the engine cowl to stop vortices lifting debris into the engine.

737-200

737-300/400/500 – Conventional Engine Instruments

This "Round Dial" 737-300 is distinguishable from a -1/200 by not having an EPR gauge and the inclusion of a vibration gauge. The hydraulic quantity and pressure gauges are still outboard of the gear lever.

Notice the Speed Brake Test and Eng Oil Qty Test buttons at the top of the panel, these are not required on EIS aircraft which have their own BITE check faciliy.

Early 737-300

737-300/400/500 – Engine Instrument System

From October 1988 the Engine Instrument System (EIS) replaced 21 individual electromechanical instruments. It comprised of a primary and secondary panel made by Smiths Industries. The primary panel had the usual N1, EGT, N2 and Fuel flow/used. The secondary panel had the oil pressure, temperature and quantity (digital only), vibration and also included TAT, hydraulic pressure and quantity (digital only) from the F/O's instrument panel.

The centre instrument panel has the Yaw Damper indicator tucked away under the glareshield panel which is easily missed on instrument checks, especially if your seating position is high. Boeing recommends that you sit so that you can see the top of the A/P–A/T–FMC lights and the bottom of the EHSI. This means that the yaw damper indicator will be out of view.

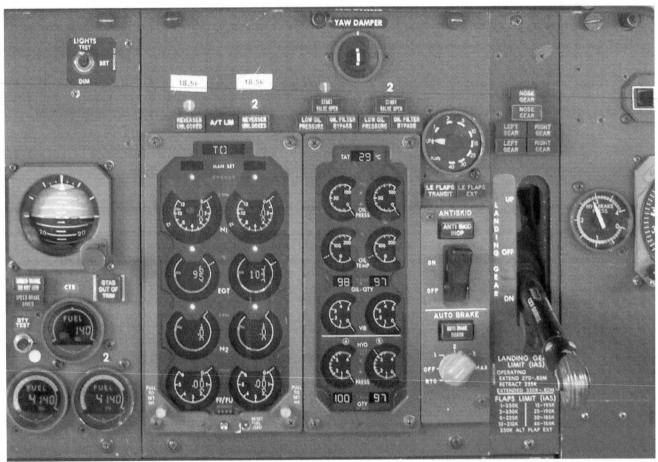

737-3/4/500

737-NG

The NG engine instruments are normally displayed on the centre forward panel upper and lower display units. The choice of gauges displayed is a customer option. Many NG operators choose to have most of the parameters hidden after engine start. If an abnormality arises, it will automatically display, thereby alerting the crew to the problem.

The fuel, oil and hydraulic gauges have gone and are displayed on the upper or lower DUs except for the brake pressure gauge which is virtually the same instrument as fitted to the original 737.

737 NG

Overhead Panel

The overhead panel (P5 panel) is the nerve centre of the aircraft systems and replaces many of the controls previously located on the flight engineers panel. the following page shows how the function and layout of this important panel has evolved from its roots in the 1950's.

The four light-grey panels are the "Primary system control panels" - fuel, electrical, hydraulics and air-conditioning.

The individual panels are usually partially covered by a 1/4in deep raised section called the light plate. This contains the backlights for any markings required for the panel. The small white cross on each light-plate shows the position of the electrical connector behind the plate. So, if the individual panel lighting is acting up then pressing on the cross will sometimes help.

The ENGINE panel complete with light-plate.

The ENGINE panel with the light-plate removed.

737-NG (opposite page)

This panel is still very similar to the original 1967 737-100. Obvious differences include:
- Digital AC & DC metering panel
- No Gen oil temp or load meters
- No APU load meter
- Independent wiper controls
- Overwing exit annunciator lights
- DCPCS pressurisation panel

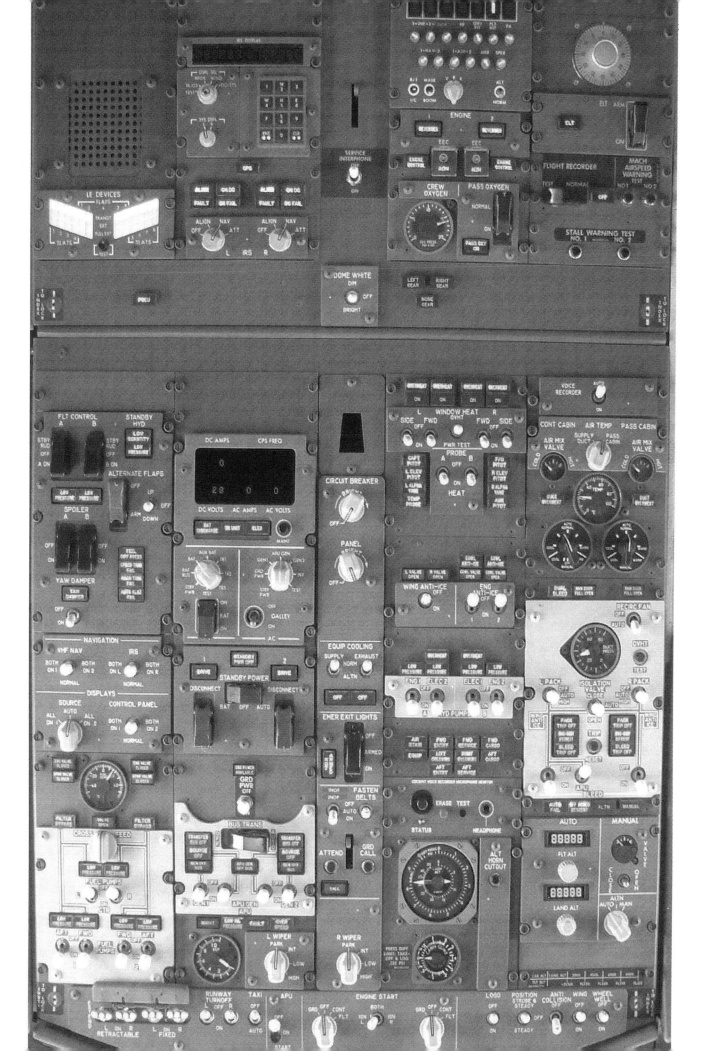

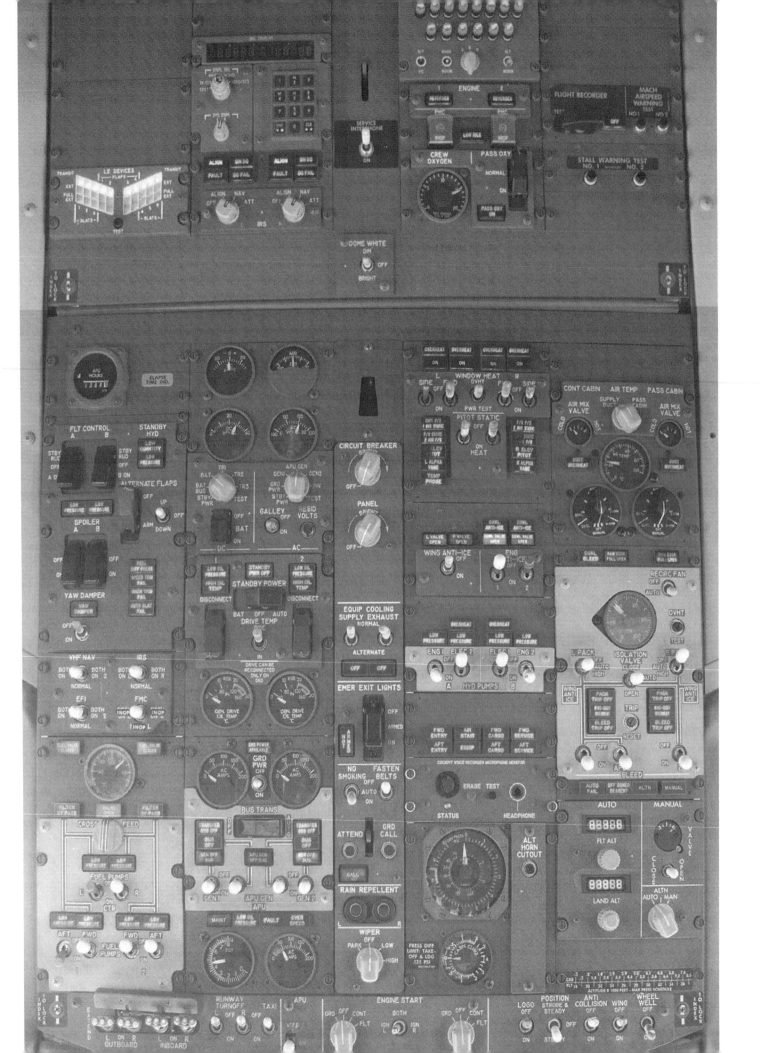

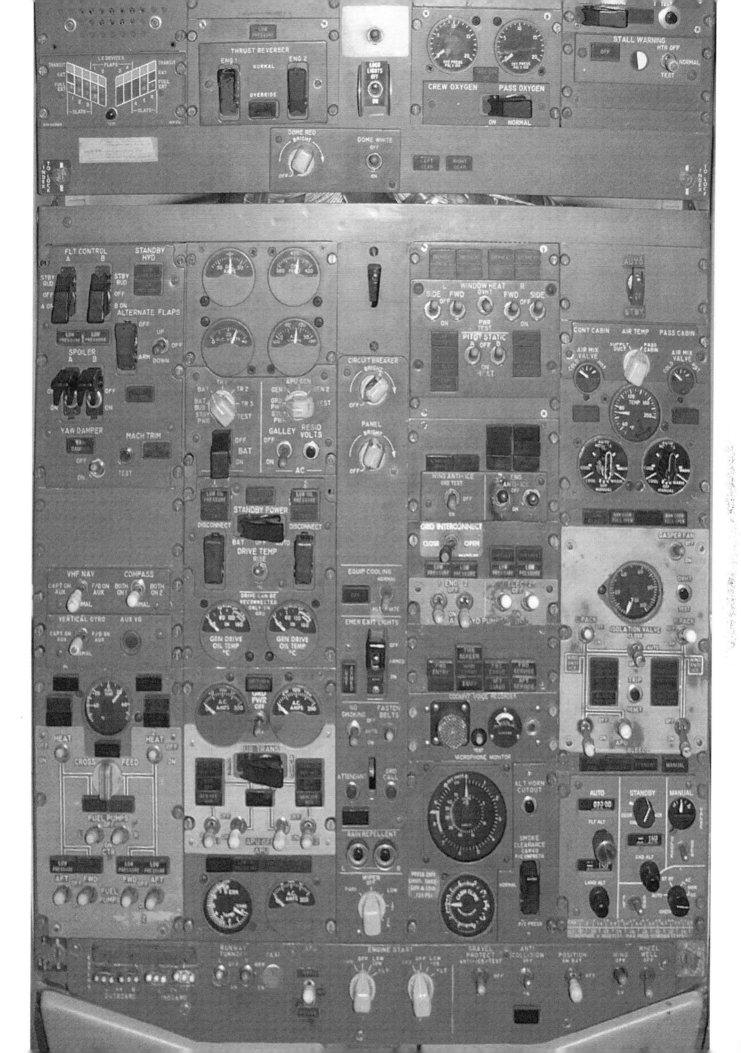

737-300/400/500 (previous opposite page)

This is a later production 737-300 overhead panel.

The air conditioning panel shows that it is not a 400 series and the digital cabin pressurisation controller shows that it is a late build aircraft.

737-200 (previous page)

Superficially very similar to the -3/4/500 panel. Differences include:

- Two extra panels (top left & right) for Compasses. DG or Slaved.
- Mach Trim test.
- Compass and Vertical Gyro transfer functions.
- Fuel filter heaters.
- Single equip cooling fan switch.
- Multi speed wipers.
- No alpha vane heat.
- Engine anti-ice indications.
- Layout of hydraulic pumps.
- Gasper fan rather than Recirc fan.
- No ignition switching.

This particular aircraft (737-200C) was built for cargo operations and some other distinguishing features:
- Extra set of main gear green lights on the aft overhead panel because the downlock viewers are not installed.
- Smoke clearance system on the pressurisation panel.
- MAIN CARGO door annunciator.

Notice also the GRAVEL PROTECT switch to the right of the ignition switches and the TYRE SCREEN light on the doors panel which are probably indicative of the rough strip kit.

727

Although difficult to identify all of the controls in this low shot photograph, it can still be seen how the 727-200 panel bridges the gap between 707 and 737, in particular, the Flight Controls panel is very recognisable to 737 pilots.

The guarded switches in the fourth column are the ignition and start switches. Notice the location of the fire switches in this version.

727-200Adv

707

707-441 Photo Valter Azevedo

This 707-441 overhead panel shows some components that are still in use on the 737 of today. For instance, the HF radios, alternate flap, emergency exit lights, rain repellent, and exterior lights are virtually identical. All credit to Boeing for living by the theory of: "If it ain't broke, don't fix it."

The 707 aft overhead (P5) panel was purely for circuit breakers.

EXTERNAL

A photographic tour of some items of interest on a 737 external check. The full list of items to check on a walkaround will be given in your company Ops manual but here are a few things that are worth looking out for.

Access Panels

This photo shows the importance of ensuring that all filler caps and hatches are correctly closed before flight. The hatches are painted Day-Glo orange on the inside to make them easier to detect and are designed so that they are difficult to close with the cap not properly in place.

In this case water had seeped through an incorrectly fitted cap and froze behind the panel forcing it open. 15 minutes into the flight the cabin crew reported that they had no water available in the rear galley even though the quantity indicator was showing half full. Shortly afterwards the forward galley also dried up.

Ice formed from an incorrectly closed water servicing panel.

After landing a 3ft icicle was observed from this hatch but it broke off before I could photograph it. During the 1hr flight, the water had frozen back up the line and into the tank. The aircraft was grounded for several hours while it was allowed to thaw out.

If this had been the toilet servicing panel, it would have formed the famous "Blue Ice", so called because of the colour of the chemicals used in the flushing agent.

Notice how the drain mast (right) from the sinks is heated to allow water to flow out without icing problems.

Birdstrike Damage

Feathers in booster blades from a birdstrike

This photo shows feathers still lodged in the booster blades of a CFM56-7 after a 2.5hr flight. Unusually, neither pilot (I was one) saw any birds during the event. There had been no impact sound, smell, vibration nor any other abnormal engine indications. In fact we were alerted only by a call from ATC telling us that a seagull had been found on the runway after our departure.

This is the same photo zoomed out and shows how easily it could have been missed on a walkaround, especially if the feathers were near the 12 o'clock position where they would be hidden from view by the fan blades.

Any birdstrike into the core, or a birdstrike where not all of the remains of the bird can be found, requires a boroscope inspection within 10 cycles.

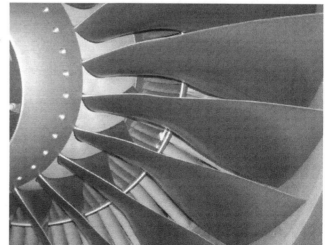

This photo shows a CFM56-3 that took a seagull at 160kts & 200ft after take-off. Again I had no abnormal engine indications but there was a strong smell of cooking bird for several minutes! You can see that blades 5, 6 & 7 have all been bent by the impact.

Bent fan blades after a birdstrike

Birdstrike remains on the Captains pitot

This photo shows the potential for trouble from a birdstrike. I caught this unfortunate bird with the pitot, if any parts had entered the tube it would have lead to unreliable airspeed indications

Brake Wear Pins

When these pins (Two on each brake unit ie eight in total) become flush with brake housing, the brakes are ready for replacement. Bring to the attention of an engineer before dispatch.

Data suggests that the average wear rate is 0.001 inch per landing. This should mean that if a pin were only 0.01" proud, it should still be good for another 10 landings.

If observed downroute then it may be comforting to know that when any pin is flush, the associated brake unit still has enough capacity for an RTO - by law.

E & E Bay

Electronics and Equipment Bay

You will have seen the EQUIP caption and know roughly where the hatch is, but if you are curious to see what is inside then I recommend that you get an engineer to show you how to open, and more importantly, close the hatch, as it is not straightforward.

Hatch Release Latch

View into the E & E bay (737 Classic)

737 Classic E & E bay

The E & E bay on the NG looks fairly similar. The large unit with the c/bs is the power distribution panel; to its left is the battery charger; above are the fire & overheat detection unit, pressurisation controller and APU start convertor unit (in grey).

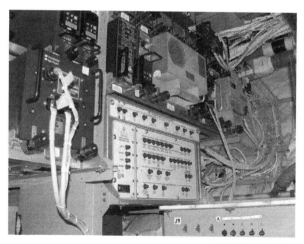

737 NG E & E bay

Fluid Ingress

On 22 October 1995, G-BGJI, a 737-200Adv experienced undemanded yaw & roll oscillations during an air test. This was put down to fluid from the cabin leaking into the E & E bay and onto the yaw damper coupler. The report stated:

"The location of the E&E Bay, beneath the cabin floor in the area of the aircraft doors, galleys and toilets made it vulnerable to fluid ingress from a variety of sources."

Only the E1 rack is vulnerable to fluid ingress because it is directly below the forward entry door, the other racks are much further aft. For protection the Classics have a carpet over the E1 rack and the NGs have drip trays over all racks. The time to be careful is when the forward doors are open on a turnaround with heavy rain coming in.

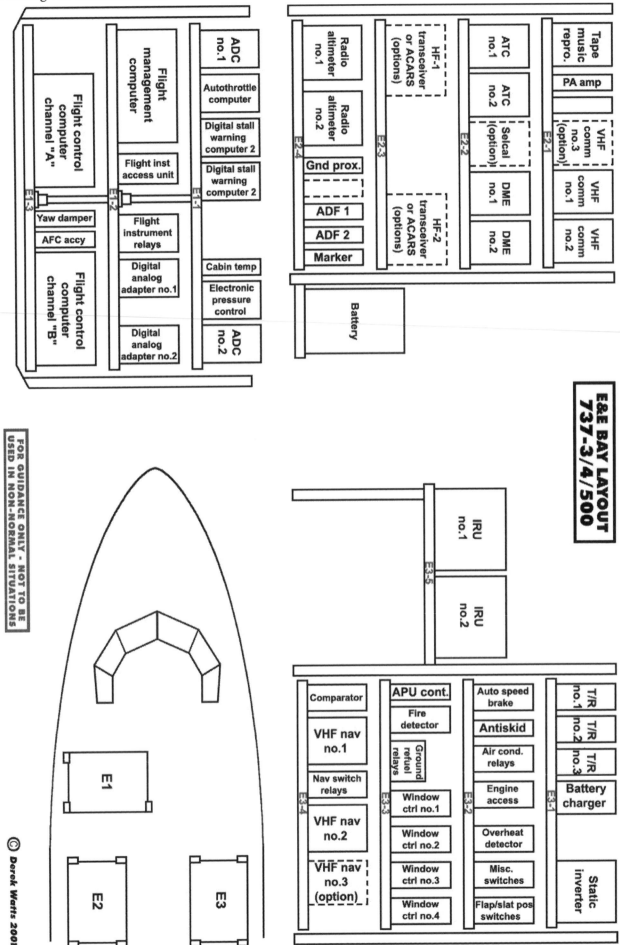

E&E BAY LAYOUT 737-3/4/500

FOR GUIDANCE ONLY - NOT TO BE USED IN NON-NORMAL SITUATIONS

© *Derek Watts 2001*

ADC no.1
Autothrottle computer
Digital stall warning computer 2
Digital stall warning computer 2
Cabin temp
Electronic pressure control
ADC no.2

Flight management computer
Flight inst access unit
Flight instrument relays
Digital analog adapter no.1
Digital analog adapter no.2

Flight control computer channel "A"
Yaw damper
AFC accy
Flight control computer channel "B"

E1-3 E1-2 E1-1

Radio altimeter no.1
Radio altimeter no.2
Gnd prox.
ADF 1
ADF 2
Marker

HF-1 transceiver or ACARS (options)
HF-2 transceiver or ACARS (options)

ATC no.1
ATC no.2
Seical (option)
DME no.1
DME no.2

Tape music repro.
PA amp
VHF comm no.3 (option)
VHF comm no.1
VHF comm no.2

E2-4 E2-3 E2-2 E2-1

Battery

IRU no.1
IRU no.2

E3-5

Comparator
VHF nav no.1
Nav switch relays
VHF nav no.2
VHF nav no.3 (option)

APU cont.
Fire detector
Ground refuel relays
Window ctrl no.1
Window ctrl no.2
Window ctrl no.3
Window ctrl no.4

Auto speed brake
Antiskid
Air cond. relays
Engine access
Overheat detector
Misc. switches
Flap/slat pos switches

T/R no.1
T/R no.2
T/R no.3
Battery charger
Static inverter

E3-4 E3-3 E3-2 E3-1

E1
E2
E3

Engine Oil Sight

Although not required for a pilot's external checks, the 737 NG engine oil sight is easily accessible. The hatch is centrally located on the starboard side of the engine and the sight glass is down to the left. Note that the #2 (right hand) engine oil sight always reads less than the #1 engine because of the angle at which it sits.

Fuel Floatstick / Dripstick

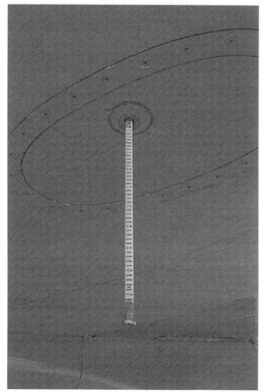

These must be used when the fuel quantity cannot be determined by the gauges. The Classics have 5 in each wing tank and none in the centre tank. The NGs have 6 in each wing tank and 4 in the centre tank.

All sticks must be read for each tank and the measurements recorded; the quantity is then calculated from tables in the maintenance manual.

Shown here is a floatstick which are fitted to later aircraft. These have the big advantage over dripsticks of not drenching you in fuel when you use them.

If any of your fuel gauges or the Fuel Summation Unit (Classics) Fuel Quantity Indication System (NGs) is u/s, the gross weight must be periodically updated in the FMC to ensure the accuracy of fuel calculations, VNAV speeds, buffet margin and max altitude etc.

Fuel Floatstick

Fuel Tank Vents

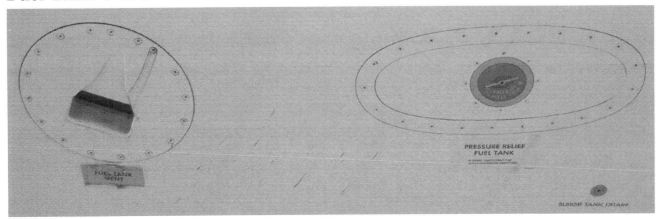

These vents, located near each wing tip, are shaped as air scoops (called NACA ducts); they provide a small positive head of pressure on the fuel in all three tanks. This prevents a vacuum forming as fuel is used, assists the fuel pumps and reduces evaporation. If you overfill the fuel tanks, then this is where the ensuing fuel spillage will come from!

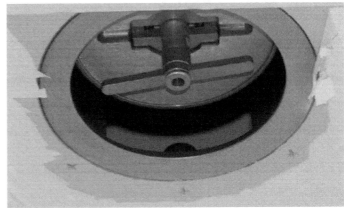

There is an option to fit a flame arrestor (SB 737-28A1131) at the vent, which has obvious safety benefits. However, there is always a chance that the flame arrestor could become clogged and block the vent, so the pressure relief valve must also be fitted. The photo to the right shows a pressure relief valve which activated during a flight test.

Activated surge tank pressure relief valve

Some auxiliary fuel tanks do not vent through the normal vent system described above, these aircraft have vents elsewhere. On this aircraft, the vent is forward of the starboard wing leading edge.

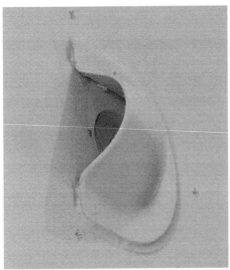

The auxiliary fuel tank vent

Location of the auxiliary fuel tank vent

Refuelling Panel

Refuelling panel with extra controls for auxiliary fuel tanks

Located under the starboard wing, near the leading edge. (This is where the F/O might find himself if it is raining during the turnaround!)

Use the three tank valve switches to direct fuel into the desired tanks. The blue lights next to the switches will illuminate when the associated tank valve switch is open and the tank is not full. They extinguish either when the tank is full or the associated switch is turned off. The gauges are repeaters of those in the flight deck - so don't press the quantity test switch while your colleague is doing the refuelling. Some gauges may be fitted with a quantity preset facility to assist the refuel process. Just visible in the centre right of the photograph are the solenoid override switches, these each have a button to override the valve switches and force the valve open - use with care, fuel spillages are expensive.

This particular panel has auxiliary fuel tanks. The guarded switch directs the fuel from normal to auxiliary and the LED display on the gauges changes from 2 to FA (Fwd Aux) and 1 to AA (Aft Aux).

There is a service interphone socket. Note: The NGs do not have a cap for the refuelling point.

Generators

Each engine has an AC generator. The constant-speed drive unit (or other system) is the link between the generator and the engine. The drive unit has its own oil system for cooling and lubrication and can be checked in the sight glass before flight. This oil system is independent of the engine oil system.

CSD - Constant Speed Drive

The CSD is driven through an in-line, variable ratio drive unit which transmits the torque to drive the generator at a constant speed from the variable speed accessory drive pad on the airplane engine. This allows the generator to produce AC at a constant frequency. Note that the CSD is only the drive, not the generator.

Check the oil quantity in the sight glass. Note the different level marks for left and right engines; this is because they are interchangeable units and can be fitted to the port side of either engine. The engines are at different angles due to the dihedral of the wing.

The reset handle is at the bottom.

Supply capacity is 45KVA.

VSCF - Variable Speed Constant Frequency

The VSCF is driven from a variable speed shaft on the accessory gearbox. The generator produces 3 phase AC at 115V between 1370 and 2545Hz. This is then converted to 400Hz in the VSCF by first converting it to DC and then back to AC.

The photograph shows an old style VSCF (two LEDs). Notice "Fault" and "Open Phase" LEDs. Press "Indicate" button (top left) with cockpit lights set bright for indication. If any LEDs illuminate, cross check the generator diagnostics panel and report to engineer. Button on top right is the LED lamp test.

In 2001 Boeing and Hamilton Sundstrand introduced modifications to the VSCF to improve its reliability which had been approx 1/3 that of CSDs. The failure rate was just over 1 every 2000hrs. You can identify a new VSCF by the three LEDs rather than two on the unit. The instruction plate for the buttons and LEDs is located just below the LEDs.

Due to poor reliability, if either generator is a VSCF there is a (UK CAA) limitation to remain within 45mins flying time of a suitable airfield.

IDG - Integrated Drive Generator (NGs only)

As the name suggests, an IDG is a drive and a generator in one unit. Unlike the CSD which is a drive and then a separate generator.

IDGs will auto-disconnect with a high oil temperature, thus no IDG oil temperature gauges on the generator drive panel. They are cooled by both fan air and a fuel-oil heat exchanger.

Supply capacity is 90KVA, with options available for up to 180KVA eg for AEW&C aircraft.

There are three coloured plugs at the top right. These are the three phase cables and are colour coded as follows:

ØA = Red
ØB = Yellow
ØC = Blue
Earth = White

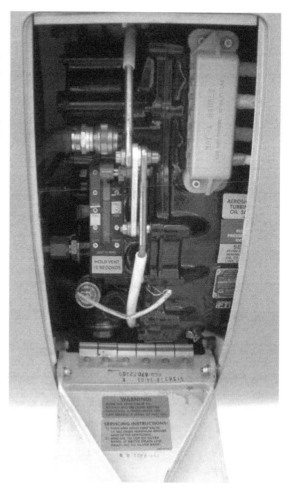

To correctly check the oil level the CSD/VSCF/IDG must first be vented for 15 seconds to release pressure. Caution: a spray of hot oil can emit from the vent during this process. The vent on the IDG is at the top of the sight glass

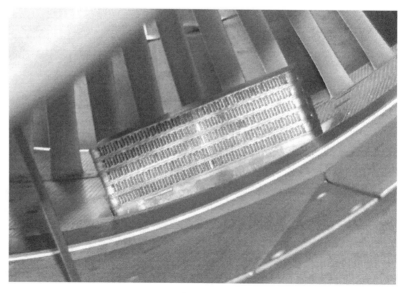

The generator oil cooler is in the bypass airflow

Instrument Probes

This photo shows how close the airbridge comes to the port instrument probes on the 1-500s. Care should be taken to park exactly in accordance with stand guidance as there is only a few inches of clearance.

I would recommend that you inspect the probes for damage if the jetty driver had any difficulty maneuvering onto your aircraft. This has greater importance these days with the introduction of RVSM airspace. Many airlines now paint a box around this "RVSM Critical Area" (Ref SRM 51-10-03) to indicate where no dents are allowable.

One of the improvements to the NG series was to move these probes further forward away from the jetty risk area.

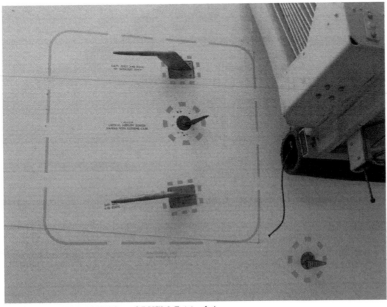

On the 1-500s, the LHS probes are from top to bottom:
- Capts pitot and static & 1st aux static.
- Alpha vane
- F/O static & 2nd aux pitot
- Temp probe

Probes on a 737-3/4/500 and RVSM Critical Area

On the NGs, the LHS probes are from top to bottom:
- Capts pitot.
- Alpha vane
- Temp probe

The pitot heat captions on the overhead panel are in a similar orientation. The elevator pitots are located on the tail-fin.

The exceptions are the freighters. These aircraft retain the pitot-static layout of the Classics because the side cargo door would interfere with the location of the static ports.

Probes on the 737-NG

Static Ports

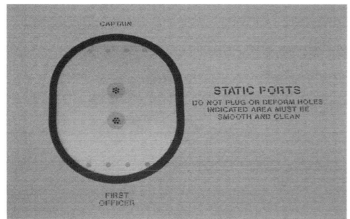

737 NG Static Ports

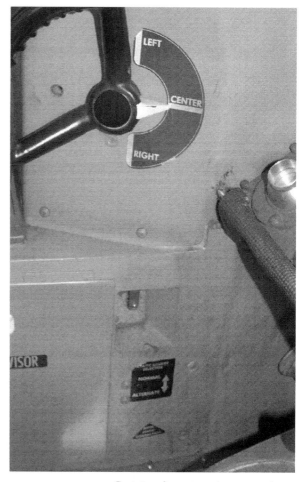

Captains alternate static source selector

In general terms, the pilots' static systems are used for the ADCs (ie flight instruments) and the aux static is used for aircraft systems. On the 737 Classics the normal and aux static sources are at the heated probes. The static sources are cross-connected for dynamic balance.

The alternate static source is the disc on the side of the aircraft. This has to be manually selected by the crew using the selectors on the cockpit wall down by the sun visors.

The static source selector is not present on all aircraft.

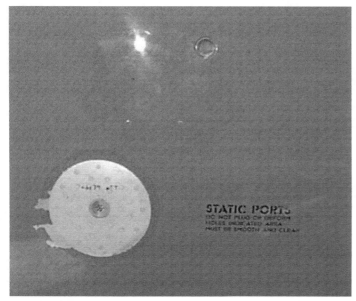

737-1/500 Static port

There are alternate static ports on both sides of the forward fuselage. This one is showing signs of paint loss around it, possibly caused by ice or airflow.

The green discharge disc above it is from the crew oxygen bottle located in the forward cargo compartment. If the disc is missing it shows that the bottle has over temp'd or over pressured. Note it does not necessarily show that the bottle is empty (the bottle could have leaked through the masks on the flight deck), although if the disc is missing it most certainly will be empty.

Test aircraft can sometimes be seen with a cone trailing behind the aircraft. This is a static cone and is aerodynamically designed to give an area of zero dynamic pressure. These are used to calibrate pressure instruments during airframe modifications and were recently used for altimeter calibration during RVSM calibration.

Static cone on 737-800 SFP prototype
(Photo Vladimir Kostritsa)

TAT Probes

TAT probes can be either aspirated or unaspirated. There are three ways to tell if you have an aspirated probe:
1. The pitot heat panel has a TAT test button.
2. The FMC TAKEOFF REF 2/2 page shows the sensed OAT in small letters.
3. Its physical appearance is unperforated (see photo below).

To get an approximate OAT indication on the ground with an aspirated probe, the left air-conditioning pack must be on. This is because aspirated TAT probes use air drawn from the left turbofan. To get an approximate OAT indication on the ground with an unaspirated probe the pitot heat must be off.

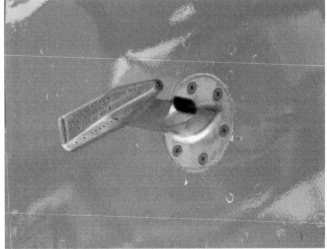

TAT Probe – Unaspirated
Perforated with large hole at rear

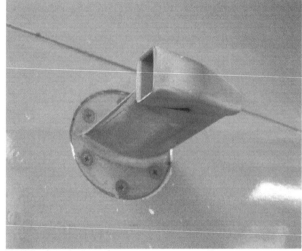

TAT Probe - Aspirated
Unperforated and no large hole at rear

Quote from the AMM: *"The TAT probe gets bleed air from the APU duct in the keel beam. Bleed air into the probe makes a negative pressure inside the probe. The negative pressure pulls outside air across the sensing elements. This permits accurate temperature measurement when the airplane is on the ground or moving at low speed."*

The 737-1/200 also has a second TAT probe on the RHS

Lightning Strike

This photograph shows damage caused by a lightning strike. These are three of the 5 entry holes, about 1 - 2 cms in diameter, there will also be some exit holes elsewhere on the airframe.

Although you are quite safe inside the aircraft when lightning strikes as the electricity is conducted away by the aircraft skin (Faradays Cage effect), it is necessary to check the compasses and radios immediately and then have the aircraft checked on the ground by an engineer.

Lightning is always located in the vicinity of CBs but is particularly likely when St Elmo's fire is observed which is defined as "Visible evidence of electrical discharge at a tolerably slow rate, this is not a problem, does not cause any form of damage, and in fact serves in a positive sense as a warning that the environment is electrified, and lightning may possibly occur."

If you do see St Elmo's fire then you should take the usual precautions for both lightning and turbulence. ie Cockpit lighting up, Start switches to FLT and fly at Turbulence speed.

Skin Dents & Buckles

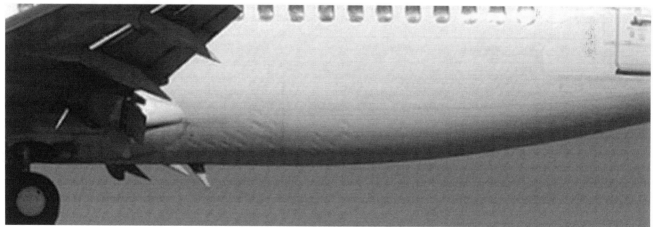

Skin wrinkles on a 737-400

These diagonal shaped wrinkles in the fuselage skin between the frames and the stringers are present in many older aircraft, particularly the longer fuselage versions. They are caused by shear loads between the tail and the centre of gravity or undercarriage due to turbulence or heavy landings. The skin stretches during the moment of tension and deforms leaving a permanent wrinkle. If the length and depth of the wrinkle is within certain limits there is no problem. If you are in any doubt, check with your engineers. They should have a record of every dent and buckle on the aircraft and I would defy any pilot, myself included, to spot even a quarter of these! Note that the wrinkles pictured above will disappear in-flight whilst the aircraft is pressurised.

Main Wheel Well

737 Originals

This photo shows the wheel well of a 1969 vintage 737-200C. The majority of components and locations are unchanged to the NG series in production today.

The tyre damage screen is to protect vital components from damage if debris is flung into the wheel well from a tyre burst or an unpaved strip. It is fitted to all 737-1/200s. The screens are hinged at their outboard edge and locking pins hold the screen in place at the inboard edge. If either pin is not secure the TYRE SCREEN caption will illuminate in the doors panel.

The standby hydraulic reservoir is much smaller on Originals than Classics & NGs. The extra hydraulic reservoir on the ceiling is to boost capacity for the side cargo door which this particular aircraft has installed.

The brake accumulator is much smaller and there is a second one just out of sight on the port aft wall.

737 Classics

At first glance the wheel-well of the classic appears the same as original 737's.

All hydraulic reservoirs have been enlarged and the standby reservoir has moved aft on the keel beam to make room for the PTU.

There is now a single, larger brake accumulator.

NG

The photographs on the following pages show some of the main components in the wheel well of the 737 NG series. One significant difference is the absence of the gear downlock viewer. The standby hydraulic reservoir has now changed shape and moved further aft to the aft wall. The standby hydraulic pump is now out of sight under an access panel aft of the RH wheel-well. The main landing gear is inclined slightly backwards.

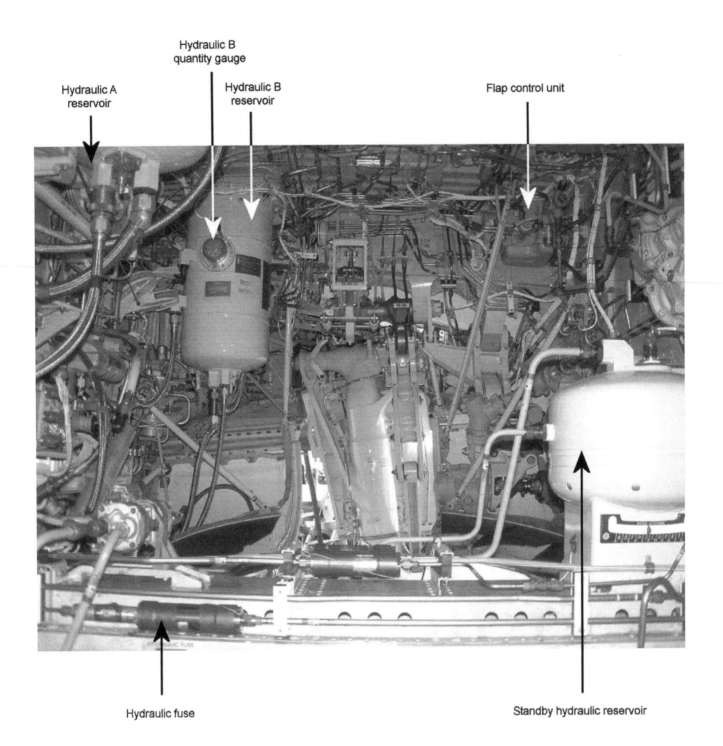

Hydraulic B
quantity gauge

Hydraulic A
reservoir

Hydraulic B
reservoir

Flap control unit

Hydraulic fuse

Standby hydraulic reservoir

NG Looking to Port

Flap torque tube

Engine fire bottles

MLG uplock mechanism

Landing gear selector valve

Aileron a/p actuator

Aileron input shaft

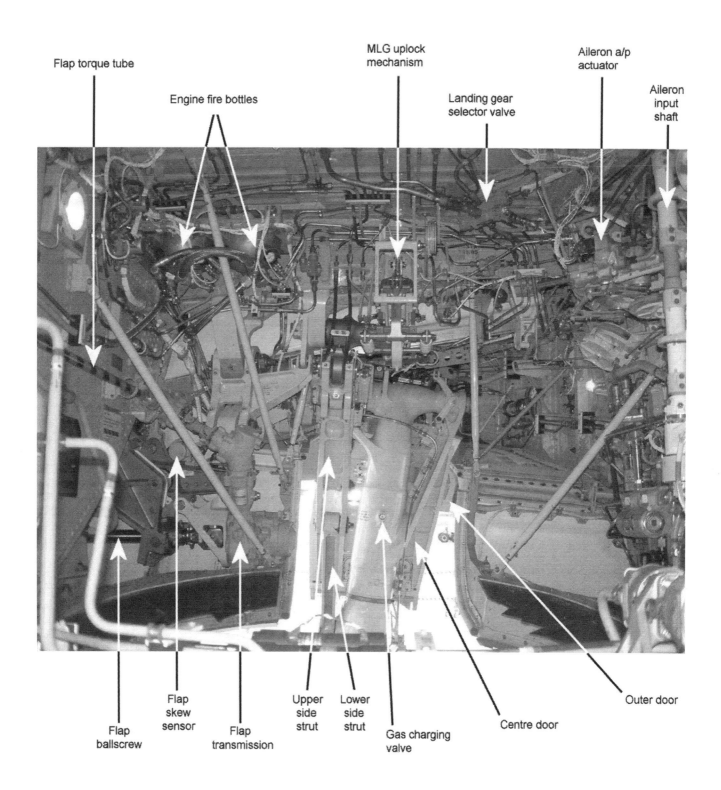

Flap ballscrew

Flap skew sensor

Flap transmission

Upper side strut

Lower side strut

Gas charging valve

Centre door

Outer door

NG Looking to Forward LHS

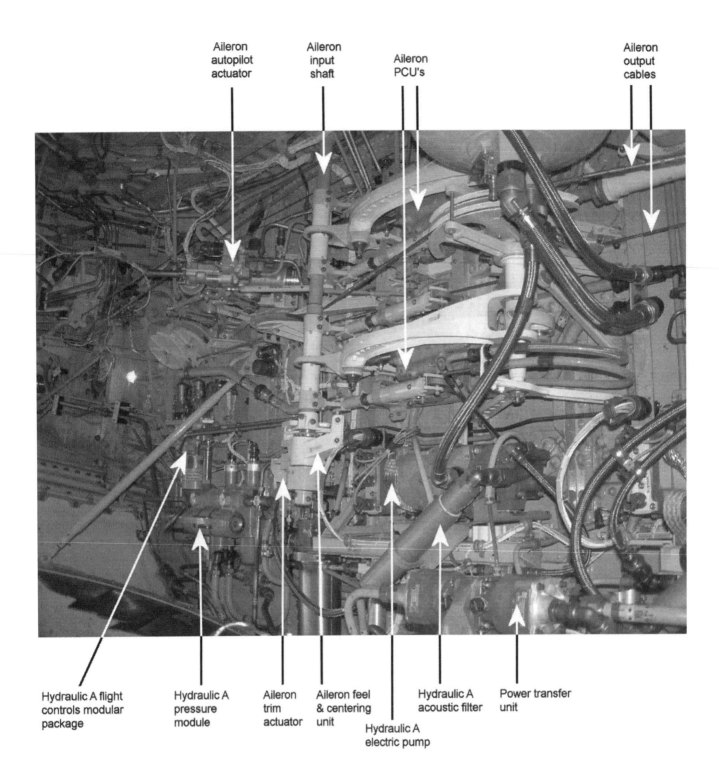

Aileron autopilot actuator

Aileron input shaft

Aileron PCU's

Aileron output cables

Hydraulic A flight controls modular package

Hydraulic A pressure module

Aileron trim actuator

Aileron feel & centering unit

Hydraulic A electric pump

Hydraulic A acoustic filter

Power transfer unit

NG Looking to Forward RHS

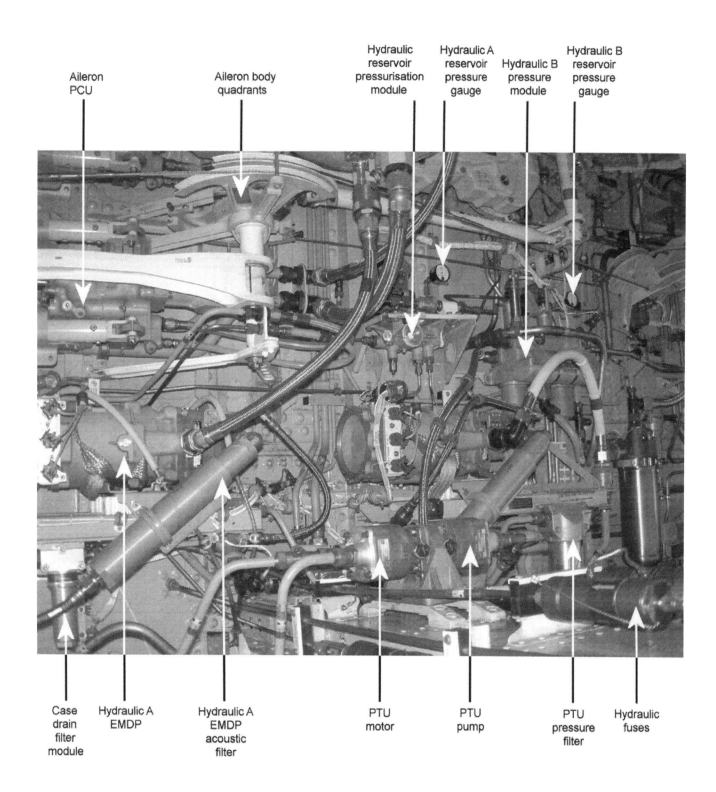

Aileron
PCU

Aileron body
quadrants

Hydraulic
reservoir
pressurisation
module

Hydraulic A
reservoir
pressure
gauge

Hydraulic B
pressure
module

Hydraulic B
reservoir
pressure
gauge

Case
drain
filter
module

Hydraulic A
EMDP

Hydraulic A
EMDP
acoustic
filter

PTU
motor

PTU
pump

PTU
pressure
filter

Hydraulic
fuses

NG Looking to Aft Wall

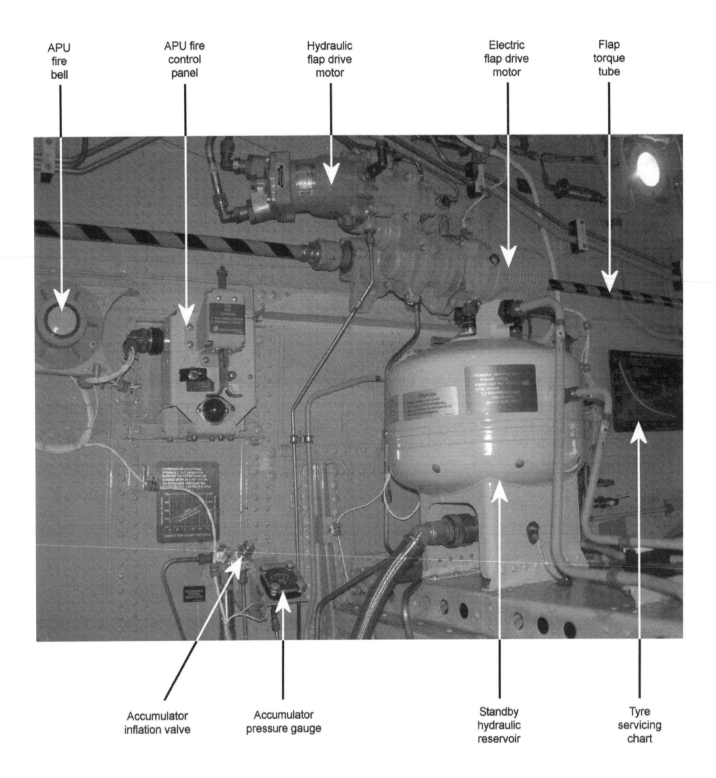

APU
fire
bell

APU fire
control
panel

Hydraulic
flap drive
motor

Electric
flap drive
motor

Flap
torque
tube

Accumulator
inflation valve

Accumulator
pressure gauge

Standby
hydraulic
reservoir

Tyre
servicing
chart

Nose Wheel

The nosewheel does not have brakes. The lines coming down the strut are for the taxi light and the downlock sensors.

The pendant is attached to the steering bypass pin and should be fitted by the groundcrew with the tow bar for pushback. There is also a hole for a nosegear downlock pin giving a total of four pins – one for each gear and one for the steering bypass valve.

The NGs have a lever on the front of the nose gear which is pulled up and held in place with the steering bypass pin. With age these levers can cease to return back to the vertical when the bypass pin is removed. The first you will know about it is when you find you have no nosewheel steering after pushback! It is worth exercising this lever on your walkaround (if the bypass pin has not yet been inserted) to check it is free.

Nosegear - Classic

NG Steering bypass valve – Correct position

NG Steering bypass valve – Sticking lever

Classics steering bypass valve - with pin inserted

The nose gear doors are moved mechanically by the control rods and bellcrank as the gear moves during extension and retraction. When the gear extends, the doors open. When the gear retracts, the doors close.

Main differences in the NG nose wheel well from the Classics include:

- Only one inspection lamp because no gear downlock viewer is installed.
- Redesigned shimmy dampers (v. ineffective on the -700 series).
- Lever on the port side to install & remove the steering bypass pin.
- Two squat switches (called "compressed sensors") for the nose gear.

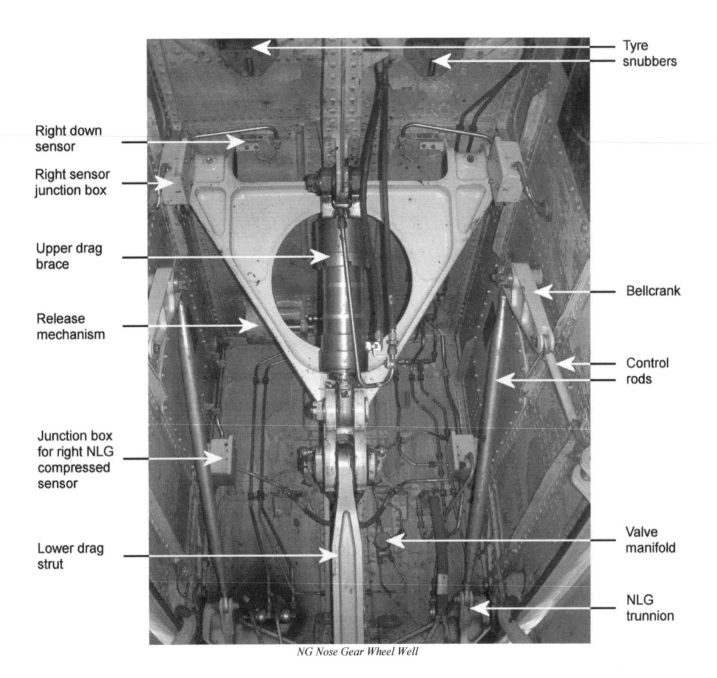

NG Nose Gear Wheel Well

Pressurisation Valves

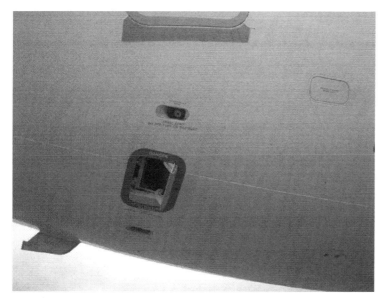

Main Outflow Valve

The main outflow valve is controlled by the pressurisation system.

It regulates the cabin pressure by adjusting the outflow of cabin air.

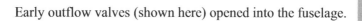

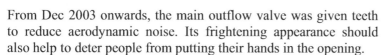

Early outflow valves (shown here) opened into the fuselage.

Later outflow valves opened out from the fuselage.

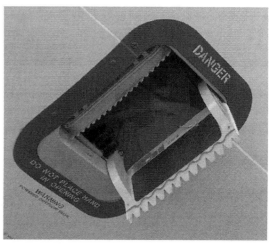

From Dec 2003 onwards, the main outflow valve was given teeth to reduce aerodynamic noise. Its frightening appearance should also help to deter people from putting their hands in the opening.

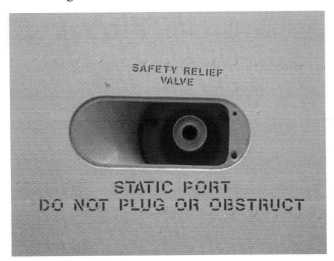

Pressure Safety Relief Valves

These two valves are located above and below the main outflow valve and protect the aircraft structure against overpressure if the pressurisation control system fails. They are set at

- Early Originals 8.5psi
- Later Originals 8.65psi
- Classics: 8.65psi
- NGs: 9.1psi.

Negative Pressure Relief Valve

This prevents vacuum damage to aircraft during a rapid descent. It is a spring loaded flapper valve that opens inwards at -1.0psid. You can check this on a walkaround by pressing it in like a letterbox.

Flow Control Valve (Classic) / Overboard Exhaust Valve (NG)

Open on the ground (check this on a walkaround) to provide E&E bay cooling and also in-flight at less than 2psi differential pressure. You can often hear this valve opening on descent when the differential pressure passes 2psi. The OEV also opens when the recirculation fan (R recirc fan on the 8/900) is switched off to assist in smoke clearance.

Strictly speaking, this is an exhaust port. The actual Flow Control Valve / Overboard Exhaust Valve is located further upstream.

Forward Outflow Valve - Classics only

This is is a vent for the E & E bay air after it has been circulated around the forward cargo compartment when in-flight (The E & E bay air is exhausted from the flow control valve when on the ground). The valve opens when the recirculation fan (R recirc fan on the 400) is off (smoke clearance mode) or when the main outflow valve is not completely closed (ie low diff pressure). It is located just below and aft of the fwd passenger door.

Note the NGs do not have a FOV. In-flight, equipment air is circulated around the forward cargo compartment and discharged from the main outflow valve.

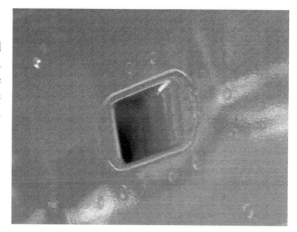

Tail Skid

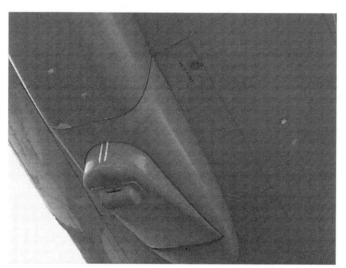

The tail skid is only fitted to the longer length aircraft – 400/800/900. It is always worth checking the tail skid, especially when accepting an aircraft if only to prove that it was not you who scraped it. Check that the decal on the skirt is showing plenty of green and that the shoe is not worn to the dimples. Some operators paint the bottom of the shoe red to help detect a tail strike.

The discs to the right of the picture are the yellow and red APU fire bottle discharge indicators.

737-400 tailskid

THE TAILSKID ONLY PROTECTS THE AIRCRAFT ON TAKEOFF, NOT ON LANDING

Data as of 2004 showed that 82% of 737-400 tail strikes and 70% of 737-800 tail strikes occurred on landing.

The 737-800SFP and 737-900ER have a two position tailskid that extends a further 5 inches during landing for tailstrike protection due to the lower landing speeds.

The two position tailskid in the landing position

271

Static Dischargers

There are static dischargers (either wicks or rods) at the tips and trailing edges of the wings, stabiliser and fin. They encourage the static build-up on the airframe to bleed off which would otherwise accumulate and cause radio interference, particularly of ADF & HF. Note that they are not for lightning protection.

They should be checked on a walkaround as they are very susceptible to damage from lightning strikes. If you notice that any are missing, you would be strongly advised to get your engineers to conduct a post lightning strike inspection in case other damage has been sustained.

You require a minimum of 2 on each wing and horizontal stabiliser and 1 on the vertical stabiliser all must include those at the tips.

There are at least two different types of static dischargers in use.

Static discharge wicks are bundles of metal needles, like stranded wire, enclosed in a rod which has a yellow conductive coating. The needles are exposed at the end to allow the static to discharge.

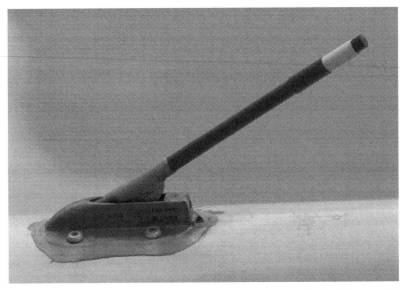

Static discharge wick

Static discharge rods each hold two tungsten needles which are bonded to the aircraft at one end and come out of the rod between the two pairs of teeth near the tip.

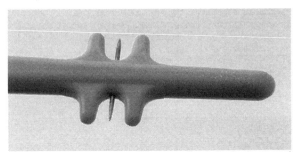

Static discharge rods

Close-up of a static discharge rod

Tyres

The Maintenance Manual gives the following guidance for tyre inspections:

Examine the tyres for air leaks, abrasions, unusual worn areas, cuts, and flat spots.

Remove the tyres that have the conditions that follow:
    1) Cuts or weather cracks in the grooves, the tread, or the sidewalls which go to the cord body.
    2) Blisters, bulges, or other signs of ply separation in the tread or the sidewall area.
    3) Tyres with a flat spot which shows the tread reinforcements/cut protector.
    4) Other types of damage which can cause tyre problems.

Examine the tyres for worn areas:
    1) Measure the tread depth at three points that are equally apart in the tyre groove.
    2) If the average tread depth on any groove is 1/32 inch (0.79 mm) or less, the tyre is not serviceable and must be replaced.
    3) If the tread belt ply (fabric) shows at any location, the tyre is not serviceable and must be replaced.
    4) If the tread reinforcement/cut protector (steel) shows at any location, the tyre is not serviceable and must be replaced.

NOTE: If the tread reinforcement/cut protector (steel) shows, the tyre may be used without safety concerns, but if the tyre is left in service you may not be able to re-tread it.

Unpaved Strip Kit

The optional Unpaved Strip Kit was made available for the 737-100/200 as early as Feb 1969. It allowed aircraft to operate from gravel, dirt or grass strips. At its peak of operation, 737s were making over 2000 movements a year from unpaved runways.

Whatever surface was to be used, certain guidelines had to be observed. The surface had to be smooth with no bumps higher than 3 inches in 100ft or 4 inches in 200ft; good drainage with no standing water or ruts; and the surface material had to be at least 6 inches thick with no areas of deep loose gravel. Boeing offered a survey service to assess the suitability of potential strips. If a surface was not particularly hard it could still be used by reducing tyre pressure down to a minimum 40psi in accordance with a chart.

Components included:
- Nose-gear gravel deflector to keep gravel off the underbelly.
- Smaller deflectors on the oversized main gear to prevent damage to the flaps.
- Protective metal shields over hydraulic tubing and brake cables on the main gear strut.
- Protective metal shields over speed brake cables.
- Glass fibre reinforced underside of the inboard flaps.
- Metal edge band on flap-to-body seals (aka "elephant ear" fairings).
- Abrasion resistant Teflon based paint on wing and fuselage under surfaces.
- Strengthened under-fuselage aerials.
- Retractable anti-collision light.
- Vortex dissipators fitted to the engine nacelles.
- Screens in the wheel well to protect components against damage.

The **vortex dissipators** prevent vortices forming at the engine intakes which could cause gravel to be ingested by engine. These consist of a small forward projecting tube which blows pressure regulated (55psi) engine bleed air down and aft from nozzles at the tip to break up the vortices.

The dissipators are operated by a solenoid held switch on the overhead panel which switches off on a squat switch so that climb performance is not affected. (Similar to the wing anti ice switch on 3-900 series).

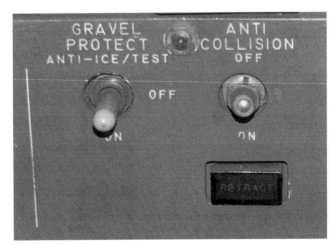

The GRAVEL PROTECT switch must be on or at the ANTI-ICE/TEST position at any time the engines are running on the ground. Take-offs and landings must be made with the engine bleeds off to ensure that there is sufficient bleed air for their effective operation. Air conditioning may be used with APU bleed air if required.

The blue caption says RETRACT for the retractable anti-collision light.

The gravel deflector on the nose gear

The nose gear gravel deflector is made of corrosion-resistant steel and has a sheet metal leading edge which acts as an aerofoil to give it aerodynamic stability.

When the gear retracts, the deflector is hydraulically rotated around the underneath of the nose wheel before seating into the fairing at the front of the nose wheel well. The rotation is programmed to maintain the deflector in a nose-up attitude during transit. No extra crew action is required to use the deflector and in the event of a manual gear extension, springs and rollers will position it correctly.

The maximum speed for gear operation (V_{LO}) is reduced considerably to 180kts and the max speed with the gear extended (V_{LE}) is only 200kts.

Note that the ground clearance of this nose-gear unit is only 3.5 inches. This is enough to allow for flat tyre clearance, but care must be taken when crossing runway arrestor cables, particularly try to avoid taxiing over the "doughnuts" that support any cables.

Operational Procedures
- Antiskid must be ON for takeoff and landing. (AFM)
- Vortex dissipators must be ON for takeoff and landing. (AFM)
- Maximum taxi EPR on gravel: 1.4. (AFM)
- Gravel Protect switch: ANTI-ICE position when using engine inlet anti-ice. (AFM)
- Use of rudder pedal steering rather than tiller is recommended to make all turns as large as possible to prevent nose gear from digging in.
- Thrust to be kept to a minimum to sustain a slow taxi speed.
- If runway is dusty try to manoeuvre so that your jet blast does not pick up loose debris that may be blown back over the runway in a crosswind. Dust should be allowed to settle before starting takeoff roll.
- Notwithstanding the above, use a rolling takeoff wherever possible to avoid debris ingestion when takeoff thrust is set. EPR should be limited to 1.4 or less before brake release.
- For landings, use of autobrake is recommended.
- When landing on gravel, use approximately idle reverse, not to exceed 1.8 EPR. Stow reversers by approx 60kts. (AFM)

PILOTS NOTES

This section is a compilation of pilots notes about all series of 737 to help the reader both fly and understand the aircraft. As always, please regard these as background information only, your company ops manual should be the definitive source of procedures and operational data.

New Normal Procedures

In 2005 Boeing took a big step and redesigned the entire Normal Procedures and checklists for all their Seattle models. This created a major upheaval for airlines, training departments and pilots, but the belief is that it was worth the effort.

Why the change?

Checklists have not always kept pace with the evolution of equipment, pilot responsibilities & operational environment. Boeing has also responded to enquiries from airlines as to why operating procedures were different from type to type and to initiatives from regulatory authorities from around the world.

The intent of the initiatives was to improve safety, particularly with regard to reducing runway incursions. They also improve efficiency and economy.

The new procedures are now standard across many Boeing models and take the best practices from each. They lead to a more even distribution of pilot workload, streamlined checklists and more flexible procedures to allow for the increased complexity of today's operational environment.

Areas of responsibility

The Captain is only responsible for those areas that relate to the control and navigation of the aircraft ie the MCP, his flight instruments, throttle quadrant and trimmers; while the F/O is responsible for all the rest including the entire overhead panel.

Once the aircraft is taxiing or flying the Pilot Flying (PF) is now responsible for those areas that relate to the control and navigation of the aircraft. The Pilot Monitoring (PM) is responsible for all the rest. Both pilots share the responsibility for engine thrust and monitoring and the fire panel.

Checklist philosophy

The checklists have been kept very short, containing only the critical items, primarily those related to safety. For instance, the number of tasks to be accomplished during taxiing is kept low to improve situational awareness, reduce missed radio calls and runway incursions. The checklist is used after doing all the procedural items. I.e. **It is used to verify that the critical items have been done.**

New Normal Checklists

PREFLIGHT

Oxygen . . . . Tested, 100%
Navigation transfer and
display switches . . . . NORMAL, AUTO
Window heat . . . . ON
Pressurization mode selector. . . . . AUTO
Flight instruments . . . . Heading___, Altimeter___
Parking brake . . . . .Set
Engine start levers . . . . CUTOFF

BEFORE START

Flight deck door . . . . Closed and locked
Fuel . . . . ___ LBS/KGS, Pumps ON
Passenger signs. . . . . ___
Windows . . . . Locked
MCP . . . . .V2___, HEADING___, ALTITUDE___
Takeoff speeds . . . . V1___, VR___, V2___
CDU preflight . . . . Completed
Rudder and aileron trim . . . . Free and 0
Taxi and takeoff briefing . . . . Completed
Anti collision light . . . .
.ON

BEFORE TAXI

Generators . . . . On
Probe heat. . . . . ON
Anti-ice . . . . ___
Isolation valve . . . . .AUTO
Engine start switches . . . . CONT
Recall . . . . Checked
Autobrake . . . . RTO
Engine start levers . . . . .IDLE detent
Flight controls . . . . .Checked
Ground equipment . . . . Clear

BEFORE TAKEOFF

Flaps . . . . ___, Green light
Stabilizer trim . . . . ___ Units

AFTER TAKEOFF

Engine bleeds. . . . . .ON
Packs. . . . . .AUTO
Landing gear . . . . .UP and OFF
Flaps . . . . .UP, No lights

DESCENT

Pressurisation. . . . . LAND ALT ___
Recall . . . . Checked
Autobrake . . . . ___
Landing data . . . . .VREF___, Minimums___
Approach briefing . . . . Completed

APPROACH

Altimeters . . . .
___

LANDING

Engine start switches . . . . CONT
Speedbrake . . . . .ARMED
Landing gear. . . . . Down
Flaps . . . . ___, Green light

SHUTDOWN

Fuel pumps. . . . . .OFF
Probe heat. . . . . .OFF
Hydraulic panel . . . . .Set
Flaps . . . . UP
Parking brake . . . . ___
Engine start levers . . . . CUTOFF
Weather radar . . . . Off

SECURE

IRSs . . . . .OFF
Emergency exit lights . . . . .OFF
Window heat . . . . OFF
Packs . . . . OFF

Rules of Thumb

Speed, Height, Distance conversion

Level flight deceleration allow 10kts/nm & 1kt/sec (deceleration is faster at lower weights)
Descending deceleration allow 5kts/nm & 0.5kt/sec
Idle descent allow 3nm/1000ft

Approach Profile Planning

Aim for 250kts, 10,000ft by 30nm out
Aim for 210kts, On ILS at 12nm

Cruise N1

N1 = (2 x Alt/1000) +10, eg at FL350 = 70+10 = 80% N1
FF = (IAS x 10)/2 - 200, eg 250kts = 2500/2 -200 = 1050 kg/hr/engine

N1s & Pitch Attitudes

Phase of flight	737-200 /-15A, 40,000kg/90,000lbs		737-300 /-20K, 48,000kg/105,000lbs		737-700 /-20K, 50,000kg/110,000lbs	
	Attitude (deg NU)	FF (/ engine)	Attitude (deg NU)	N1 (%)	Attitude (deg NU)	N1 (%)
Level Flight:						
250kts	4	1250	4	63	4	59
210kts	6	1100	6	58	6	54
Flap 1, 190kts	6	1200	6	60	6	56
Flap 5, 180kts	7	1300	6	61	6	57
Gear down, flap 15, 150kts	7	1600	7	68	6	65
On glideslope:						
Flap 15, 150kts	4	900	4	52	3	46
Flap 30, Vref + 5	2	1050	2	58	2	51
Flap 40, Vref + 5	0	1200	1	65	1	58

N1 may vary by 5% and attitude by 2° at other weights.
Add 2% N1 in turns.
For single engine add 15% N1 + 5% N1 in turns. For 737-200 double the fuel flow on the remaining engine.

Climb Speeds

If ECON info not available, use 250KIAS until 10,000ft then 280KIAS/M0.74 thereafter.
Best Angle = V2 + 80
Best Rate = V2 + 120

Kinetic Heating

Is approximately 1°C per 10kts IAS

Driftdown

Driftdown speed and level off altitude are for the **terrain critical** case; if terrain is not critical you may accelerate to Long Range Cruise (LRC), this will cost approximately 3000ft. Otherwise slowly accelerate to LRC at the level off altitude as weight reduces with fuel burn. If anti ice is required, the altitude penalties are severe. See table below for figures (QRH PI.13.7).

| | Altitude penalty for engine bleed requirements | | | |
| | 737-300 | | 737-700 | |
Bleed requirements	Terrain critical	LRC	Terrain critical	LRC
Eng Anti-ice ON	-1500ft	-4000ft	-5600ft	-5900ft
Eng & Wing Anti-ice ON	-4800ft	-7600ft	-12500ft	-13000ft

Fuel Consumption Formulae

Optimum FL   (FPPM p2.1.1)

| Altitude away from optimum | Fuel Mileage Penalty % | |
	737-300 M0.74	737-700 M0.78
2000ft above	1	2
Optimum	0	0
2000ft below	2	2
4000ft below	4	5
8000ft below	11	14
12000ft below	20	24

Step climb & wind/altitude trade   (FPPM p3.2.16)

Step climb under consideration	Break even wind 737-300 M0.74	Break even wind 737-700 M0.78
FL290 $\Rightarrow$ FL330	< 34Kts	< 75Kts
FL310 $\Rightarrow$ FL350	< 25Kts	< 69Kts
FL330 $\Rightarrow$ FL370	< 12Kts	< 55Kts
FL370 $\Rightarrow$ FL410	N/A	< 24Kts

The 737 burns approx 30kg(70lbs)/min. Hence subtract (30kg/70lbs x reduced trip time in mins) from the trip fuel at the proposed level. If this figure is less than the trip fuel for the planned flight level, the lower level is justified.

Brake release to ToC: 1700kg (3800lbs), 120nm, 20mins.
ToC landing: 400kg (1000lbs), 120nm, 20mins.

Trip Fuel Reduction = Weight reduction x Flight time in hrs x 3.5%
Eg: 10 pax less over a 2hr flight = 1000kg x 2 x 3.5% = 70kg lower trip fuel. (2200lbs x 2 x 3.5% = 155lbs)

Anti Ice

Engine Anti-ice burns 90kg (200lbs) per hour.
Engine + Wing Anti-ice burns 250kg (550lbs) per hour.

Non-Normal Configurations

Compared to 2 Engine LRC at Optimum Altitude for any given weight:
* Engine-out LRC burns 20% more fuel & reduces speed by 15%.
* Depressurised LRC (2 Engines at 10,000ft) burns 50% more fuel & reduces speed by 20%.
* Gear Down burns 90% more fuel! & reduces speed by 30%.

Limitations

Note: Systems limitations are listed with the system description. Not all limitations given here are AFM

Operational

Runway slope limits	**+/-2%**
Max takeoff / landing tailwind component	**10kts (Or 15kts as customer option)** **No tailwind component allowed on contaminated runways.**
Max windspeed for taxying	**65kts**
Turbulent airspeed	**1/200: 280kts/.70M** **3-500: 280kts/.73M** **6-900: 280kts/.76M**
Max operating altitude	**1/200 (Unmodified): 35,000ft** **1/200 (Most a/c): 37,000ft** **3-500: 37,000ft** **6-900: 41,000ft**

Flight Manoeuvring Load Acceleration Limits (AFM)

Flaps up	**+2.5g to -1.0g**
Flaps down	**+2.0g to 0.0g**

Crosswind Guidelines (FCTM)

		Takeoff		Landing
Runway Condition	**1/200**	**3/4/500**	**NG**	**All series**
Dry	40kt	40kt	36kt*	40kt
Wet	25kt	40kt	25kt	40kt
Snow – no melting	21kt	35kt	25kt	35kt
Standing water / Slush	16kt	20kt	15kt	20kt
Ice – no melting	7kt	17kt	15kt	17kt

Do not land on untreated ice or snow if melting is present.
*Crosswind limit reduced to 34kt with winglets

Contaminated Runways (FCTM)

Max depths as follows		
Dry snow	**102mm (4in)**	**NB Above -5°C snow is to be considered as slush.**
Standing water, Wet snow, Compacted snow or Slush	**13mm (0.5in)**	

Minimum Width for Snow Clearance

Width	**Depth**
Central 30m	**13 mm**
Next 8m (38)	**23 cm**
Next 16m (54m width)	**38 cm**
Beyond 54m	**120 cm**

Landing Technique

During intermediate approach - before glideslope capture

Speed is controlled by pitch
Rate of descent is controlled by thrust

During final approach - after glideslope capture

Speed is controlled by thrust
Rate of descent is controlled by pitch

Pitch & Power Settings on Final Approach

Use 5deg nose up for initial flap settings.
Use 2.5deg nose up for flap 30.
For flap 30, start with 55% N1, then adjust as required.

Short finals at LHR R/W 27R

Stabilise the aircraft at the selected approach speed with a constant rate of descent, between approx 600 to 800 fpm, on a desired glide path, in trim.
Descent rates above 1000fpm should be avoided.

Visual Aiming Point

Aim for the aiming point markers or your desired gear touchdown point if no markers are available. Now adjust the final approach glide path until the selected point is stationary in relation to the aircraft. ie it does not appear to move up or down the windscreen.
The approach lights & runway centreline should run between your legs until touchdown, then keep the centreline running down your inside leg.

Flare and Touchdown

After the threshold goes out of sight under the nose, shift the visual sighting point to a point approximately 3/4 down the runway while maintaining descent, this will assist in determining the flare point. Initiate the flare when the main gear is approx 15 feet above the runway by increasing the pitch attitude by about 3deg and smoothly bring the thrust levers back to idle. Do not float, but fly the aircraft onto the runway and accomplish the landing roll procedure.

Instructors Notes

- The importance & necessity of achieving a stabilised approach.
- Use of all available clues - visual and instrument.
- Do not wait until "Decision" before taking in the visual picture.
- Below 200ft, the landing is primarily a visual manoeuvre backed up by instruments.
- The best way to judge the flare near the ground is to fix your eyes on a point near the far end of the runway.
- A firm landing in the TDZ is a good one; a smooth landing outside the TDZ is bad - despite any comments from the cabin crew.

Assumed Temperature Thrust Reduction

An assumed temperature thrust reduction is a way of reducing the take-off thrust to the minimum required for a safe take-off, thereby conserving engine life and hence reducing your chances of an engine failure.

The CFM56-3 and -7 are flat rated at ISA+15C ie 30C. This means that they are guaranteed to give (at least) the rated thrust at the full throttle position when the OAT is below this temperature. Above this temperature, they will give less thrust because the air is less dense.

On occasions when full thrust would be more than is safely required, eg light weight, long runway, headwind etc. we can choose a thrust setting below full thrust by telling the engines (via the FMC) that the OAT is much higher than it actually is. This higher temperature is called the assumed temperature.

If we fool the engines into thinking that the temperature is much higher than it actually is, by entering an assumed temperature into the FMC, they will use a correspondingly lower N1 to give the rated thrust for the higher temperature when TOGA is pressed. You may, at any stage after TOGA is pressed, advance the thrust levers further to give the full rated thrust again.

In practice, we find the assumed temperature by entering the take-off tables with the actual takeoff weight, and then determining the hottest outside air temperature at which the take-off could be performed. This temperature is called the "Assumed Temperature" and is entered into the FMC TAKEOFF REF 1/2 page as "SEL TEMP". The ambient temperature is also entered here (as "OAT") and the reduced thrust take-off N1s will be computed by the FMC and displayed on the CDU. On the NG series, the temperatures are entered in the N1 LIMIT page.

Takeoff Ref FMC U7.x

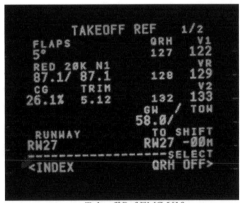

Takeoff Ref FMC U10.x

Some Facts about Assumed Temperature Thrust Reduction

1. **It also known as**: "Flex" (Airbus & Fokker), "Graduation", "Reduced Take-off Thrust (RTOT)" or "Factored Take-off Thrust (FTOT)".

2. **It is not the same as "De-rate"**
A de-rate is a semi-permanent engine fix, used to reduce the maximum thrust available; for instance down to 20k from 22k on -3/700s. It is also used to equalise the thrust where B2 & C1 engines are mixed on the same airframe. When an engine is de-rated, the full (un-de-rated) thrust is no longer available because this would require changes to the EEC, HMU, fuel pump, engine ID plug and the loadable software; none of which can be done by the pilot in-flight.

A temporary form of de-rating known as a "T/O de-rate" is accessible through the FMC on TAKEOFF REF 2/2 or N1 LIMIT (NGs), but this is prohibited by some operators. The T/O de-rates (TO-1 & TO-2) can be 10 to 20%. It follows that an engine may be de-rated and also be using reduced thrust, in which case you could be taking off at Full power -20% -25% = 60% of the full power of the engine - scary thought! Note that a T/O de-rate can be overridden by firewalling the thrust levers; this action will give the thrust rating shown on the IDENT page.

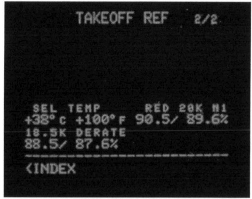
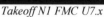

Takeoff N1 FMC U7.x

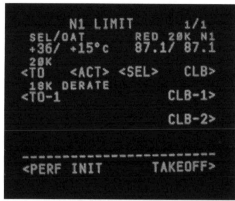

Takeoff N1 FMC U10.x

3. The "Double De-rate"
Using a de-rate and an assumed temperature is known as a "Double De-rate" or "Common N1" option on FMC U10.5 & CDS BP02.

4. Max & Min Temperatures
The normal range of assumed temperatures is from 30 to 55C. However the QRH shows that temperatures from 16 to 75C may be used. The maximum temperature is set by the maximum amount of thrust reduction allowable i.e. 25%; and the minimum is where the engine becomes "flat rated" and no further performance gain can be achieved.

5. It Reduces Vmcg.
Because there will be less swing from the remaining engine if it is at a reduced thrust. Some airlines/aircraft use this (when done as an FMC de-rate) to increase the RTOW on contaminated runways. This must be done by a de-rate rather than an assumed temp, so that the pilot cannot accidentally reapply full power.

6. Saves Engine life.
Most engine wear comes from operation at high internal temperatures, so even a slightly reduced thrust (32C) can make a significant difference to engine life. Increasing engine life will not only save the company money but it will also reduce the chance of you, or the crew after you, having to practice their EFATO technique. Every 1C of assumed temperature will reduce the EGT by approx 3.5C.

7. EFATO
It is not necessary (although it may be good practice) to set full power on the remaining engine after an EFATO, the performance figures allow for this. Doing so will increase your climb gradient but also the asymmetric thrust.

8. Increases TODR.
Therefore you may come to rest on the stopway after a stop from V1.

9. Inherently Safe.
This procedure does have a built in safety factor. For example, if you used an assumed temp of 50C to enable you to use a lower N1, all the figures are done for 50C but the engine will be operating at ambient (which is cooler) so will deliver more thrust than the calculations allowed for. Also, if you do lose an engine you can still increase the thrust to the ambient temperature limit.

10. Increases fuel burn.
Strange, but true. This is because:
    a) Assuming an uninterrupted climb, you will take longer to reach the more economical cruise altitude.
    b) Engines are less efficient when not at full thrust.

11. After a reduced thrust take-off, the climb thrust is also reduced.
This is then gradually increased until the engine is back to full climb thrust at about 15,000ft. The cut-off between CLB-1 & CLB-2 is at about 45C depending upon the amount of take-off thrust reduction. (Effectively, whether TO-1 or TO-2 was used.) CLB-1 or CLB-2 may be either pre-armed before departure or selected manually during the climb after a full thrust take-off if desired.

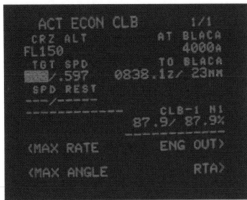

Climb N1 FMC U7.x   *Climb N1 FMC U10.x*

12. Tailwinds.
Reduced thrust is permissible for tailwind take-offs, subject to the normal takeoff flight planning considerations.

13. Provable Numbers.
If you really want to, you can calculate the reduced N1 by using the following formula:

$$\text{Reduced N1} = \text{Full N1} \times \sqrt{(\text{Ambient temp (°K) / Assumed temp (°K)})}$$

This calculation is not an approved airline procedure!

14. Take-off Bump Thrust
This is a customer option to allow an increased take-off thrust **above** the normal max take-off thrust setting. Note the bump only applies to take-off; not to climb, max cont or TOGA thrust. It uses a reserve thrust capability which is only intended for emergency use.

Note: In the "bump" configuration the highest takeoff thrust level is an FMC selectable thrust bump and the use of assumed temperature thrust reduction is not permitted in combination with the bump rating.

Effect of Centre of Gravity

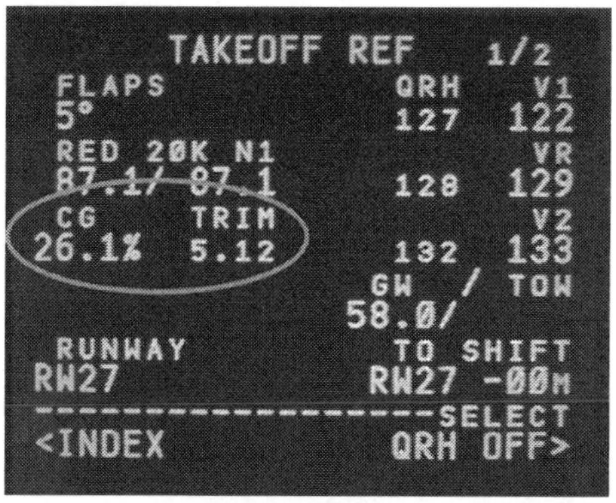

Most loadsheets show both the stab trim and C of G, usually as a Mean Aerodynamic Chord at Take-Off Weight (MACTOW); but how should we use this information?

The stab trim is set before departure to give an in-trim take-off; the corresponding MACTOW can be read from the markings on the stab trim indicator. Unfortunately these markings were removed on the NG series, but they can still be cross checked by entering the CG (MACTOW) in the TAKEOFF REF 1/2 page of the FMC.

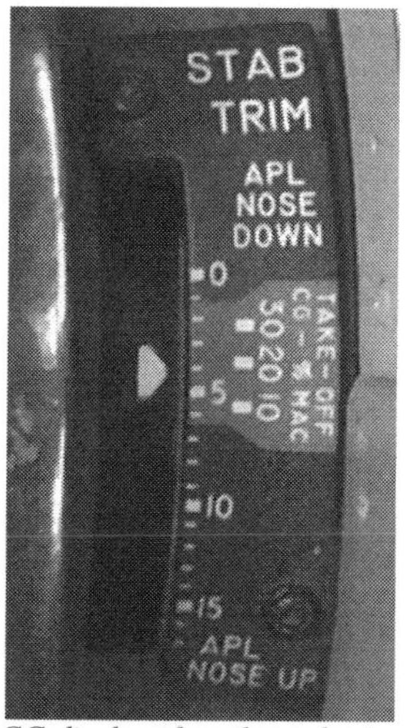

CG displayed on the stab trim

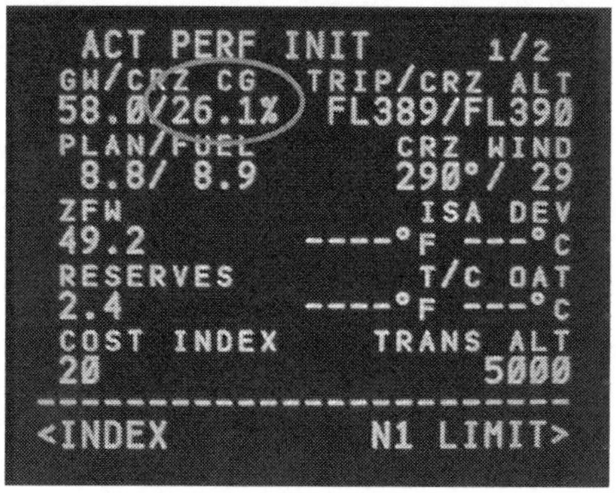

The cruise CG should also be entered into the PERF INIT 1/2 page to give an accurate estimate of the altitude capability. For most purposes the cruise CG can be assumed to be the take-off CG (ie the MACTOW) however, inspection of a manual loadsheet will show a trim correction with fuel burn. Since the fuel tanks are centrally located the CG change will be minimal. In general terms if only wing tank fuel is used, there will be no trim change; if centre tank fuel is used, subtracting 4% from the MACTOW will cover the worst case trim change. If your aircraft has aux tanks the trim change due to fuel burn will be more significant.

The further aft the CofG (ie higher MAC), the higher the altitude capability. This is because an aft CG reduces the amount of downforce the stabiliser has to produce, thereby giving more net lift between the wings & stabiliser. This can alter the altitude capability of the 737 by 2-300ft within the allowable CG range.

The minimum manoeuvre speed also varies with MAC for the same reason. From top to bottom these photos taken on a 737-700 at GW 60,000Kg (132,277Lbs) show CRZ CG 10, 15, 20, 25, 30%. The low speed limit can be seen changing from 219, 217, 215, 213, 211kts.

The default CRZ CG is 4% on CAA/JAR rules aircraft, ie the most forward and limiting. The FAA defaults are quite close to a typical mid-cruise CG; 18.5% for a -300 and 22% for a -700.

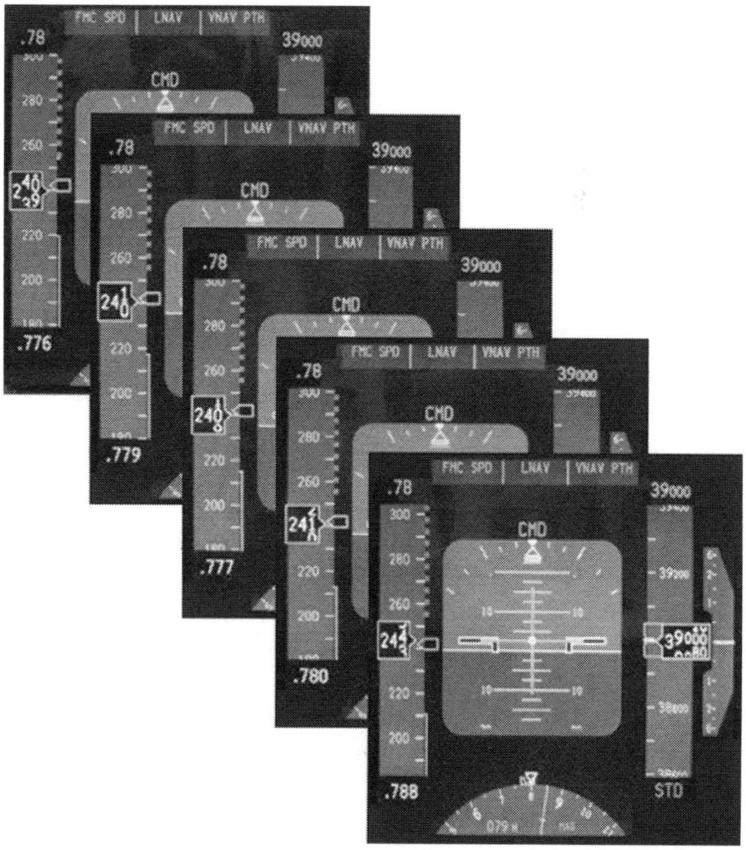

Variation of minimum manoeuvre speed with CG

ATA Codes

Air Transport Association (ATA) codes are used by maintenance to label parts of the aircraft, such as systems, components, placards, documents and software. I have listed some of the chapters, but there are others which are not applicable to the 737, eg "18 - Helicopter Vibration". ATAs 46 & 47 are new categories for the 737 with the introduction of Electronic Flight Bag and Nitrogen Generation System.

As aircrew your only involvement with ATA codes will be in the Tech Log, where defects & rectification work are written with reference to these codes. Also, if your aircraft is carrying a defect and you wish to check it in the MEL/CDL, you will need the ATA code to locate it. Eg ATA 21-15 = Air conditioning – Main outflow valve.

Unfortunately the codes do not keep their meaning across different books. Eg In the maintenance manual the recirculation system is 21-25-00 but in the MEL 21-25 is the water separator anti-icing system and the recirc fans are at 21-31 ! The only thing that remains constant are these chapter codes.

ATA Code	Chapter
	Aircraft
05	Time Limits
06	Dimensions and Areas
07	Lifting and shoring
10	Storage & return to service
11	Placards and markings
12	Servicing
	Airframe systems
21	Air conditioning
22	Auto flight
23	Communications
24	Electrical power
25	Equipment/furnishings
26	Fire protection
27	Flight controls
28	Fuel
29	Hydraulic power
30	Ice and rain protection
31	Indicating/recording systems
32	Landing gear
33	Lights
34	Navigation
35	Oxygen
36	Pneumatic
38	Water/waste
46	Information systems
47	Inert gas system
49	Airborne/auxiliary power
	Structure
51	Structures - General
52	Doors
53	Fuselage
54	Nacelles/pylons
55	Stabilizers
56	Windows
57	Wings
	Power plant
71	Power plant
72	Engine
73	Engine fuel and control
74	Ignition
75	Air
76	Engine controls
77	Engine indicating
78	Exhaust
79	Oil
80	Starting

Airtesting

An airtest, also known as an "airworthiness flight test" by ICAO or "maintenance check flight" in Part M, is now officially called a "check flight" by EASA. It is not an experimental or certification test flight. It is merely a thorough examination of the aircraft to ascertain that it handles normally, its performance is within limits and that its systems are fully serviceable.

I have been airtesting for almost 20 years on types from light aircraft to airliners and I have conducted over 100 on the 737 Classics and NG series. This guide is for pilots who may be new to airtesting or those who are experienced at airtesting other types, but are new to the 737. It should not in any way replace a full briefing or training course before conducting these types of flights.

Admin & Responsibilities

An airtest may be required for any of the following reasons.
- For the issue or renewal of the Certificate of Airworthiness. Note that EU member states only require a Check Flight Report if the aircraft is being imported from a non-EU country.
- As required by the Maintenance Manual; eg after certain work on the primary flying controls.
- For customer demonstration; eg when handing back an aircraft to the leasing company or selling to another airline.
- If the handling or performance of the aircraft is in doubt.

To conduct airtests on UK registered aircraft, the pilot in command must be "approved" by the CAA. This comprises of a briefing and a full airtest with the CAA test pilot and flight test engineer. This ensures that the crew are aware of the risks, potential problems & recovery techniques and helps to maintain a common standard for data collection. Any company insurance for the flight is likely to be invalid if the pilot is not CAA approved as he will be deemed to be unqualified. These days "Duty of Care" legislation in the UK means that you personally will also be liable for any accident – not just your airline.

The regulations in other countries vary. EASA under many circumstances set the "conditions for flight" and that may include the flight crew and persons who may be on board. If your authority does not have any approval or training requirements, I would recommend simulator training and to observe/fly at least two full airtests before attempting an airtest as PIC.

You should inform your life insurance company, in writing, that you are conducting "check flights". Explain that this is not high risk test pilot work.

Surprisingly, if the aircraft has a C of A it is not illegal to carry passengers on airtests, although it is not recommended. Limit any observers to essential persons and ascertain why each one needs to be on board. If you do allow them to fly, brief them that you will be manoeuvring and that the flight carries a higher degree of risk than a normal flight. Consider devising a proforma for them to sign acknowledging that they have been briefed, accept the risk and stating why they need to be on board.

If you need to conduct an airtest in another country to which the aircraft is registered, you must get approval from the authority of that country. For airtests where the aircraft has a C of A this is just a formality and is unlikely to be refused. But again, non-compliance may invalidate any insurance etc. An EASA Permit to Fly is now valid throughout the EU, but not elsewhere, when special permission must be sought to fly an aircraft without a C of A.

All AFM or company limitations are binding. However aircraft with an AFM to UKCAA standard have an airtesting supplement in which Mmo & Vmo are increased to 360kts/M0.84.

Performance climbs need not be done on every airtest but be aware that you are also signing off the aircraft as having satisfactory performance. If you suspect the performance of the aircraft, a performance climb should be done.

Planning

Find out beforehand what the reason for the airtest is (see possibilities above) and what needs to be done, ie a full airtest or just certain items. If it is a customer demonstration flight, they (the customer or leasing company) may well ask for some tests which are not on your usual schedule. If so then first read carefully what the test is and assess whether or not you are happy to do it. I always carry a copy of the Boeing Production Flight Test Procedures and check if and how Boeing does it. I have been presented with company created airtest schedules which have contained some poorly thought out procedures which posed an unnecessary risk. As the airtest Captain you have the final say of what tests you are prepared to conduct. If you accept the test then you should choose the most suitable place to fit it into the existing schedule and append your working copy accordingly.

The profile of the flight will usually be pre-determined by the particular airtest schedule that you are following. This should have been written in such a way that one test follows logically on from the next by virtue of your height, speed or configuration.

The UK CAA profile is a typical example and runs as follows:

1) Full thrust takeoff followed by an optional 5 minute takeoff performance climb.
2) Engine slam acceleration tests between 5000 & 8000ft at 150kts.
3) Climb to FL150, 150kts for engine number one shutdown and starter assist relight.
4) Climb to FL180 for depressurisation checks including cabin altitude warning and automatic passenger oxygen dropout.
5) Climb to FL350 at 280/M0.74 checking control response & trim and spoiler isolation.
6) At FL350 check cabin differential pressure, safety valve operation, yaw damper, manual reversion and APU relight.
7) FL350 to FL300: High Mach runs.
8) FL300 drifting down to FL240 for number two engine shutdown and windmill relight.
9) FL200 to FL150: High IAS runs.
10) FL150: Stick shaker at flap 5 and flap 30, PTU, timed flap and gear extensions/retractions, alternate flap extension, manual gear extension.

It is possible that you may have to change the sequence for weather, airspace or other reasons; if so then don't forget them later.

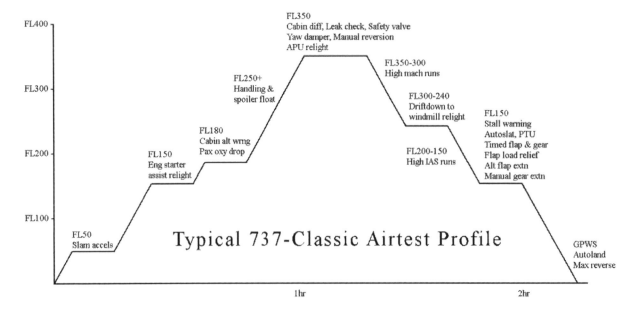

You will need to get clear of controlled airspace to have the flexibility to manoeuvre and ideally have a straight run of 100nm so that you do not have to turn about too often which can be awkward if you are in manual reversion. Have a couple of areas in mind and choose the best for weather on the day. Try to speak to the person who is going to file your flight plan as he may not be used to non-scheduled flights. He will need the points and EET's where you want to leave and rejoin controlled airspace. If necessary discuss your requirements with ATC by landline beforehand to save time in the air.

All airtests must be conducted in weather conditions that would allow for the safe conduct of the test, should anything unexpected occur. In practice this means day, VMC. This is primarily to make the manoeuvres and any recovery action easier to fly. But also to allow you to make a day landing if you encounter any problems. You should allow three hours from brakes-off to sunset for a full airtest. Set a deadline for pushback and stick to it. Smooth air aloft is also required both to ensure accuracy of results and to allow the high speed dives without risk of overstressing in turbulence.

Most airtest schedules require the aircraft to be within certain weight and CofG limits. Typically 40-46,000kg, 12-18% MAC for the 737-3/4/500 and 44-50,000kg, 15-20% MAC for an NG. A full airtest will use about 6,000kgs (13,000lbs) of fuel, so start with 12,000kg to stay within these limits. If the aircraft has been repainted it will have a new weight schedule, make sure that you know how to make a loadsheet from this raw information.

Defects – Pre Flight

Be very careful about what defects you accept before an airtest. The MEL, whilst still a useful guide, is not applicable. Under normal operations, the MEL states whether or not an unserviceable item is acceptable, and if so for how long it may be carried. The answer or time period is derived by considering your unserviceable item combined with the worst case other single failure to give an acceptable probability of getting into the second failure condition. For example, the probability of an engine failure under normal circumstances is around 10^{-5} but on an airtest the probability of an engine failure is 1 because you will be intentionally shutting down an engine.

Your First Officer and the Observer

If your F/O is new to airtesting then you should allow 2 hours to brief. It is important that you put them at their ease. It may be the first time that they have (almost) stalled since their PPL; they may be expecting something more dramatic during a depressurisation and it will possibly be the first time that they have ever shut down an engine. All of which could be worrying them which will detract from their performance and increase your workload. You should take time to explain that the airtest schedule is well proven, takes no unnecessary risks and that they should enjoy experiencing some of the more unusual procedures for real which others only get to see in the simulator.

Brief your F/O and observer that your accuracy will not always appear to be exact compared to airline operations. This is because you may be prioritising speed, height, loading or power etc. This will help dispel any loss of confidence.

Always brief your intentions and expectations of the F/O for:
- Engine shut downs
- Manual reversion
- High speed phase
- Low speed phase

PF (you) will probably do the R/T and navigation, especially if PNF is configuring the aircraft or reading ahead the schedule. If possible ask ATC if they can give you vectors to keep you clear of controlled airspace, danger areas etc and always ask for block heights to work within.

If conditions permit, allow your F/O to handle the aircraft when in manual reversion. This will give him experience and an appreciation of the handling and you may need him to share the flying if you cannot restore the hydraulics and have to reland in that configuration.

It is possible, particularly on customer demonstration airtests, that you will not know the observer. You should (tactfully) assess his ability to record the data required. He may not be a pilot, or even an engineer and there may be language barriers. If necessary you may have to ask the F/O to also record data so that the results can be pooled; if so be prepared to take the test more slowly so that everybody can keep up.

Pre-Flight Checks

If the aircraft is completely de-powered this gives you an opportunity to check some of the hot battery bus items. Select BAT on the DC metering panel (on NG you will need the battery switch ON and standby power OFF) and observe a small discharge on the ammeter as you move each start lever to IDLE. Now pull each engine fire switch and again observe the battery discharge as the fuel valves close.

Switch the battery ON. The crossfeed valve and fire panel should now be checked for operation. Start the APU and put it on line to save the battery.

External Checks

Put the hydraulic pumps on for the external checks as this will help identify any leaks. Be meticulous with your walkaround. Stand back from the aircraft and see that it looks right. Inspect the ground beneath the aircraft for evidence of leaks. Check that all probes, ports & vents are uncovered, especially if the aircraft has been repainted. Look inside the holds to see that they are empty; if anything is carried in the hold (not recommended) it must be well secured to ensure that it does not shift. Check every fastener and panel by eye to see that they are secured, if not they will probably detach during the high speed runs. I have found panels missing both before and after airtests. If you have any doubt, check with an engineer. Note what gear pins are in place and if you are chocked.

Get your F/O to do the flight control checks whilst you stand behind the tail of the aircraft, from this position you will be able to see the correct deflection of the spoilers and flight controls.

Internal Checks

Check that you have all the manuals, plates, charts, checklists and necessary aircraft certificates on board, these may have been removed by engineering for safe keeping. I now carry spares in case any are missing. For security's sake, check that the galley is secure and the overhead bins and toilets are empty. Check that the gear downlock viewers are accessible and clean enough to be used.

Ensure that the PSU's are latched up but leave about three to drop. I usually leave one passenger PSU on each side near the front and rear eg row 1DEF & 18ABC and one flight attendant or toilet PSU at the rear because they are less in the way after dropping.

It may sound ridiculous but make sure that you both know how to operate the airstairs, since this is normally done by the cabin crew. Check that the handles are fully stowed before retracting and continue retracting until the AIRSTAIRS OPERATING light extinguishes. You should arm the forward door slides but don't forget them after landing!

Flight Deck Preparation

Before you sit down check that all the circuit breakers are in, if any are not check with the engineers why before you go any further, it could mean that maintenance work is incomplete or faulty. Locate all c/b's that you will pull during the test and mark them with a piece of masking tape so that you can find them easily in-flight. Whilst you have the masking tape out, you may want to put a strip on the throttle quadrant so that you can mark it for the slam acceleration tests. Check also that the oxygen valve (not NG's) is open; you will need this during flight.

Now, take your seat and check the escape rope, there should be a coloured mark at the knot nearest to the securing point. Check your oxygen mask and interphone in the normal way. For the light test, take your time and replace any dead bulbs, record the positions of any that you can't replace or have an engineer do it. Make sure that you still have some spares for the flight.

Now you can begin the panel scan. Done thoroughly, this can take 15 minutes and can uncover many faults. I find it best to follow a systematic cockpit panel scan checking every panel you meet along the way; this ensures that everything is checked which is a useful technique if you have a varied fleet.

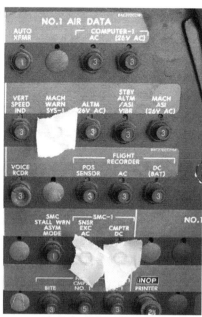

Mark the C/Bs to be pulled in-flight

This is my list of pre-flight airtest checks. It is not exhaustive but it is sufficient for a check flight. For the full definitive list see the Boeing Production Flight Test Procedures.

LE Devices Panel: With the hydraulic pumps off this may show some LED's in transit which is normal for older aircraft. If the droop is unusually large or fast after hydraulic depressurisation you may want to record it for info. Put the B system pump on and check that all devices retract. Press test button for light test.

Flight Controls Panel: With the hydraulics off both flight control low pressure lights should be illuminated. Switch the flight control A switch to STBY RUD and check that the standby hydraulics LOW PRESSURE light briefly illuminates and then extinguishes. The standby LOW QUANTITY light and flight controls A LOW PRESSURE light should be extinguished. With the flight deck door open you will also be able to hear the standby hydraulic pump running. Return the flight control switch to "A ON" and check that the standby pump stops. Repeat for system B.

With the flaps up select the alternate flap switch to ARM; again you should now hear the standby hydraulic pump running. Move the flap lever to 1 and check that the flaps do not move. Run the alternate flaps down for 10 seconds and see that the trailing edge flaps and leading edge devices extend. The LE slats should continue to run to the FULL EXT position. Bring the trailing edge flaps up with alternate flap and check that the leading edge devices remain extended. Now move the flap lever up and switch off the alternate flaps, this will retract the leading edge devices.

Select either electric hydraulic pump on and check that the FEEL DIFF PRESS light illuminates.

Transfer Switches:
VHF NAV transfer: Select an ILS frequency on Nav 1 and a VOR frequency on Nav 2 and VOR/ILS on the EHSI/ND, selecting BOTH ON 1 will give Nav 1 freq on the F/O's EHSI (with G/S & LOC dev scales on the PFD) and vice versa. Return to NORMAL.

IRS transfer: Select SYS DSPL to L on the IRS panel, turn the DSPL SEL to TEST and observe that the Captains dispiays test. Switch the IRS transfer to BOTH ON R, the Captains displays should now be restored. Repeat the procedure for the other side and return to NORMAL.

EFI / Control Panel transfer: Select different EHSI/ND modes on Capt & F/O's EHSI/ND, observe the displays transfer when selected. Return to NORMAL.

Fuel Panel: If not done earlier on, take any AC power off the buses (leave battery on) and check that the crossfeed valve still works, restore AC power. Switch all fuel pumps on and check that the LP lights extinguish.

IRS Panel: This will have been tested during the IRS transfer checks. You can now start the IRS alignment, but be aware that the subsequent electrical tests will briefly remove power to the IRS's.

Electrical Panels: Check the metering panel for the correct indications. With the APU on the busses, all positions should be powered except the engine generators and GPU if not connected. Note that the correct frequency for the old Garrett 85-129 APU is 415Hz, unlike the others which is 400Hz. Do not press the RESID VOLTS button when a generator is on line as it uses the 30V scale and will overload the meter. Galley power should be selected ON to find any problems in these circuits.

Standby power check: Select standby power on the AC & DC meters. Remove any AC power, AC & DC standby busses will be unpowered and STANDBY PWR OFF light illuminates. Select BAT on the standby power switch and check that all standby instruments are powered ie Standby ASI, altimeter & horizon; Fuel gauges, primary EIS, VHF 1, NAV 1, ADF 1, ADI 1, RMI 1, HSI 1, left IRS, interphone & dome light. Select standby power OFF and these should switch off. Return to AUTO and restore AC power.

Bus switching checks: If you have ground power connected, check that putting the APU on a bus disconnects the ground power from that side. Remove AC power from one side and check that the BUS OFF light illuminates. Now switch the bus transfer OFF and check that the TRANSFER BUS OFF light illuminates on the side without AC power. Return the bus transfer to AUTO and repeat on the other bus. Restore electrics to normal.

Service Interphone: Check communications to the attendant stations and check that ground engineer can be heard through service interphone with switch in ON position.

Equipment Cooling: Switch over to alternate for a while and check that you hear the fans restart. You may find that the supply fan OFF light illuminates during the depressurisation checks, this is normal and changing to alternate should restore the flow.

Emergency Exit lights: Switch ON and check the operation of all interior and exterior emergency lights. Return to AUTO. This check can be left until after flight if desired.

Signs & Calls: If you wish, the no-smoking and fasten belts signs can be left in AUTO to check their operation with the gear & flap. Once you have checked this after departure you would be advised to return them to OFF or ON due to the number of configuration changes on the test. Check the operation of the attendant & ground call buttons and the blue CALL light when called from the cabin.

Wipers: Don't check these unless you have a wet windscreen but do record if they travel up from the park position at high IAS. The rain repellent should have been disconnected years ago.

Engine Panel: Check that both PMC's / EEC's are switched ON.

Crew Oxygen: Check minimum quantity for despatch against how many people you will have on the flightdeck. Re-check the shut-off valve is open on the P6 panel (Not NG's).

Window Heat: Put all window heat ON then put the test switch to OVHT, all overheat lights should illuminate. Extinguish these by cycling the window heat OFF then ON. Remember the PWR TEST will only work when the heat is switched on but power has been removed because it is at temperature. This is usually only seen on very hot days.

Pitot Heat: Put all pitot heat ON and check that the lights extinguish then switch OFF again. If a TAT TEST is installed, press and observe that the TEMP PROBE light extinguishes and the TAT increases.

Wing Anti-ice: With APU bleed air on, select wing anti-ice ON, check the VALVE OPEN lights go bright then dim blue and duct pressure has reduced. Advance each thrust lever beyond 30 degrees and check that the lights go bright again, duct pressure should increase as the valves close. Close the thrust levers and the reverse occurs. If the VALVE OPEN lights illuminate bright again this is because of an overheat trip which is normal with the aircraft stationary, especially in warm conditions. Finally pull the AIR/GND RELAY & LTS c/b (P6-2) to simulate take-off and the wing anti-ice switch should move OFF. Reset the c/b.

Hydraulic Pumps: When switched on, normal hydraulic pressure is 3000+/-200psi. This can be taken as 3000+/- 1 segment on EIS displays.

Doors: Check that the lights agree with the doors that are open. If your EQUIP light is on it could be either the E&E bay or radar bay. If the AIR STAIR light is on after you have closed up, run the airstairs up further to ensure that the airstair door is fully closed.

CVR: If your CVR has an AUTO/ON switch, select ON. Plug your headset into the unit and press TEST, you should hear a tone and see the status light illuminate. Speak into the area microphone and check that you can hear it through the headphone, there should be a slight delay. The switch should trip back to AUTO during the first engine start.

FDR: Move the guarded switch to TEST and check that the OFF light extinguishes.

Mach Airspeed Warning: Test both systems by pressing the buttons.

Stall Warning Test: Again, test both systems by pressing the buttons. If the stick shaker is not strong enough it can be adjusted by engineers.

Air Conditioning: Select engine and APU bleeds ON and the recirculation fan OFF for these checks.

Select both packs on and both temperature selectors to AUTO COOL. Move CONT CABIN temp selector to AUTO WARM and check that the air mix valve moves towards hot and the supply temperature increases without a duct overheat. Return to AUTO COOL and repeat for PASS CABIN. Maximum supply duct temp should be 70C.

Switch off the left pack and gradually select MANUAL WARM on the right pack. Duct overheat should occur between 78 & 95C. Check the DUCT OVERHEAT light illuminates, the air mix valve moves closed and the TRIP RESET function. Repeat for the left pack.
Note: The air conditioning checks for the 737-4/8/900 are significantly different.

Press the OVHT TEST button and check that both WING-BODY OVERHEAT lights illuminate after a short delay.

The following bleeds & pneumatic checks can be monitored by selecting FMCS Discretes 1/4 on the FMC.

Check the DUAL BLEED light illuminates in the appropriate bleed configurations. Switch all bleeds ON, isolation valve to AUTO, left pack OFF and right pack AUTO. Check that the L & R duct pressures are within 3psi. Close the isolation valve and check that the right duct pressure falls to zero and that the right pack shuts down. Put the isolation valve and left pack to AUTO and check that left pack operates. Open the isolation valve check the right duct pressure rises and right pack operates.

Check each pack in AUTO and HIGH flow. Then switch one pack OFF and put the other in AUTO and pull the AIR/GND RELAY & LTS c/b (P6-2), check that the pack goes to high flow (duct pressure will decrease and APU EGT will increase). Now select flap 1 and the pack should return to low flow. Retract flaps and reset the c/b.

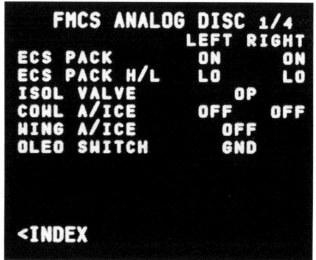

Use FMCS Discretes 1/4 to check valve positions

Pressurisation: Do these checks with the packs off and a door or flight deck window open.

For CPCS aircraft, perform the Auto Trip & Standby Check from the supplementary procedures. Check also manual AC & DC outflow valve operation. For flight set 35000ft in the Auto window, 8000ft in Standby and a high rate.

For DCPCS aircraft, check manual outflow valve operation. Set 35000ft in the Auto window for the flight.

MCP: Engage either autopilot and disengage with the following methods: Disengage bar/switches on MCP, switches on yoke, electric stab trim, autopilot stab trim cutout switch, instrument transfer switches & by switching off the corresponding electrical hydraulic pump (expect 5 sec delay). Set up MCP for departure.

Instrument Panel: Systematically check each instrument from left to right. Check the operation of the orange ASI bugs and N1 bugs (Classics) or N1 SET & SPD REF controls (NG).

Fuel Gauges: Press the test button and hold to test gauges. Note any error codes. As a general rule if the fuel gauge is still indicating, the error code is allowable under the MEL. Beware of error code 7 on Simmonds gauges as this could indicate water in the tanks which is possible if the aircraft has not been flown for a long time.

The N1 SET & SPD REF controls should be checked

EIS Panels: Both of these have built in BITE test features. Use a ball point pen to press the recessed button at the bottom of the panels; this will start an LED check during which the various checks are conducted. If any of the checks fail, the appropriate code will be shown in place of the affected parameters readout.

Upper/Lower DUs: With the start levers at CUT OFF, place the start switches to CONT and observe that the EGT, FF/FU, Oil pressure & Oil temp displays unblank. The ENGINE CONTROL & FILTER BYPASS lights should remain extinguished. Return start switches to OFF. Check also that pressing MFD SYS will display the appropriate systems and that the DU selectors operate normally on all DUs

GPWS: Ensure that the weather radar is on in TEST mode and displayed on the EHSI. Pressing SYS TEST quickly will give a short confidence test, pressing for 10 seconds will give a full vocabulary test; the latter should be done.

Wheel-Well Fire: As you go past the fire panel, do a wheel-well fire test now that you have AC power.

Electronics Panel: These tests will vary depending upon the specific avionics installed. Include Nav 1 & 2 test function, transponder test and TCAS test (when IRS has aligned).

Trimmer Checks:
- Rudder and aileron trim should be moved several units each side of neutral to see that the pedals/yoke follow the trim, then leave centered.
- With aileron trim at zero and hydraulic pumps on, rotate the control wheel to 6 units then back to 2 units and release the wheel. Check that the control wheel centers to within 1/2 unit. Repeat in other direction.
- Run the electric stab trim then move control column in the opposite direction, trim should stop. Now select STAB TRIM to OVERRIDE and trim should resume. Reselect NORMAL.
- Run the electric stab trim then select STAB TRIM - ELEC - CUTOUT, trim should stop. Reselect NORMAL.
- Engage an autopilot, select STAB TRIM - AUTOPILOT - CUTOUT, autopilot should disengage. Reselect NORMAL.

Takeoff Configuration Warning: Check individually, by advancing either thrust lever, the five takeoff config warning modes: Parking brake set, Trim not in greenband, Speedbrakes not down, LE flaps not extended, TE flaps not at takeoff (assume flap 5 to 15 positions). Note that some early aircraft do not have the parking brake mode and some aircraft have takeoff TE flap allowable from 1 to 15.

In-Flight Tests

Engine Start

Crank the first engine to its max motoring before introducing the fuel to get a feel for the condition of the APU / engine starter. Ideally each starter should be able to reach 25% N1 but make allowances if the engine has not been started for a long time or conditions are unusually cold. Start each engine on a different igniter.

After Start

Check that the engine driven hydraulic pumps can be switched off. These are not switched very often and the solenoids can stick in the ON position. If so, you will not be able to complete some of the in-flight test items.

Steering & Brake Checks

Ensure that you have sufficient space and that nobody is taxying behind you. Advise ATC of your intentions. Beware any slope or parked aircraft as you will be without brakes and/or steering for a short period.

Check normal operation of brakes (L & R pedals) whilst watching for hydraulic pressure fluctuations on B but not A. Then depressurise system B (pumps off and exercise the control column) and try the brakes again to test the alternate brakes, you should get fluctuations on A but not the brake pressure gauge. Now if the area is clear, depressurise system A and check the accumulator braking down to the pre-charge pressure.

With no hydraulics, no steering will be available. Select system B hydraulics ON and alternate nose-wheel steering, check steering with both tiller and rudder pedals. Restore all hydraulics.

Thrust Reversers

Move the thrust levers out of idle and check that reverse cannot be selected.
Select idle reverse and check that the thrust levers cannot be moved forward.

Take-Off

Do a full thrust take-off from the full length of the runway. For DCPCS aircraft check that the outflow valve starts to close at 60% N1.

Engine Slam Accelerations

This is to check that the engines can give go-around thrust from flight idle at Vref in the certified time of 8s.

There are two techniques for this. The first is to set the go-around N1 and mark the position of the thrust levers on the throttle quadrant (put some masking tape down first). Then you can set the thrust levers to these marks in the timed test. The second technique is not to mark the quadrant, but to simply firewall each thrust lever from idle and slightly retard the thrust lever as the target N1 is approached, then quickly bring it back to the middle range as the target N1 is seen. The logic is that this gives only one max N1 event per engine whereas the marking method gives two.

Pre-marking the throttle quadrant can make setting G/A thrust easier

I believe that the marking method is probably better if you are new to this test since there is potential for a large exceedance with the firewalling method. Both techniques have their plusses and minuses; if you go with the firewall technique, you must react quickly and not get distracted. It is important that you use the technique that you are most comfortable with, particularly if your co-pilot is new to it.

Performance Climbs

If you are doing performance climbs (not always necessary) then you should do them at the beginning of the flight when you have centre tank fuel. This eliminates the need for fuel balancing and the associated risks of crossfeeding errors and possibly flaming out the live engine. It is acceptable to keep the other engine running at idle rather than shutting it down, although it may be shutdown when at a safe altitude (say for the en-route perf climb) in preparation for the starter assist relight.

Two 5 minute performance climbs are normally done; a second segment climb at V2 & flap 5, and an en-route climb clean at en-route climb (clean) speed. You should try and find an area which is clear of cloud, has smooth air and is away from hills which may give mountain waves. Speed should be held to within 2kts of target and turns should be avoided, but if necessary must be restricted to 10 degrees AoB to ensure that the results are accurate. The rudder forces should be trimmed off and the aileron no more than 2 units to ensure that the spoilers are down.

The charts for these climbs can be found in the AFM, ensure that you use the gross data charts which are specifically for the purpose. The tolerances are: V2 climb -80fpm, en-route climb -120fpm.

Engine Shutdown & Starter Assist Relight

This is done at the worst case conditions allowable in the QRH; ie FL150, 150kts, 15% N2 for Classics or FL 200, 140kts, 11% N2 for SAC NG's, 13% N2 for DAC NG's.

737 IN-FLIGHT START ENVELOPE

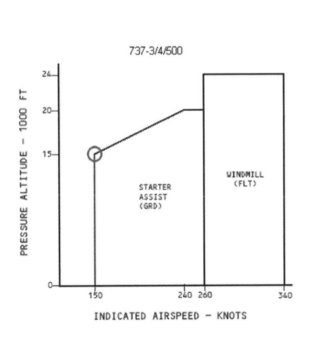
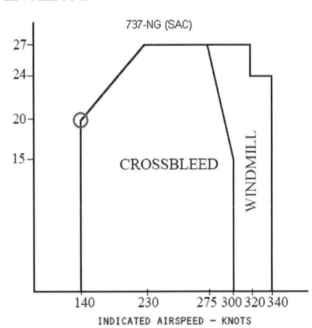

I would advise you to do the QRH "Engine failure/shutdown" immediately followed by the "In-flight engine start" drills up to the point were you actually start the engine. This will enable you to restart the engine quickly if the other were to fail during this exercise. If this happens, you can continue with a starter assist relight using APU air. It is good practice to operate the engine at idle for 3 minutes before shutting it down and leave it idling for 2 minutes after the relight before applying thrust.

If required, this is a suitable place to do a check of the LGTU, ie No1 engine shut down and low speed. Extend the gear, switch off system A pumps, deplete the hydraulic pressure and then retract the gear.

De-Pressurisation Checks

These are best done at about FL180. This is sufficiently above FL140 to allow depressurisation, but also low enough such that an emergency descent to FL100 will not take too long.

Checking the cabin alt warning horn at 10,000ft (DCPCS)

Checking the passenger oxygen mask drop at 14,000ft (CPCS)

Before you start, take the radio and get your F/O onto oxygen first. This is so that you can help him sort out his mask, headphones, glasses etc and you can explain the mask and ASP selections required without having to speak through your mask. After donning your own mask I always then do a radio check to ensure both that the mask microphone works and that I have made the correct ASP selections and am not going to be out of radio contact.

Tips: Check the EMERG mode before you put on your mask. This will blow away any dust and stale air that may have lain in the mask since it was last used. Locate the NORM/100% switch with the mask facing you because it will be difficult to find when you are wearing it.

Make sure that you are both on oxygen before going above 10,000ft cabin altitude. You may need to turn one or both packs off to get the cabin to climb above 14,000ft. Don't forget to stop the oxygen flow with the reset slider after you have finished with it otherwise you may not have any left for later! Don't bother checking that the PSU's have all dropped in-flight; this can be done when you are back on the ground.

You cannot see your oxygen mask controls when you are wearing it!

Handling in the Climb

Use the climb for more accurate control & trimmer checks. Ensure that the fuel is in balance, and N1's are equal and use rudder trim with the autopilot engaged in HDG SEL to simply level the yoke. If the aircraft appears not to be rigged correctly you can then refine the trim with the following method from the FCTM:

- Engage the autopilot in HDG SEL & set zero rudder trim.
- Move the rudder trim toward the low side of the yoke until the bank angle is zero. The rudder trim should not exceed ½ unit at M0.74 or ¾ unit at 250kts.
- The control wheel deflection now indicates the true aileron trim. This should not exceed ¾ unit at 250kts/M0.74.
- If after the above steps, the autopilot is disengaged and the aircraft rolls, hold the wings level and use more aileron trim.
- When properly trimmed, the aircraft will maintain a constant heading (check for this against the IRS heading display) and the aileron trim will be the same with the autopilot engaged or disengaged.

Spoiler Float

This test is done to ensure that if hydraulic pressure is lost, the spoilers do not float up such that they could cause loss of airspeed, pitch or roll. The AMM states *"When hydraulic pressure is lost, springs should reseat the spoiler extension check valve to prevent spoiler float."* (AMM 27-61-00)

Spoiler float – Classic – CAA method   *Spoiler float – NG – FAA method*

On the CAA schedule this is done clean, in the climb to FL350 at 280/M0.74; On others it is done with flap 40 at FL170. For the CAA method, set the speedbrake to flight detent and then switch off spoilers A & B. Check for any roll and that the spoiler float is both symmetrical and not excessive. The amount of spoiler float should be recorded by the observer, but as you can see from the photo this can be difficult to judge at a distance. **Ensure that you put the speedbrake lever down again before switching the spoilers back on.** Note that for the flap 40 method, the speedbrake lever should be left down for this test.

Functioning at FL350

Safety Valve Operation

Manually close the outflow valve to raise the cabin differential pressure. The valve does not always open abruptly so keep a careful eye on the cabin differential and VSI for any changes which may indicate safety valve operation. If the aircraft is old and leaky this may require both packs to be in HIGH to trigger the safety valve. I have had aircraft were the safety valve did not operate. Do not exceed the maximum cabin differential pressure.

Leak Check

Start by switching off each pack individually and checking that cabin pressure is maintained. Then switch off both packs and record the stabilised cabin rate of climb. The Boeing new build aircraft limit is 2000fpm, or 2250fpm on the shorter -500 & -600 series aircraft.

This is not in the UK AFTS because it is considered to be a purely commercial check. If the aircraft has a high leak rate the packs have to work harder to pressurise the aircraft and so the fuel consumption is increased.

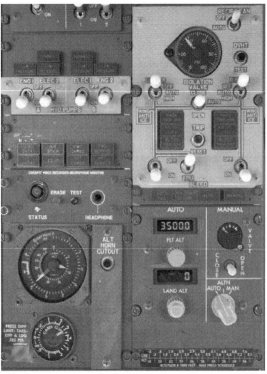

The pressurisation panel during a Leak Check at 35,000ft

Yaw Damper

Switch the yaw damper off and squeeze in rudder to yaw the aircraft until it starts to roll then gently reverse until wings are level. The aircraft should roll/yaw past level. Switch on the yaw damper and after a short delay you will feel it kick in and the Dutch Roll will stop within 2 cycles.

Dutch Roll is quite docile on the 737 and left alone it will naturally stop within about 8 cycles. I am told that this was not the case with its predecessors, the 727 & 707. These aircraft were fitted with two yaw dampers as one was required for despatch. The prototype 737's were also built with two yaw dampers until flight testing demonstrated that this was unnecessary.

Manual Reversion

This exercise is to test the function of the primary flight controls with flight control switches A and/or B switched off. With A and B system off, movement of the ailerons and elevators is assisted by balance tabs and balance panels. The gearing of the elevator balance tabs is changed when hydraulic power is off to reduce their deflection in this condition. A noticeable deadband exists when the elevator system is controlled manually (approx 1.5 inches at the control column). This is because during manual operation, an input to the control system moves the elevator PCU input crank until the crank hits the stops and then the PCU is used as a link to drive the elevator.

These tests are done at 250kts/5,000ft+ (Originals), M0.74/FL350 (Classics) and 250kts/FL150 for the NG series. Try to have a straight run for about 10 minutes and an altitude block of +/-1000ft from ATC before you start. Take all the time you need to ensure that the aircraft is accurately in trim with no turbulence or mountain wave. The tolerances for this test are very tight and the 737 has a tendency to take up a long period phugoid oscillation which can also affect the results. Remember to have the fuel in balance and N1's equal.

Take extreme care when switching off the hydraulics to the flight controls. Have the F/O do this on your command whilst you cover the controls. I have had some significantly out of trim aircraft with dramatic results! If the aircraft is too far out of trim to hold it level then abandon the test. A simple way to restore all flight control and spoiler switches quickly is to sweep down the flight controls panel with your hand, this will close all of the top hinged guarded switches thereby restoring them to their normal positions.

The allowable limits of stab trim for Classics vary depending upon whether aluminium or titanium rods are fitted. The tolerances are greater with titanium rods due to the better thermal matching of the materials. The limits on the 737NG are much larger; this is because of differences in design which makes the aircraft lighter to control in manual reversion. The limits are based upon a maximum stick force of 50lbs at elevator power off.

To ease handling under manual reversion you may enter & exit the turns with a small gentle rudder input, but remember that the rudder will still be powered. The NG will do this for you with the Wheel To Rudder Interconnect System (WTRIS). Beware, unlike the Classics, the yoke does not centralise but will stay deflected almost like CWS roll. Give your F/O some handling practice in this condition, he will appreciate it and one day you may need his assistance for a manual reversion recovery.

The test is also a check of the standby rudder actuator. This is activated whenever either flight control switch is moved to STBY RUD. If the standby actuator is out of rig, this will cause a rudder deflection which should be recorded for rectification. My worst case is 3 units of rudder trim required to bring the aircraft straight. Take care not to over-control in standby rudder, as unlike the elevators & ailerons it will still have hydraulic power. Be very careful not to perform a reversal of the rudder input too quickly as this will place large sideways loads on the fin.

APU Relight

Monitor the EGT carefully during the start; it will be higher than a ground start. Typical start times are 1 min +/- 20secs. Don't forget that if you have an old Garrett 85-129 APU (identifiable by having colour bands on the EGT gauge and its 415Hz running frequency), the third attempted restart altitude is FL300 (not FL350). The NG APU's should be tried at FL410. Two start attempts are allowed for all APU's.

High Mach Number Handling

With all the exercises in the cruise complete, you can use your initial descent to do a high mach number handling check. The recommended altitude block is FL350-FL300 this ensures that you will not be near the Vmo/Mmo crossover point which could cause an inadvertent exceedance.

The CAA Schedule calls for a handling check at M0.84, which is above Mmo (M0.82) but safely below Mdive (M0.89). This is only permissible because British registered 737's have an AFM supplement permitting flight at this speed for the purpose of airworthiness flight testing. If you are testing an aircraft registered in another country you should check if the AFM of that aircraft has a similar allowance otherwise conducting this test could be illegal.

The high speed runs must only be conducted in smooth air and clear of cloud. Trip the MACH WARN SYS-2 c/b and accelerate to about M0.80 in level flight. You can now reduce thrust slightly and put the aircraft into a gentle descent. Allow the speed to increase slowly and record the indicated Mach number at buffet onset and of the number 1 high speed warning. The first run is also a handling check. Put very small control inputs in all three axies to check there are no unusual vibrations; be particularly careful not to make large inputs with rudder. Gently select the speedbrake at high speed and observe that the aircraft pitches up with a normal amount of buffet and little roll. Now you can reduce thrust and return to normal speeds.

The ASI at high Mach. Notice the slow acceleration and shallow dive

You should find on the Classics that the onset of Mach buffet occurs at about M0.815-0.825, there is little or no Mach buffet on the NG models. If you experience any unusual vibrations or problems then knock it off and do not repeat. If all went well, then repeat for the number 2 high speed warning check.

Engine Shutdown & Windmill Relight

Again, you would be well advised to do the QRH shutdown procedure immediately followed by the restart drills up to where you would raise the start lever. This is so that if you subsequently need to restart the engine quickly you are ready to do so. If you do lose the other engine, you should note that the LOSS OF THRUST ON BOTH ENGINES relight drills assume the worst and don't show a relight envelope. Attempted relights outside the normal restart envelope will have a reduced chance of success. So, if you have deliberately shut down one engine and the other fails you should, if possible, continue with the normal drill for a relight on that engine. If you can't get the minimum N2 for a windmill relight then go for a starter assist using the APU if available.

The test calls for a relight at exactly FL240, 260kts, 15% N2 for Classics or FL270, 275kts, 11% N2 for a SAC NG or FL200, 290kts, 13%. It can be difficult to achieve all three parameters, the following technique is recommended: Shutdown the engine as high as possible after your 3 minutes cooling from the high Mach runs and set 80%N1 on the remaining engine. Descent at 290kts will stop the N2 decreasing. As you get to within 1,000ft of the restart altitude adjust the power and attitude to decelerate to target speed, using speedbrake if necessary, and raise the start lever as you hit the three targets. Don't allow the N2 to drop below 15% as it will take a lot of speed (altitude) to get it back up and do not attempt a start below min N2 as you may damage the engine. Thereafter maintain target speed with MCT and accept any slight height loss.

Flight instruments during driftdown to a windmill relight

High IAS Handling

The CAA CFS calls for a handling check at 360kts, which is above Vmo (340kts) but safely below Vdive (400kts). This is only permissible because British registered 737's have an AFM supplement permitting flight at this speed for the purpose of airworthiness flight testing. If you are testing an aircraft registered in another country you should check if the AFM of that aircraft has a similar allowance otherwise conducting this test could be illegal.

Before starting this test you should ensure that you have smooth air and be clear of cloud. Do not do this test above FL200 to ensure margin from the Vmo/Mmo crossover point which could cause an inadvertent exceedance. Trip the MACH WARN SYS-2 c/b and accelerate to about 320kts in level flight. You can now reduce thrust slightly and put the aircraft into a gentle descent. Allow the speed to increase slowly and record the indicated airspeed of the Number 1 high speed warning. The first run is also a handling check. Put very small control inputs in all three axies to check there are no unusual vibrations; be particularly careful not to make large inputs with rudder. Gently select the speedbrake at high speed and observe that the aircraft pitches up with a normal amount of buffet and little roll. Now you can reduce thrust and return to normal speeds. If you encounter any unusual vibration during the high speed run, decelerate quickly. If the vibration is or was severe, consider a diversion to the nearest suitable airfield and follow the procedure below (from the Mayday call). If all went well, then repeat for the Number 2 high speed warning check.

Detached panel during an airtest

This test, more so than the high Mach runs, carries the risk of panel separation and possible subsequent airframe damage. If this happens, abandon the test and decelerate slowly as the structural integrity is in doubt. Put out a Mayday or a PAN and consider a diversion to the nearest suitable airfield, especially if your intended destination is short. At about 5000ft slowly configure towards the landing configuration and back to the appropriate Vref and assess the handling at each stage of flap/gear. Choose your landing configuration based upon this assessment. If you get to a speed at which you feel uncomfortable above the Vref speed you should note this new speed, add 10 knots to it and this will become the new Vref. You may need to reconsider your landing distance based on the final configuration and Vref. Do not change the configuration once your landing configuration has been reached unless fuel considerations warrant it.

Low Speed Handling

The standard stall tests are conducted at the take-off configuration, gear up flap 5 & 15 and landing configuration, gear down flap 30. If other configurations are requested, be aware that some schedules do not give a trim speed, in which case 1.23 x stick shake may be used.

Remember that NG's have a TAI bias which raises the stick shaker speeds by approximately 6kts if wing anti-ice has been selected on for more than 5 sec. Note these biased speeds assume there is no ice on the airframe.

The test technique is to decelerate to the trim speed whereupon it should be held and trimmed at that speed. Eventually you should be able to take your hands off the column without the speed changing with the throttles fully closed. The aircraft will be descending gently but this is to be expected at idle thrust and is acceptable. You then decelerate at approx 1kt/sec using the elevator only, with both hands on the control column. At the stick shake you gently unload the aircraft allowing the nose to drop and the speed to increase. When the airspeed is back above stick shake you can move your right hand to the throttles and gently apply thrust by "walking" the thrust levers back up to the required power.

Increasing thrust slowly will help to prevent blade stresses or tip rub which tends to occur at high power and high AoA. It will reduce the pitch up tendency of increasing power which could cause a secondary stall and it will also help prevent any sideslip problems if the engines do not spool up symmetrically - not good if you are still below Vmca.

Stall warning / PTU check at flap 5. The correct deceleration rate can be attained by keeping the speed trend vector at the bottom of the box.

Stall warning check at flap 30. Notice the missing low speed warning bar and PLI due to the No1 SMC c/bs being pulled.

Notice how the airtest stall recovery technique is different to conventional stall recovery technique. This is because we are conducting the test at a safe altitude and terrain contact is not an issue. On the test our priorities are not to overstress the airframe or lose an engine.

If you get wing drop, or any other unusual attitude; unload, roll the wings level and ease out of the dive. Do not pull out too hard. It is far more desirable to pull out gently and overspeed rather than pull quickly and 'g' overstress. You will survive the flaps falling off, but not the tail! Make sure that you have briefed your F/O that he may retract the gear and flap without prompt if he sees an overspeed condition but if the overspeed is more than 20-30 knots it may be better to leave them down to prevent asymmetric retraction or further damage.

PTU check

Switch off the hydraulic system B pumps and repeat the flap 5 stall again, have your observer watch the LED panel for autoslat deployment. Be careful with this test, I have seen these fail a number of times. Be aware of the change in stall speed if the auto-slats do not deploy or undemanded roll if they extend asymmetrically. The maximum acceptable roll-off is 15°AOB.

Flap Load Relief

Be careful not to exceed the flap 40 limiting speed as this only functions at 2kts before Vfe. The most accurate way to control speed is with pitch rather than power. Be aware of the different flap limiting speeds and flap load relief speeds for the different series of 737.

Alternate Flap Extension

This check is similar to that which was done in the cockpit preparation, but with timings since we now have air loads. Some schedules require this check to be done to flap 40, others only to flap 10. Either way, remember the duty cycle limits.

This check should be done out in the area, but close to base, because if you cannot retract the flaps or LED's it will take a long time and a lot of fuel to get back at this configuration.

Landing Gear Manual Extension

Do this out in the area rather than when rejoining the circuit as it would be too busy for troubleshooting if any abnormalities arose. The nose rope will pull approx 8 inches and the mains approx 18 inches.

Pulling the nose gear manual extension handle

Approach

If there have been no control or autopilot problems in-flight, and the airport has the facilities, then an autoland may be conducted.

The GPWS TOO LOW GEAR mode may be checked by selecting the landing gear lever to the OFF position after a normal extension. If the gear is accidentally raised I would suggest a go-around rather than recycling the gear at low altitude. The gear indicators will now show three reds (gear lever in disagreement with gear position) & three greens (down and locked). At 500ft radio altitude the TOO LOW GEAR warning will sound; switch the GPWS flap/gear to INHIBIT and verify that the warning is silenced; now move the gear lever to DOWN and restore the flap/gear inhibit.

Touch and Go

One is almost always asked to demonstrate an autoland, but not all maintenance bases have such facilities. An autoland to a touch and go at somewhere suitable can expedite matters considerably over a full stop and taxy back. This is not a problem if the runway is long, but be aware of configuration differences for landing (no autobrake and do not arm the speedbrake) and landing roll procedure (no reverse thrust, retract flap to 15 & retrim). I would recommend a simulator check-out before doing this. It should be remembered that the configuration warning may sound if the flaps and trim are not reset quickly, which is why good technique and prior thought is required to prevent this happening.

Landing

Use full reverse thrust; there is a table for minimum N1 at maximum reverse thrust for a given temperature. The differences between the engines should be less than 5% N1.

Rejected Take Off

Treat an RTO test with a great deal of respect. Only use a long runway, with ATC permission and the fire vehicles standing by. Accelerate to 90kts then retard the thrust levers to idle; the autobrake should operate at maximum. Check that the brakes will not release by pedal pressure in the first two seconds but will release after two seconds. Select idle reverse and check that the speedbrakes deploy.

A production RTO test, note the tyre smoke and no reverse thrust.

Taxi in & Shutdown

There are several items which can be safely done after landing:

Fuel Suction Feed

After vacating the runway, switch off all fuel boost pumps and ensure that the engines continue to operate normally until shutdown.

Generator Disconnect

When the APU is on the buses, disconnect the generators checking that the voltage & frequency drop to zero and the LOW OIL PRESSURE or DRIVE lights illuminate. Don't forget to make a tech log entry to get them re-connected.

Engine Fire Warning Switch Shutdown

Before this test pull the hydraulic shut-off valve c/b's on the P6 panel to avoid EDP cavitation. Override and pull the two engine fire switches but do not rotate unless you want to fire the bottles. It takes about 25 seconds for the engines to run down because the fuel is cut off at the spar rather than the MEC. Ensure that you put the start levers to CUTOFF before restoring the fire switches and hydraulic c/b's.

A mask which failed to drop from a PSU

IRS Drift

After shutdown you should note the residual groundspeed of both IRS's available from FMC POS 2/3. Anything more than 6kts per hour of flight should be investigated.

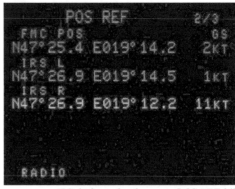

IRS drift can be shown on POS REF 2/3

PSU's

While you are walking down the cabin, check that the PSU's have dropped. This photograph shows a PSU that has been allowed to drop for the pressurisation checks. Notice how even though the door has opened, only one of the two pairs of masks has dropped out. This should be noted on your report.

Emergency Exit Lighting

Switch on the emergency exit lighting and check that all lights (inside and outside) are illuminated and that any captions are legible. Switch them off as soon as you have finished checked them so as not to deplete the batteries.

Post Flight Duties

All defects should be entered both in the Tech Log and on the test schedule. Defects can be classified in three ways:

1) Items requiring a re-fly eg elevator power off limits, poor performance climb etc. In general terms, any items that cannot be adequately tested on the ground due to air loads require a re-fly, eg landing-gear, flight controls or pressurisation defects.
2) Items requiring rectification before C of A can be issued, eg VHF1 U/S.
3) Items for engineering info, eg leak rate check excessive.

Depending upon the reason for the airtest, you may find that there are pressures on you from various sources to record more or less defects than you consider necessary, this can start as early as after landing. If this is the case find yourself somewhere private where you can write your report until you are ready to debrief.

Be detailed but objective with your tech log entries. Try to discuss any subjective feelings or theories with the engineers face to face.

The test schedule should then be signed, dated and copies given to the interested parties eg engineers, leasing company, customer, CAA etc and keep a copy for yourself in case there are any queries at a future date.

And Finally…

If you are fortunate enough to be involved in airtesting, either as a Captain or an F/O, then you have one of the best jobs in aviation. You will be returned to the hands-on, non-radar controlled way of flying that you used to enjoy when you learnt to fly – except that you will be doing it in a 737. You will also learn a great deal about the aircraft. There is nothing like doing something for real to learn about it, rather than reading or even seeing it in the simulator.

Obviously there are responsibilities and a greater margin for error than routine airline operations. The best advice I can offer is to know your aircraft and the test schedule thoroughly; don't be rushed by anybody and if you are uncomfortable with something, don't do it.

The aircraft that required 19 turns of stab trim during its manual reversion test looking innocent in the morning sun

ACCIDENT REPORTS

This is a complete list of all Boeing 737 write-offs. It should be said that there have been other accidents with more serious damage than some of these listed here, but if the aircraft was repaired they do not appear. Similarly, some less serious accidents have resulted in a write-off because the operator could not, or did not wish to, pay for a repair. This is particularly true of 737-200s in recent years.

Breaking the accidents down by phase of flight shows the following results:

Phase of Flight	Number of Occurrences
Ground	10
Take-off – RTO	12
Take-off – Initial climb	13
Climb	4
Cruise	7
Descent	4
Approach - Non Precision	28
Approach - Other	10
Landing	59
Go-Around	3

More controversially, listing the main factor for each accident shows the following:

Main Accident Factor	Number of Occurrences
Unstable approach	34
Weather	24
Mechanical / structural failure	17
Engine failure	15
Controlled flight into terrain	14
Mis-configuration	11
Hijack / bomb, etc	8
Stall	7
Collision	6
Crew training	6
Double engine failure	3
Rudder problem	2
Unknown	3

It should be remembered that the above table is an oversimplification of the cause of the accidents. In almost every case there were multiple factors, I have simply tried to indicate the main one. I am sure that if the reader compiled his own table from these same reports he would get a different set of results.

There have been 153 737 hull-losses, including 8 hijackings/bombings and 10 ground accidents. This may sound high, but remember that over 7500 737s have been built since 1967. This gives a 2% accident rate or approx 3 per year or one every 2.5 million flight hours. Furthermore, over 40% of occupants survive fatal 737 accidents.

The next table shows that more accidents have befallen the older aircraft. This is to be expected because they have amassed more flying hours & cycles and because later generations have 40 years of design and technology improvements built in to them. A fairer comparison across the generations would be rate per flying hours or cycles but I do not have that data.

Series	W/Os	No Built	W/O Rate / A/C
737-1/200	108	1144	1 in 11
737-3/4/500	36	1990	1 in 57
737-NG	9	4000+	1 in 444+

The following summaries have been compiled using data from a variety of sources including aviation journals, books, news reports, internet sites and of course the official accident reports, some of which is contradictory. Any additional information or corrections about any of these incidents would be gratefully received.

The Complete Boeing 737 Write-off Listing

19 Jul 1970, N9005U, 737-200, 19043/18, FF 29 Feb 68, United Airlines; Philadelphia, USA.

The First Officer (PF), initiated the take-off roll. Shortly after rotation, at a speed above V2, a loud bang was heard and the aircraft veered to the right. The Captain moved both throttles forward but there seemed to be no response. The Captain then made the instantaneous decision to land back on the runway. The aircraft touched down 1075ft before the end of the runway; it overran and continued across the blast pad. It crossed a field, passed through a 6-foot high aluminium chain link fence into an area covered with high grass, weeds and brush. The aircraft came to rest 1634ft past the end of the runway.
The NTSB report states the cause as: "The termination of the take-off, after the No.1 engine failed, at a speed above V2 at a height of approximately 50 feet, with insufficient runway remaining to affect a safe landing. The Captain's decision and his action to terminate the take-off were based on the erroneous judgment that both engines had failed." There were no fatalities.

8 Dec 1972; N9031U, 737-200, 19069/75, FF 24 Sep 68, United Airlines; Chicago Midway, USA:

Aircraft crashed in a residential area about 1.5 miles from Runway 31L during a non-precision approach. The aircraft was observed below the overcast in a nose-high attitude and with the sound of high engine power just before it crashed into structures on the ground. It is believed that the aircraft was in the landing configuration but with the flight spoilers still deployed after a deceleration which caused the aircraft to stall.
The NTSB determined that the cause was "the Captains failure to exercise positive flight management during the execution of the non-precision approach, which culminated in a critical deterioration of airspeed into the stall regime were level flight could no longer be maintained."
3 of the 6 crew, 40 of the 55 passengers and 2 people on the ground were killed.

31 May 1973; VT-EAM, 737-200, 20486/279, FF 26 Mar 71, Indian Airlines; near New Delhi, India:

The aircraft was making an NDB approach with visibility below minima, but the crew continued the approach and descended below minimum descent altitude without seeing the runway. The aircraft collided with high tension wires, crashed and caught fire. The crash killed five of the seven crew members and 43 of the 58 passengers.

31 Mar 1975; N4527W, 737-200, 20131/165, FF 7 May 69, Western Airlines; Casper, WY, USA.

The Captain was flying a back-course localizer approach to Runway 25, even though a full ILS was available on runway 07. The weather was 500ft cloud base with visibility less than 1 mile in light snow and a slight tailwind of 030/08. The runway had been ploughed earlier but there was now an estimated 2-3 inches of snow on the runway. The aircraft levelled off at MDA with flap 25 set. After 4 sec the F/O became visual and selected flap 30 at the Captains request who descended toward the runway. The aircraft remained high and fast until touchdown and no advisory/warning callouts were made by the F/O. He touched down 2375ft into the 8681ft runway at Vref+20 and ran off the end by 800ft hitting the approach lights at the far end. Aircraft sustained substantial damage but there were no fatalities.

4 Dec 1977; 9M-MBD, 737-200 Adv, 20585/306, FF 12 Sep 72, MAS; near Johor Strait, Malaysia:

The flight was approaching Kuala Lumpur when the pilot radioed that a hijacker had taken control of the aircraft. The aircraft continued to Singapore. While descending from FL210 to FL070 the nose suddenly pitched up. Control was lost and the aircraft crashed into a swamp and disintegrated. All seven crew members and 93 passengers were killed. Both pilots had been shot.

11 Feb 1978; CF-PWC, 737-200, 20142/253, FF 24 Apr 70, Pacific Western Airlines; Cranbrook, Canada:

The aircraft touched down just as the crew noticed a snow blower on the runway. A go-around was initiated, but the thrust reversers did not stow away properly because hydraulic power was automatically cut off at lift-off. The aircraft missed the vehicle, overran the runway, crashed and burned. Estimated time of arrival given by Calgary ATC was considerably in error. Crew did not report over the final approach beacon. The crash killed four of the crew members and 38 of the 44 passengers.

3 Apr 1978; PP-SMX, 737-200 Adv, 20969/369, FF 14 Aug 74, Vasp; Congonhas, Brazil:

During landing in darkness and fog, on their last flight of the day, the crew forgot to lower the gear. The aircraft landed on the runway and skidded for about 2000 ft before it veered off the side and came to rest in grass. The aircraft collided with small trees on its path sustaining damage to both engines, forward and mid fuselage and horizontal stabilizers. No fatalities.

4 Apr 1978; OO-SDH, 737-200C Adv, 20914/396, FF 12 Feb 75, Sabena; Charleroi, Belgium:

The aircraft was performing circuits on a training flight. During a touch-and-go on Runway 25, the aircraft struck a flock of birds. The take-off was aborted, but the aircraft overran, struck the localiser antennas and skidded. The right main gear collapsed and the no.2 engine was torn off in the slide. The aircraft came to rest 300m past the runway end and was destroyed by fire. There were no fatalities.

17 Dec 1978; VT-EAL, 737-200, 20485/277, FF 25 Feb 71, Indian Airlines; Hyderabad-Begumpet, India:

The aircraft lifted off from Runway 09, but could not climb because the leading edge devices did not deploy and as a result the aircraft became aerodynamically unstable. The take-off was aborted and the aircraft was flared for a belly landing with undercarriage retracted. The aircraft belly landed in a nose up, left wing low attitude, on the centre line of the runway. It slid for 3080 feet, hit a boundary fence, crossed a drain and ploughed into rough terrain hit some small boulders and came to rest. Fire broke out on impact. One passenger and 3 maintenance workers cutting grass were killed.

26 Apr 1979; VT-ECR, 737-200 Adv, 20962/380, FF 16 Oct 74, Indian Airlines; Madras, India:

On its way from Trivandrum to Madras, the aircraft was cleared to descend from FL270. Shortly afterwards an explosion took place in the forward lavatory, causing a complete instrument and electrical failure. The aircraft had to make a flapless landing at Madras. The aircraft touched down 2500ft past the Runway 25 threshold and overran. The right side of the plane caught fire. PROBABLE CAUSE: "Detonation of an explosive device in the forward lavatory of the aircraft. The aircraft overshot the runway due to high speed of touchdown, non-availability of reverse thrust and anti-skid system due to systems failure from explosion."

5 Nov 1980; D2-TAA, 737-200, 21172/439, FF 7 Nov 75, TAAG Angola Airlines; Benguala, Angola:

The aircraft landed 4m short, causing the gear to collapse. The aircraft slid 900m and came to rest 20m to the side of the runway. The no.1 engine and right wing caught fire. The aircraft was destroyed during recovery attempts. PROBABLE CAUSE: Lack of VASIs and threshold markings caused the pilot's inability to follow the correct approach slope.

17 Feb 1981; N468AC, 737-200, 20334/232, FF 26 Aug 70, Air California; Santa Ana, CA, USA:

The crew had received a clearance for a visual approach to Runway 19R. Meanwhile the controller cleared another flight for take-off from the same runway. When recognising the hazard, the controller ordered the landing aircraft to go-around and the departing aircraft to abort its take-off. The departing aircraft rejected its take-off, but the Captain of the landing aircraft delayed the go-around by approximately 12 seconds and then selected the gear UP before achieving a positive rate of climb. The 737 left the runway surface at 900ft past the threshold and skidded another 1170ft before coming to rest 115ft to the right of the centreline. No fatalities.
PROBABLE CAUSE: "The captain's failure to immediately initiate a go-around when instructed to do so by the tower's air traffic controller and his subsequent failure to correctly execute the specified go-around procedure which resulted in the retraction of the landing gear after the aircraft touched down on the runway."

22 Aug 1981; B-2603, 737-200, 19939/151, FF 30 Apr 69, Far Eastern Air Transport; near Sanyi, Taiwan:

The aircraft experienced in-flight structural failure at 22,000ft. The crash killed all six crew members and 104 passengers. The aircraft had experienced rapid depressurisations 2 weeks previously and that morning. The aircraft was the one built immediately before the Aloha 737 which ripped open in-flight.

The cause was found to be serious belly corrosion, exacerbated by the fact that the aircraft had been frequently used to carry fish in the hold. The first 418 737s were susceptible to structural problems as they were built with a production method of stiffening with bonded doublers (cold bonding). This was changed to chemical milling.

13 Jan 1982; N62AF, 737-200, 19556/130, FF 15 Feb 69, Air Florida; Washington, DC:

The crew used reverse thrust to assist the push-back off stand and taxied close behind another aircraft for heat from the exhaust, both of which may have contributed to the airframe snow/ice accretion. They also did not use engine anti-ice whilst taxiing out during snow causing the PT2 (EPR) probes to misread. Although anomalous engine instrument readings were called during the takeoff, the Captain neither aborted nor adjusted the thrust levers to allow sufficient thrust for takeoff and the aircraft stalled and crashed into the frozen Potomac River. Take-off was almost an hour after de-icing had been completed and the CVR shows that the crew were aware of 10-20mm of snow on the wings. Four of the five crew members and 70 of the 74 passengers were killed.

25 May 1982; PP-SMY, 737-200 Adv, 20970/376, FF 24 Sep 74, VASP; Brasilia, Brazil:

The aircraft landed heavily in a rainstorm and broke in two. One report stated that "The pilots' misuse of rain repellent caused an optical illusion". The crash killed two of the 112 passengers.

26 Aug 1982; JA8444, 737-200 Adv, 21477/545, FF 1 Dec 78, Southwest (Later JTA); Ishigaki, Japan:

The crew made a crosswind landing on Runway 22 with a wind of 300/12, the speed at touchdown was Vref+6kts. The aircraft bounced and landed back. When the spoilers and reversers didn't seem to operate, the crew shut down both engines, thus making it impossible to use the anti-skid system. The inner tyres on both main gear legs burst almost simultaneously at 125m short of the runway end. The aircraft skidded to the left, overran the runway and came to rest 145m further on. The aircraft caught fire after the evacuation. No fatalities.

22 Feb 1983; PP-SNC, 737-200C Adv, 21187/443, FF 12 Dec 75, VASP; Manaus, Brazil:

The First Officer lost control of a simulated engine failure after take-off during a training flight. The retarded thrust lever was advanced, but the engine stalled. The Boeing then rolled left and crashed.
PROBABLE CAUSE: Failure to follow procedures, attempted operation with known deficiencies and inadequate supervision were factors. Both pilots were killed.

28 Mar 1983; C9-BAB, 737-200, 20281/228, FF 16 Dec 69, LAM-Mozambique; Quelimane, Mozambique:

While on final approach at night, the pilot reduced engine power to flight idle. At 230ft (70m) the pilot added some power, but the airspeed had decayed below Vref. The stick-shaker activated and full power was applied. The 737 touched down 400m short of the runway, causing the undercarriage to fail.
PROBABLE CAUSE: Misjudged speed, distance and altitude. Contributing factors were the inadequate crew supervision and the non-use of VASIs. There were 110 people on board but no fatalities.

11 Jul 1983; HC-BIG, 737-200 Adv, 22607/775, FF 11 Jun 81, TAME; near Cuenca, Azway, Ecuador:

The aircraft struck a ridge with its tail section during an approach in fog and then crashed into hilly terrain approx 1 mile short of the runway. The crash killed all eight crew members and 111 passengers. The pilot was reportedly "under-qualified".

23 Sep 1983; A40-BK, 737-200 Adv, 21734/566, FF 4 Apr 79, Gulf Air; Mina Jebel Ali, UAE:

Crashed during approach into Abu Dhabi after a bomb had exploded in the baggage compartment. The crash killed all six crew members and 105 of 111 passengers. There were indications of a pre-impact explosion in a cargo hold, with resultant structural damage and an uncontrollable fire producing toxic fumes that rapidly overcame the occupants. The evidence pointed away from a blaze of either electrical or fuel origin, and it was later concluded that the 737 had been sabotaged. Some checked-in luggage was carried by a ticket holder who did not board the aircraft.

8 Nov 1983; D2-TBN, 737-200 Adv, 22775/869, FF 29 Apr 82, TAAG Angolan; Lubango, Angola:

The aircraft took off and climbed to 200ft before turning steeply to the left and crashing about 800m from the end of the runway. The crash killed all five crew members and 121 of 126 passengers. Angolan authorities blamed technical failure; however, anti-government guerrillas claimed to have shot the aircraft with a surface-to-air missile.

22 Mar 1984, C-GQPW, 737-200 Adv, 22265/755, FF 3 Apr 81, Pacific Western; Calgary, Canada:

About 20 seconds into the take-off roll, at 70 knots, the flightcrew heard a loud bang which was accompanied by a slight veer to the left. The No1 engine 13th stage compressor disc had failed 1,300 feet into the take-off roll. A piece of the disc punctured a fuel cell which ignited instantaneously and the fire engulfed the left wing and aft section of the aircraft.

The Captain immediately rejected the take-off using brakes and reverse thrust. Both the pilots suspected a tyre on the left main landing gear had blown. While the aircraft was vacating the runway, the crew noted that left engine low pressure unit rpm was indicating 0 per cent. The purser then entered the flight deck and reported a fire on the left wing. It took approximately 1min 30secs for the crew to initiate the evacuation. There were 119 people on board, but no casualties.

9 Feb 1984; D2-TBV, 737-200 Adv, 22626/802, FF 21 Sep 81, TAAG Angola Airlines; Huambo, Angola:

An explosion occurred in the rear cabin whilst climbing through 8000ft after take-off. The crew retuned to Huambo for an emergency landing but were unable to extend the flaps because of damage to the hydraulic systems. The aircraft landed flapless and fast and overran the runway by 180m. No fatalities

30 Aug 1984; TJ-CBD, 737-200C Adv, 21295/484, FF 15 Feb 77, Cameroon Airlines; Douala, Cameroon:

When taxiing out for a flight to Garoua via Yaound, a compressor disc in the No.2 JT8D-15 engine failed. Debris punctured the wing and fuel tank, causing a fire. The aircraft burned out. Two of the 108 passengers were killed.

15 Apr 1985; HS-TBB, 737-200 Adv, 21810/604, FF 24 Sep 79, Thai Airways; Phuket, Thailand:

Shortly after starting a descent to 3000ft the crew reported that they were not receiving the DME. ATC asked if they would prefer a VOR approach to Runway 09 instead of a visual to 27, although it was night (23:25 local), the visual approach to 27 was continued. Later the crew reported that they had lost both engines and were descending through 3400ft and had nearly hit a mountain. The aircraft eventually hit high ground at 800ft and was destroyed by the impact and subsequent fire. The crash killed all four passengers and seven crew members.

22 Aug 1985; G-BGJL, 737-200 Adv, 22033/743, FF 26 Feb 81, British Airtours; Manchester, UK:

The aircraft began its take-off from Runway 24 at Manchester with the First Officer handling. About 36 seconds later, as the airspeed passed 125 knots, the left engine suffered an uncontained failure, which punctured a wing fuel tank access panel. Fuel leaking from the wing ignited and burnt as a large plume of fire trailing directly behind the engine. The crew heard a thud, and believing that they had suffered a tyre-burst or bird-strike, abandoned the take-off immediately, intending to clear the runway to the right. They had no indication of fire until 9 seconds later, when the left engine fire warning occurred. After an exchange with ATC, during which the fire was confirmed, the Captain warned his crew of an evacuation from the right side of the aircraft, by making a broadcast over the cabin address system, and brought the aircraft to a halt.

As the aircraft turned off, a wind of 7 knots from 250° carried the fire onto and around the rear fuselage. After the aircraft stopped the hull was penetrated rapidly and smoke, possibly with some flame transients, entered the cabin through the aft right door which was opened shortly before the aircraft came to a halt. Subsequently fire developed within the cabin. Despite the prompt attendance of the airport fire service, the aircraft was destroyed and 55 of the 137 persons on board lost their lives.

The cause of the accident was an uncontained failure of the left engine, initiated by a failure of the No 9 combustor can which had been the subject of a repair. A section of the combustor can, which was ejected forcibly from the engine, struck and fractured an under wing fuel tank access panel. The fire which resulted developed catastrophically, primarily because of adverse orientation of the parked aircraft relative to the wind, even though the wind was light.

The major cause of the fatalities was rapid incapacitation due to the inhalation of the dense toxic/irritant smoke atmosphere within the cabin, aggravated by evacuation delays caused by a door malfunction and restricted access to the exits.

23 Nov 1985; SU-AYH, 737-200 Adv, 21191/450, FF 26 Feb 76, Egyptair; Luqa, Valletta, Malta:

The aircraft was hijacked to Malta. After several hours of negotiations, Egyptian troops stormed the aircraft. During the ensuing battle, the hijackers threw several hand grenades. The aircraft was severely damaged by the explosions and fire. Two of the six crew members and 58 of the 90 passengers were killed.

28 Jan 1986; PP-SME, 737-200, 20096/190, FF 16 Jul 69, VASP; Guarulhos, Sao Paulo, Brazil:

The crew unknowingly tried to take-off from a taxiway in fog. The take-off was aborted, but the aircraft overran, collided with a dyke and broke in two. One of the 60 passengers was killed.

16 Feb 1986; B-1870, 737-200, 20226/168, FF 12 May 69, China Airlines; Makung, Taiwan:

A nose gear tyre had reportedly burst during the first attempt to land. The aircraft went around and crashed into the sea 12 miles North of Makung. All six passengers and seven crew members were killed.

15 Oct 1986; EP-IRG, 737-200 Adv, 20499/284, FF 8 Aug 71, Iran Air; Shiraz, Iran:

Passengers were disembarking when the airport came under attack from Iraqi aircraft and the aircraft was hit. Three passengers were killed.

25 Oct 1986; N752N, 737-200, 19073/90, FF 27 Oct 68, Piedmont Airlines; Charlotte, NC.

Flight 467 landed on Runway 36R after an ILS approach. About 24 seconds after touchdown, the aircraft overran the runway, struck a localizer antenna array, a concrete culvert, continued through a chain link fence and came to rest upon the edge of railroad tracks, 440ft past the runway end. Three passengers sustained serious injuries, but there were no fatalities.
PROBABLE CAUSE: "The Captain's failure to stabilize the approach and his failure to discontinue the approach to a landing that was conducted at an excessive speed beyond the normal touchdown point on a wet runway. Contributing to the accident was the Captain's failure to optimally use the airplane deceleration devices. Also contributing to the accident was the lack of effective crew co-ordination during the approach. Contributing to the severity of the accident was the poor frictional quality of the last 1500ft of the runway and the obstruction presented by a concrete culvert located 318ft beyond the departure end of the runway."

25 Dec 1986; YI-AGJ, 737-200C Adv, 21183/446, FF 15 Jan 76, Iraqi Airways; Over Saudi Arabia:

The aircraft was en route between Baghdad, Iraq and Amman, Jordan when hijackers set off a hand grenade in the passenger cabin and started a gunfight with security forces on board the aircraft. An emergency descent was initiated immediately. Descending through FL160, a hand grenade in the cockpit exploded. The aircraft crashed trying to land at a small airfield near Arar Saudi Arabia; it broke in two and caught fire, killing 67 of the 107 passengers.

4 Aug 1987; CC-CHJ, 737-200 Adv, 22602/711, FF 17 Oct 80, LAN Chile; Santiago/Calama, Chile:

The aircraft was approaching Runway 27 which had an 880m displaced threshold due to construction work. The pilot was landing into sun and touched down 520m short of the displaced threshold. The nose-gear collapsed as it hit obstacles and the aircraft broke in two. A fire broke out 30mins later and destroyed the aircraft. One of the 27 passengers was killed.

31 Aug 1987; HS-TBC, 737-200 Adv, 22267/685, FF 25 Jul 80, Thai Airways; near Ko Phuket, Thailand:

While descending during a daylight approach in good weather, the crew lost control of the aircraft and crashed into the Andanan Sea. All of the nine crew members and 74 passengers were killed.
The airport had no radar and the crew were concerned that another aircraft was behind them flying 500ft lower, on a different VOR radial, and was cleared to land. During this distraction the aircraft stalled and was unable to recover before hitting the sea. The controllers were re-assigned and their supervisor disciplined.

4 Jan 1988; D-ABHD, 737-200 Adv, 22635/774, FF 15 Jun 81, Condor; Izmir, Turkey:

The flight was cleared to the CU NDB for an ILS approach to Runway 35. After passing the NDB the pilots switched to ILS and thus couldn't verify their position in the procedure turn. The aircraft was outside the 35deg sector of the ILS centreline and the crew followed the wrong side beam. The crew descended to Outer Marker altitude and the 737 struck a hill. PROBABLE CAUSE: Presumed NDB passage, wrong use of VOR and ILS, overconfidence of the Captain and considerable inactivity of the First Officer (PF). All five crew members and 11 passengers were killed.

28 Apr 1988; N73711, 737-200, 20209/152, FF 28 Mar 69, Aloha; near Maui, HI:

The aircraft experienced an explosive decompression at 24,000ft due to metal fatigue in upper cabin area. The crew was able to execute a successful emergency landing with an 8ft x 12ft section of the upper fuselage missing from aft of the forward entrance door to the leading edge of the wing. One of the flight attendants was swept overboard and killed.

The aircraft was the second highest cycle 737 in the world at 89,193; the highest also being in the Aloha fleet. The original design life of the 737 was 75,000 cycles but this had been increased to 130,000 in 1987. However Boeing had been expressing concern at the condition of three high cycle Aloha aircraft since 1987. The NTSB blamed Aloha for failure to detect fatigue damage and an inadequate maintenance policy. Aloha voluntarily scrapped the other two aircraft in July 1988.

15 Sep 1988; ET-AJA, 737-200 Adv, 23914/1456, FF 2 Oct 87, Ethiopian Airlines; Bahar Dar, Ethiopia:

During take-off, just past VR, at a speed of 146kts, the 737 suffered a multiple birdstrike. Both engines ingested 10-16 Columba Guinea birds (approx. 320 grams each), which resulted in a rise in engine temperature on climb-out and the engines backfiring at 100-200ft. The crew made a right turn to return to the airport with the EGT at limits. On base leg at 7100ft amsl with 190kts, both JT8D-17A engines backfired & flamed out. A wheels-up crash landing was made and the aircraft caught fire. The aircraft had been delivered on October 29, 1987 and had accumulated just 1377hrs flying time and 1870 cycles.

PROBABLE CAUSE: "The accident occurred because the airplane could not be safely returned to the runway after the internal destruction and subsequent failure of both engines to operate arising from multiple bird ingestion by both engines during take-off." As a result of the crash landing, 31 of the 105 passengers were killed.

26 Sep 1988; LV-LIU, 737-200 Adv, 20964/379, FF 10 Oct 74, Aero Argentinas; Ushuaia, Argentina:

The runway in use at Ushuaia was 16 with a wind of 230/12. However during the approach the wind changed to 360/20 so the crew elected to use Runway 34, despite being warned of possible windshear for that runway. The aircraft touched down hard (1.89G) 12kts fast, bounced and landed back 3/4 of the way down the 1400m runway. It then veered off and went down a slope into 2m deep water. No fatalities.

15 Oct 1988; 5N-ANW, 737-200 Adv, 22771/866, FF 20 Apr 82, Nigeria Airways; Port Harcourt, Nigeria:

The aircraft landed in heavy rain and overran the runway. The nose and right main gear both collapsed. No fatalities.

19 Oct 1988; VT-EAH, 737-200, 20481/271, FF 24 Nov 70, Indian Airlines; Ahmedabad, India:

The aircraft was approaching Runway 23 in fog when it hit trees and an electric pylon three miles (5km) out on approach. All six crew members and 124 of 129 passengers were killed.

8 Jan 1989; G-OBME, 737-400, 23867/1603, FF 6 Oct 88, British Midland Airways; Kegworth, UK:

13 minutes after take-off from LHR, while climbing through FL283, moderate to severe vibration was felt, accompanied by a smell of fire in the cockpit. The outer panel of one of the No.1 engine fan-blades detached, causing compressor stalls and airframe shuddering. Believing the No.2 engine had been damaged the crew throttled it back. The shuddering stopped and the No 2 engine was shut down. The crew then decided to divert to East Midlands. The flight was cleared for a Runway 27 approach. At 900ft, 2.4nm from the runway, engine No.1 power suddenly decreased. As the speed fell below 125kts, the stick shaker activated and the aircraft struck trees at a speed of 115kts. The aircraft continued and impacted the embankment of the M1 motorway and came to rest against the wooded embankment, 900m short of the runway. 47 people were killed.

The engine was a CFM56-3C1 at 23,500lbs and failed after only 500hrs, this was followed by several other failures of the 3C1 which ran at higher rotational speeds than previous CFM56s. The engines were then derated down to 22,000lbs until a permanent cure was found.

9 Feb 1989; C9-BAD, 737-200 Adv, 20786/323, FF 24 Oct 73, LAM; Lichinga, Mozambique:

Overran runway landing in a rainstorm. One engine separated from the wing. No fatalities.

3 Apr 1989; OB-R-1314, 737-200, 19425/153, FF 2 Apr 69, Faucett Peru: Iquitos, Peru:

Aircraft landed in stormy weather and the gear collapsed. Aircraft slid off the runway and the No.2 engine separated. No fatalities.

3 Sep 1989; PP-VMK, 737-200 Adv, 21006/398, FF 7 Feb 75, Varig; near Sao Jose do Xingu, Brazil:

The pilot set a heading of 270 instead of 027 and ended up 600 miles off course. The error led to fuel exhaustion and a forced landing in jungle, 12 of the 48 passengers were killed in the crash. It took two days for the survivors to be found.
The heading mistake went unnoticed because the crew was reportedly listening to the Brazil v Chile World Cup qualification football match.

20 Sep 1989; N416US, 737-400, 23884/1643, FF 9 Dec 88, USAir; La Guardia Airport, New York:

The aircraft swung to the left during the F/O's takeoff, the Captain aborted at Vr+5. Auto-throttle was not used and autobrake RTO mode had not been selected both of which increased the accelerate-stop distance. The aircraft overran the wet runway and dropped onto the wooden approach light pier, which collapsed causing the aircraft to break into three and drop into 7-12m deep East River. Two of the 55 passengers were killed.
The swing was caused by the rudder trim having been placed unintentionally at 16 degrees (full) left position. This could have been caused by the foot of the jumpseat passenger on the centre console or the knob could have stuck whilst being moved. Boeing have since made the rudder trim knob round rather than blade shaped and fitted a guard rail around the console.

26 Oct 1989; B-180, 737-200 Adv, 23795/1319, FF 3 Dec 86, China Airlines; near Hualien, Taiwan:

The crew were using the wrong SID, causing the aircraft to make a left instead of right turn and hit cloud shrouded high ground at 7000 feet. All 7 crew members and 49 passengers were killed.

30 Dec 1989; N198AW, 737-200, 19710/54, FF 2 Aug 68, America West Airlines; Tucson:

While descending to Tucson, a 115Volt AC wire of the No.2 B hydraulic pump shorted and punctured a hydraulic system A line. As the aircraft was approaching the airport a fire erupted and burned through to the electrical power wires to the standby hydraulic pump. After landing at Tucson, the aircraft overran the runway, collided with an abandoned concrete arresting gear structure, shearing off the nose-gear and continued to slide for 3,803ft. The aircraft had flown 62,466hrs and 38,827 cycles.
PROBABLE CAUSE: Mechanically worn thrust reverse accumulator check valve and inboard brake isolation check valve.

2 Jun 1990; N670MA, 737-200C Adv, 23121/1025, FF 2 May 84, MarkAir; Unalakleet, Alaska:

The aircraft crashed about 7.5 miles short of Runway 14, Unalakleet, Alaska, while executing a localizer approach to that runway in IMC. There were no passengers on board but the crew sustained injuries and the airplane was destroyed.
The NTSB determines that the probable cause of this accident was deficiencies in flightcrew coordination, their failure to adequately prepare for and properly execute the UNK LOC Rwy 14 non-precision approach, and their subsequent premature descent. The Safety Board issued a safety recommendation on approach chart standardization to the FAA.

22 Jul 1990; N210US, 737-200, 19555/129, FF 9 Feb 69, US Air; Kinston, NC, USA:

The No.1 engine accelerated beyond target EPR on take-off. The crew aborted the take-off, but the No.1 engine didn't respond to the retarded power lever, so had to be shut down with the fuel shut-off lever. Asymmetric thrust was controlled with nose wheel steering. The nose-gear wheels separated from the gear before the aircraft was brought to a halt. No fatalities.
PROBABLE CAUSE: "Failure of the fuel pump control shaft because of improper machining by the repair facility during maintenance modification of the pump and improper procedures during overhaul of the nose landing gear."

2 Oct 1990; B-2510, 737-200 Adv, 23189/1072, FF 7 Dec 84, Xiamen Airlines; Baiyun, Canton, China:

A hijacker detonated a bomb during approach, causing the 737 to hit parked aircraft on the ground. Seven of the nine crew members and 75 of the 93 passengers were killed. The hijacker had ordered the flight crew out of the cockpit except for the Captain. He refused an offer from the Captain to fly to Hong Kong and the dispute continued until the fuel was nearly exhausted, necessitating the landing. Shouts and sounds of a struggle were heard from the cockpit during the approach. The aircraft landed hard and veered off the runway and clipped a 707 and a 757 before coming to a halt upside down in a grassy area.

5 Nov 1990; EI-BZG, 737-300, 24466/1771, FF 24 Aug 89, Philippine Airlines; Manila, Philippines:

The centre fuel tank exploded while the aircraft was taxiing for departure. 8 of the 113 passengers were killed.

The airline had fitted logo lights after delivery which involved additional wires to be passed through vapour seals in the fuel tanks. The NTSB recommended to the FAA that an AD be issued requiring inspections of the fuel boost pumps, float switch and wiring looms as signs of chafing had been found. The FAA declined to issue the AD.

1 Feb 1991; N388US, 737-300, 23310/1145, FF 23 Aug 85, USAir; Los Angeles, CA, USA:

US Air flight 1493 entered LAX airspace around 17.57 and was cleared for a profile descent and ILS Runway 24R approach. At 17.59 this was changed to a Runway 24L approach clearance. At about the same time a SkyWest Metro II aircraft (Flight 5569 to Fresno) taxied to Runway 24L. At 18.03 the crew were advised to, "taxi up to and hold short of 24L" because of other traffic. At 18.04:49 the flight was cleared to taxi into position and hold. Immediately thereafter, the controller became preoccupied with instructing WingsWest Flight 5006 who had unintentionally departed the tower frequency. The WingsWest 5072 reporting ready for takeoff caused some confusion because the controller didn't have a flight progress strip in front of her. The strip appeared to have been misfiled at the clearance delivery position. Meanwhile, Flight 5569 was still on the runway at the intersection with taxiway 45, awaiting takeoff clearance. At 18.07 Flight 1493 touched down. Simultaneous to the nose-gear touchdown, the US Air B737 collided with the SkyWest Metro. Both aircraft caught fire and slid to the left into an unoccupied fire station.
Two of the six crew members and 20 of the 83 passengers on the USAir jet were killed. All 10 passengers and two crew members on the Metro III were killed.

3 Mar 1991; N999UA, 737-200 Adv, 22742/875, FF 11 May 82, United Airlines; nr Colorado Springs, CO:

The aircraft departed from controlled flight approximately 1,000 feet above the ground and struck an open field on approach to Colorado Springs. All 25 people on board were killed.
After a 21-month investigation, the Board issued a report on the crash in December 1992. In that report, the NTSB said it "could not identify conclusive evidence to explain the loss of the aircraft", but indicated that the two most likely explanations were a malfunction of the airplane's directional control system or an encounter with an unusually severe atmospheric disturbance.
The NTSB has since adopted a revised final report on this crash. The Board said that the most likely cause of the accident was the movement of the rudder in the direction opposite that commanded by the flight crew. The decision tracks information learned from the investigation of two fatal 737 accidents - including this one - and a non-fatal incident.

16 Aug 1991; VT-EFL, 737-200 Adv, 21497/504, FF 17 Nov 77, Indian Airlines; near Imphal, India:

The aircraft crashed into a hill about 30km from the airport while positioning for a Runway 04 ILS approach. The pilot had extended the outbound leg too far, flying over mountainous terrain. All six crew members and 63 passengers were killed.

17 Nov 1991; EI-CBL, 737-200 Adv, 20957/377, FF 26 Sep 74, SAHSA; San Hose, Costa Rica:

The aircraft appeared to be left of the centerline during the final stages of the ILS approach. The crew made a correcting manoeuvre, which resulted in the aircraft landing on the right main gear first. The right main gear collapsed, followed by the left gear.

6 Jun 1992; HP-1205CMP, 737-200 Adv, 22059/631, FF 15 Jan 80, COPA Panama; near Ticuti, Panama:

Twenty minutes after leaving Panama City at FL 250, the crew became disorientated when the artificial horizon failed. The aircraft rolled through 90deg, entered a steep dive and broke up at approximately FL130. The aircraft entered an area with thunderstorms, but it is not known if it sustained a lightning strike. However, a wire from the gyroscopes to the instruments had frayed, creating a short circuit and was giving erroneous attitude indications. The aircraft, although in VMC, was flying over featureless woodland at night and the crew were unable to determine the aircrafts attitude. All seven crew members and 40 passengers were killed.

22 Jun 1992; PP-SND, 737-200C Adv, 21188/444, FF 22 Dec 75, Vasp; Cruzero Do Sol, Brazil:

During the descent to Cruzeiro do Sol, the crew's attention was distracted by the cargo compartment warning system, which began to activate intermittently. The aircraft crashed in the jungle while performing a Delta 1 arrival. All 3 crew on board were killed.

20 Nov 1992; LV-JNE, 737-200C Adv, 20408/265, FF 11 Sep 70, Aerolineas Argentinas; San Luis:

The aircraft had previously made a heavy landing which was inspected by an engineer on the turnaround. The aircraft back-tracked Runway 18 for departure, but tyre marks on the runway showed that the number 3 & 4 tyres had locked up during the line-up. During the take-off run, No3 tyre lost pressure and No4 tyre burst just before V1. The aircraft began vibrating and pulling to the right. The take-off was aborted and the aircraft overran by 125m and caught fire. No fatalities.

24 Nov 1992; B-2523, B737-300, 24913/2052, FF 10 May 91, China Southern Airlines; Guilin, China:

Aircraft hit high ground at 7000ft (500ft below MSA) during approach to Runway 36. A serious vibration had reportedly occurred in the No 2 engine before the aircraft crashed in a steep right hand turn. FDR data indicates that the aircraft levelled off at 7200ft during a descent towards the airport with the autopilot and autothrottle engaged, the left throttle lever advanced but the right did not. The FDR also indicated that the same problem had occurred earlier in the same flight and had been corrected manually. All 8 crew members and 133 passengers were killed.

30 Mar 1993; 33-333, 737-300, 24480/1773, FF 28 Aug 89, Royal Thai Air Force; Bangkok, Thailand:

On approach the aircraft pitched violently up and down several times, appeared to go out of control and crashed. The stabilizer was said to have been mis-trimmed, but the cause is not known. The aircraft tech log showed recent problems with horizontal stabilizer control, which were being investigated at the time by Boeing. All 6 crew killed.

26 Apr 1993; VT-ECQ, 737-200 Adv, 20961/375, FF 18 Sep 74, Indian Airlines; Aurangabad, India:

The aircraft failed to climb after take-off with weight about 2000kg (4400lbs) above RTOW for the conditions (OAT 40C). It struck a large vehicle on a road just outside the airport with its landing gear. The vehicle strike also damaged one engine and the aircraft later hit power lines and crashed 7km from the airport. Four of the six crew members and 52 of the 112 passengers were killed. Maintenance deficiencies were found including the fact that the aircraft was despatched with an unserviceable FDR. Both pilots were charged with negligence. The administrators of the airport were also cited for failing to regulate traffic on that same road.

18 Jul 1993; N401SH, 737-200 Adv, 20584/305, FF 17 Aug 72, SAHSA; Managua, Nicaragua:

The aircraft landed heavily on Runway 09, skidded to the right off the runway. The nose-gear collapsed and both engines were torn off. The aircraft came to rest 200ft right of the runway. The weather was bad at the time of the accident with heavy rain and lightning. No fatalities.

26 Jul 1993; HL7229, 737-500, 24805/1878, FF 14 Jun 90, Asiana Airlines; near Mokpo, Korea:

The aircraft collided with a ridge of the Mount Ungeo (1050ft high) at an altitude of 800ft, in strong winds & heavy rain about 4nm from the runway while it was making its third attempt at a VOR/DME approach into Mokpo. Four of the six crew members and 64 of the 104 passengers were killed.

19 Nov 1993; HP-873CMP, 737-100, 19768/184, FF 26 Jun 69, COPA Panama; Panama City, Panama:

The aircraft touched down in a crosswind following an ILS approach onto Runway 03R, but wasn't properly aligned with the runway. It then departed the runway 2500ft past the threshold and crossed a taxiway. The nose-gear collapsed. Weather was bad with low clouds, turbulence and rain. No fatalities.
This was the only 737-100 ever to be written off.

8 Mar 1994; VT-SIA, 737-200C Adv, 21763/571, FF 25 Apr 79, Sahara India Airlines; Delhi, India:

A Training Captain and three new pilots were performing circuits for crew training. The aircraft took off, attained a steep nose-up attitude and the left wing dropped. The aircraft crashed and slid into a parked Ilyushin 86 airliner. Both aircraft burned out. All 4 occupants and 5 persons on the ground were killed. Whilst this accident looked like pilot error, there was evidence suggesting the rudder had reversed to the left when the pilot had correctly commanded it to move right. A rudder PCU jam remains a possibility.

29 Jul 1994; B-2540, 737-300, 27139/2400, FF 17 Nov 92, Yunnan Provincial; Kunming, China:

The aircraft landed fast and long on Runway 03 in thunderstorms and rain. It overran across soft ground and struck approach lights and the ILS antenna. The nose-gear collapsed and was pushed into the avionics bay. No fatalities.

8 Sep 1994; N513AU, 737-300, 23699/1452, FF 24 Sep 87, USAir; near Pittsburgh, PA:

Flight 427 was approaching Pittsburgh Runway 28R; the aircraft was levelling off at 6000ft & 190kts and rolling out of a 15deg left turn (roll rate 2deg/sec) with flaps at 1, the gear still retracted and autopilot and auto-throttle systems engaged. The aircraft then suddenly entered the wake vortex of a Delta Airlines Boeing 727 that preceded it by approx. 69 seconds (4.2nm). Over the next 3 seconds the aircraft rolled left to approx. 18deg of bank. The autopilot attempted to initiate a roll back to the right as the aircraft went in and out of a wake vortex core, resulting in two loud "thumps". The First Officer then manually overrode the autopilot without disengaging it by putting in a large right-wheel command at a rate of 150deg/sec. The airplane started rolling back to the right at an acceleration that peaked 36deg/sec, but the aircraft never reached a wings level attitude. At 19.03:01 the aircraft's heading slewed suddenly and dramatically to the left (full left rudder deflection). Within a second of the yaw onset the roll attitude suddenly began to increase to the left, reaching 30deg. The aircraft pitched down, continuing to roll through 55deg left bank. At 19.03:07 the pitch attitude approached -20deg, the left bank increased to 70deg and the descent rate reached 3600fpm. At this point, the aircraft stalled. Left roll and yaw continued, and the aircraft rolled through inverted flight as the nose reached 90deg down, approx. 3600ft above the ground. The 737 continued to roll, but the nose began to rise. At 2000ft above the ground the aircraft's attitude passed 40deg nose low and 15deg left bank. The left roll hesitated briefly, but continued and the nose again dropped. The plane descended fast and impacted the ground nose first at 261kts in an 80deg nose down, 60deg left bank attitude and with significant sideslip. All 132 aboard died.

PROBABLE CAUSE (US Airways): "An uncommanded, full rudder deflection or rudder reversal that placed the aircraft in a flight regime from which recovery was not possible using the known recovery procedures. A contributing cause of this accident was the manufacturer's failure to advise operators that there was a speed below which the aircraft's lateral control authority was insufficient to counteract a full rudder deflection."

26 Nov 1994; N11244, 737-200, 20073/142, FF 6 Mar 69, Continental Airlines; Houston, USA:

Two mechanics were repositioning the aircraft to Continental departure gate 41. Simultaneously, Flight 1176, a Boeing 737-300, from gate 44 was under the control of the pushback team consisting of a tug driver and a wing walker when the right wing of N11244 contacted the left outboard flap. The cockpit crew, who were in the process of starting engine No1, felt the impact and aborted the engine start. N11244 was damaged substantially and considered a loss. PROBABLE CAUSE: "The failure of maintenance personnel to follow the taxi checklist resulting in the hydraulic pumps not being turned on."

21 Dec 1994; 7T-VEE, 737-200C Adv, 20758/322, FF 17 May 73, Air Algerie; Coventry, UK:

The aircraft had left Amsterdam at 07.42h but had to divert to East Midlands due to bad weather at Coventry. After awaiting better weather, the flight took off again at 09.32h. While approaching Runway 23 the aircraft descended below MDA, clipped the roofs of two houses, rolled and crashed inverted into a wood.

CAUSAL FACTORS: "i) The crew allowed the aircraft to descend significantly below the normal approach glidepath during a Surveillance Radar Approach to Runway 23 at Coventry Airport, in conditions of patchy lifting fog. The descent was continued below the promulgated Minimum Descent Height without the appropriate visual reference to the approach lighting or the runway threshold.; ii) The standard company operating procedure of cross-checking altimeter height indications during the approach was not observed and the appropriate Minimum Descent Height was not called by the non handling pilot.; iii) The performance of the flight crew was impaired by the effects of tiredness, having completed over 10 hours of flight duty through the night during five flight sectors which included a total of six approaches to land."

29 Dec 1994; TC-JES, 737-400, 26074/2376, FF 25 Sep 92, THY Turkish Airlines; near Van, Turkey:

The aircraft hit a hill 4km from the airport during the fourth VOR/DME approach in snow (900m visibility, reducing to 300m in driving snow). Six of the seven crew members and 49 of the 55 passengers were killed.

2 Jan 1995; 9Q-CNI, 737-200C Adv, 20793/333, FF 13 Nov 73, Air Zaire; Kinshasa:

The aircraft landed heavily in bad weather. The crew lost control and the aircraft ran off the side of the runway. This resulted in a nose-gear collapse and both engines were torn away. No fatalities.

16 Jan 1995; PK-JHF, 737-200 Adv, 20508/287, FF 23 Sep 71, Sempati Air; Jakarta:

Flight SG416 landed on Runway 09 (1858m long) with flap 30 at Yokyakarta, which was wet with pools of standing water after a thunderstorm. It overran the runway by 100m and the nose-gear collapsed. No fatalities.

2 Feb 1995; PP-SMV, 737-200 Adv, 20968/367, FF 19 Jul 74, Vasp; Sao Paulo, Brazil:

Aircraft departed Sao Paulo-Guarulhos for a flight to Buenos Aires. Following flap retraction the No.3 flap IN TRANSIT light remained on and the crew noticed some other problems: They were not able to reduce No.2 engine thrust below 1.15 EPR and the hydraulic system A suffered a pressure loss. An emergency return was made and the aircraft touched down on Runway 09L at 185kts, flaps 15. The 737 overran the runway by 200m and came to rest following a collapse of the nose-gear and right hand main gear. It appeared that the No.3 leading edge flap actuator attachment fitting on the wing front spar had fractured due to corrosion. The actuator came away and caused the failure of some hydraulic lines and damage to the thrust control cables. Some 1981 Boeing Service Bulletins had not been complied with. One of these included the replacement of the aluminium leading edge flap actuator attachment fitting with a steel one; this had not been done. No fatalities.

9 Aug 1995; N125GU, 737-200 Adv, 23849/1453, FF 26 Sep 87, Aviateca; nr San Salvador, El Salvador:

Flight GUG901 encountered heavy rain & thunderstorms while approaching San Salvador at night. The crew diverted off Airway G346 to avoid the thunderstorms but the accident report states that the aircraft's DME had been damaged by a lightning strike. The aircraft should then have passed overhead the airport and turn right downwind for an ILS approach to Runway 07. There seemed to be some confusion as to the position of the aircraft. The aircraft was at 5000ft, as cleared by ATC, when the GPWS sounded. Full power was applied, but the aircraft struck Mt. Chinchontepec volcano (2181m high) at an altitude of 1800m. All 7 crew members and 58 passengers were killed. The accident report attributes the let-down error to the pilots' failure to realise that the DME readouts were incorrect.

13 Nov 1995; 5N-AUA, 737-200 Adv, 22985/920, FF 14 Oct 82, Nigeria Airways; Kaduna, Nigeria:

Flight WT 357 touched down more than half way down Kaduna's Runway 23 in good, dry weather but with a 10-15kt tailwind. The plane veered off the left side, skidded sideways and came to rest 35m beyond the end of the runway. A fire broke out on the dry grass under the aircraft on the right hand side and destroyed it. Nine of the 129 passengers were killed.

2 Dec 1995; VT-ECS, 737-200 Adv, 20963/383, FF 4 Nov 74, Indian Airlines; Delhi, India:

At 12.56h LT flight IC492 arrived from Bombay and Jaipur, but touched down just 600m before the end of the runway and overran by 450m. Both engines, all undercarriage and the wings sustained major damage. No fatalities.

3 Dec 1995; TJ-CBE, 737-200 Adv, 23386/1143, FF 14 Aug 85, Cameroon Airlines; Douala, Cameroon:

The aircraft was on a flight from Cotonou, Benin and crashed in darkness in a steep dive about three miles (4.8 km) short of the runway in a mangrove swamp. The crew members had reportedly aborted the first approach due to landing gear problems and they were on a go-around from their second approach when the accident occurred. It appeared that the No.2 engine was operating at high power while the No.1 engine was not developing power. Four of the six crew members and 68 of the 72 passengers were killed.

29 Feb 1996; OB-1451, 737-200, 19072/86, FF 26 Oct 68, Faucett Airlines; Arequipa, Peru:

The aircraft was on a scheduled domestic night flight from Lima to Arequipa. When doing a VOR/DME approach to Runway 09, the aircraft crashed into a hillside at 3 miles out, at an altitude of 8015ft - almost 400ft below the airfield elevation of 8404ft. It appeared that the pilot reported flying at 9500ft, but was actually at 8644ft. Visibility was given as between 2000 and 4000m and the FDR showed that the aircraft had been well below the published approach path for some distance before impact. All 117 passengers and six crew members were killed.

3 Apr 1996; 73-1149, 737-200/CT-43A, 20696/347, FF 27 Mar 74, U.S. Air Force; nr Dubrovnik, Croatia:

The aircraft crashed into a hill at 2300ft while making an NDB approach to Runway 12 in IMC conditions. It was 1.7nm left of the extended centerline and 1.8nm North of the runway, at a speed of 133kts and a 118 degree right bank. It appeared that the aircraft had strayed off course because the aircraft flew a 110 bearing instead of 119, after passing the KLP beacon (final approach fix). Weather at the time was 8km in rain; wind 120/12kts; cloud base 120m broken and 600m overcast; temp 12C. All 6 crew members and 29 passengers were killed, including the U.S. Secretary of Commerce, Ron Brown.

2 Aug 1996; 7T-VED, 737-200C Adv, 20650/311, FF 24 Oct 72, Air Algerie; Tiemcen, Algeria:

Take-off aborted due to difference in engine N1 readings. The aircraft overran by approx 40m. The nose-gear collapsed. No fatalities.

14 Feb 1997; PP-CJO, 737-200 Adv, 21013/393, FF 10 Jan 75, Varig; Carajas, Brazil:

The aircraft touched down at Carajas' Runway 10 in bad weather (thunderstorm, bad visibility) following a VOR approach. The right main-gear collapsed rearwards, causing the plane to veer off the right side of the runway, 700m from the point of touchdown. The aircraft ended up in a forest. The First Officer was the only fatality.

8 May 1997; B-2925, 737-300, 27288/2577, FF 28 Jan 94, China Southern Airlines; Shenzhen, China:

The aircraft made a heavy landing at night during a rainstorm at Shenzhen-Huangtian airport, pushing the nose-gear up into the fuselage. The crew performed a go-around and tried to land 9 minutes later. After the 2nd touchdown the aircraft broke up in three pieces. It veered off the runway and caught fire. Two of the nine crew members and 33 of the 65 passengers were killed.

3 Aug 1997; TU-TAV, 737-200C, 19848/157, FF 15 Apr 69, Air Afrique; Douala, Cameroon:

The take-off from Runway 30 was abandoned following a tyre burst at 110kts. The aircraft skidded off the runway into shrubs and was engulfed in smoke when coming to rest 130m past the runway. No fatalities.

6 Sep 1997; HZ-AGM, 737-200 Adv, 21282/476, FF 20 Oct 76, Saudia; Najran, Saudi Arabia:

At 95kts on the take-off run, the crew noted that the No2 engine thrust suddenly increased. At 120kts the EGT warning light illuminated. The Captain attempted to reduce thrust of the No2 engine but was unable to do so; he aborted the take-off. The thrust reversers did not deploy and the aircraft overran the runway. The gear and No2 engine detached as the aircraft ground looped and caught fire. No fatalities.

19 Dec 1997; 9V-TRF, 737-300, 28556/2851, FF 27 Jan 97, Silk Air; near Palembang, Indonesia:

SilkAir Flight 185 was en-route to Singapore from Jakarta. While cruising at FL350, the aircraft disappeared from radar screens and was seen crashing nose-down into the river bed of the River Musi. One of the wings is understood to have broken off during the dive. The FDR was retrieved 27 December and the CVR January 4, both were buried in the mud of the river bed. Sections of the aircraft's empennage were found on land, away from the main wreckage. Investigators found more than 20 screws missing on the top and bottom of the right-hand horizontal stabilizer where the leading edge attaches to the front spar. It appeared that the fasteners were never installed. The stabilizer may have separated in flight, causing the plane to lose control. The FAA issued an AD January 8, 1998 requiring operators of Boeing 737s to check the horizontal stabilizers to make sure that all fasteners and elevator attachment fitting bolts are properly in place. Due to other circumstances, many people also suspect suicide by the Captain. All 7 crew members and 97 passengers were killed.

2 Feb 1998; N737RD, 737-200, 20365/220, FF 24 Nov 69, Ram Air Sales; Miami:

Damaged beyond economical repair by a 104mph tornado whilst on ground at Miami airport. No fatalities.

26 Feb 1998; YU-ANU, 737-200 Adv, 24139/1530, FF 17 Mar 88, Chanchangi Airlines; Lagos, Nigeria:

Burst a tyre on landing during circuit training. Aircraft taxied to stand and parking brake was set. Brakes subsequently caught fire and the aircraft was written off. No fatalities.

12 Apr 1998; P4-NEN, 737-200 Adv, 20925/373, FF 9 Sep 74, Orient Eagle Airways; Almaty, Kazakstan:

Overran runway at 80kts after a heavy landing on a wet runway. No fatalities.

5 May 1998; FAP-351, 737-200 Adv, 23041/962, FF 25 Apr 83, Peruvian AF; near Andoas, Peru:

The aircraft crashed near the Andoas airport during an NDB approach, in a rainstorm, after a flight from Iquitos. The aircraft had been leased by Occidental Petroleum from the Peruvian Air Force in order to ferry its workers to the Andoas area. Five of the seven crew members and 69 of the 80 passengers were killed.

16 Jul 1998; ST-AFL, 737-200 Adv, 21170/430, FF 11 Sep 75, Sudan Airways; Khartoum, Sudan:

The Boeing 737 suffered hydraulic problems shortly after takeoff. The crew elected to return to Khartoum. Upon landing one of the tires burst. The crew, hearing the bang thought it was an engine malfunction and deactivated the thrust-reversers. The 737 overran the runway and came to rest in a ditch. No fatalities.

16 Sep 1998; N20643, 737-500, 28904/2933, FF 12 Sep 97, Continental Airlines; Ghadalahara, Mexico:

After executing a missed approach on their first ILS approach to Runway 28, the flight was vectored for a second approach to the same runway. The second approach was reported by both pilots to be uneventful; however, after touchdown, the aircraft drifted to the left side of the runway. The left main landing gear exited the hard surface of the runway approximately 2,700 feet from the landing threshold and eventually all 3 landing gears exited the 197 foot wide asphalt runway. The First Officer, who was flying the airplane, stated that he never felt any anti-skid cycling during the landing roll and did not feel any "radical braking" which was expected with the autobrake 3 setting.

The airplane's nose landing gear collapsed resulting in significant structural damage. A total of 15 runway lights on the southern edge of runway 28 were found either sheared or knocked down. There were no injuries.

The tower operator reported that intermittent heavy rain showers accompanied with downdrafts and strong winds associated with a thunderstorm northeast of the airport prevailed throughout the area at the time of the accident. The two transport category airplanes that landed prior to Continental flight 475 reported windshear on final approach. The winds issued to Continental 475 by the tower while on short final were from 360 degrees at 20kts, gusting to 40kts.

1 Nov 1998; EI-CJW, 737-200 Adv, 21355/493, FF 10 Jun 77, Air Tran Airways; Atlanta, USA:

Shortly after takeoff from Runway 8R, the crew reported a hydraulic problem and declared an emergency. The aircraft was vectored to a visual approach and landing on Runway 9L. During the landing, while decelerating through 100 knots, the aircraft's steering system failed due to hydraulic pressure, and the aircraft departed the left side of the runway, coming to rest between Runway 9L and Taxiway L. Mechanical failure. No injuries.

1 Jan 1999; 9Q-CNK, 737-200C Adv, 20795/348, FF 29 Mar 74, LAC; Kilimanjaro, Tanzania:

Damaged after landing with engine failure at Kilimanjaro Airport. No injuries. Aircraft ferried to Harare for D check, but considered damaged beyond economical repair.

4 Mar 1999; F-GBYA, 737-200 Adv, 23000/930, FF 27 Nov 82, Air France; Biarritz, France:

The aircraft was approaching Runway 27 at night with reported wind 280/15G30kt and turbulence. The approach was stable but a gust from the right was encountered below 100ft RA. The aircraft touched down softly (1.1G) 5m left of centreline, with its left main gear which delayed deployment of the spoilers by two seconds until the right MLG touched down. Directional control was lost and the aircraft ran off the side of Runway 27. The nose-gear dug into the ground and collapsed. No fatalities.

6 Apr 1999; TC-JEP, 737-400, 25378/2732, FF 9 Jun 95, Turkish Airlines; Ceyhan, Turkey:

The Boeing 737 departed Adana at 00.36h for a ferry flight to Jeddah to pick up Turkish pilgrims. Weather was poor when the plane crashed, nine minutes after takeoff. All 6 occupants were killed.

The accident report concluded that severe weather conditions probably contributed to the cause of the accident; The pitot static anti-ice system was probably not activated during preparations for flight because of missed checklist items; The crew failed to recognize the cause of an erratic airspeed indication; The crew failed to use other cockpit indications for control and recovery of the airplane; The presence of cabin crew in the cockpit probably distracted the attention of the cockpit crew..

10 May 1999; XC-IJI, 737-200, 20127/144, FF 8 Mar 69, Mexican Air Force; Loma Bonita, Mexico:

The aircraft was on a training flight when it overran the runway. The nose-gear collapsed. Grass near the aircraft caught fire causing the airplane to burn out. No fatalities.

17 May 1999; CC-CYR, 737-200, 20195/205, FF 9 Sep 69, Ladeco Airlines; Santiago de Chile:

One of the fuel tanks burst open during fuelling of the aircraft. The aircraft was damaged beyond repair.

9 Jun 1999; B-2525, 737-300, 24918/2087, FF 2 Jul 91, China Southern Airlines; Zhanjiang, China:

The aircraft skidded off the runway at Zhanjiang Airport, probably due to the bad weather and heavy rain. The landing gear collapsed after exiting the runway. No fatalities.

31 Aug 1999; LV-WRZ, 737-200C, 20389/251, FF 14 Apr 70, LAPA; Buenos Aires, Argentina:

The aircraft settled back onto the runway just after take-off, overran the runway, hit two cars on a nearby road, and caught fire. The crew had not selected take-off flap and had continued the takeoff despite the take-off configuration warning horn sounding for the entire 37 second take-off run. There were 65 fatalities among the 98 passengers and five crew members. Two of the occupants in the cars were also killed.

5 Mar 2000; N668SW, 737-300, 23060/1069, FF 21 Dec 84, Southwest Airlines; Burbank, USA:

Aircraft had been held high by ATC and eventually touched down at 181kts deep into an 1840m runway. The aircraft overran and went through the perimeter fence at 32kt coming to a halt at a petrol station.
The NTSB determined that the probable cause of the accident was the flight crew's excessive airspeed and flight path angle during the approach and landing at Burbank, California. The Board also attributed the cause of the accident to the crew's failure to abort the approach when stabilized approach criteria were not met.
Contributing to the accident was the air traffic controller's positioning of the airplane, which was too high, too fast, and too close to the runway threshold. As a result, no safe options existed for the flight crew other than a go-around manoeuvre. Furthermore, the Board found that had the flight crew applied maximum manual brakes immediately upon touchdown, the aircraft would likely have stopped before impacting the blast fence,

19 Apr 2000; RP-C3010, 737-200 Adv, 21447/508, FF 19 Jan 78, Air Philippines; nr Davao, Philippines:

The aircraft had gone around from an ILS approach onto Runway 05 at Francisco Bangoy Airport in Davao because of an aircraft on the runway. The crew requested a VOR/DME onto Runway 23 and was cleared to do so. The aircraft hit a hill on Samal Island at 570ftamsl at 7dme on the non-precision approach and was destroyed by impact forces and a post-accident fire. All 7 crew and 124 passengers were killed.

17 Jul 2000; VT-EGD, 737-200 Adv, 22280/671, FF 29 May 80, Alliance Air; Patna, India:

The aircraft was inbound to Patna. It was cleared for the VOR/DME arc to ILS Runway 25. The crew took a direct track to the intercept and ended up high on the approach. 30 seconds before impact they requested an orbit to lose height and hit the ground in the orbit. The CVR recorded the stick-shaker in the orbit suggesting that the aircraft stalled. The aircraft crashed into a residential area about 2 km from the airport. All six crew members and 49 passengers were killed. Five people on the ground were also killed.
"The court of enquiry determined that the cause of the accident was loss of control of the aircraft due to aircrew error. The crew had not followed the correct approach procedure which resulted in the aircraft being high on approach. They had kept the engines at idle thrust and allowed the airspeed to reduce to a lower than normally permissible value on approach. They then manoeuvred the aircraft with high pitch attitude and executed rapid roll reversals. This resulted in actuation of the stick shaker warning indicating an approach to stall. At this stage the crew initiated a go-around procedure instead of an approach to stall recovery procedure resulting in an actual stall of the aircraft, loss of control and subsequent impact with the ground."

3 Mar 2001; HS-TDC, 737-400, 25321/2113, FF 22 Aug 91, Thai Airways; Bangkok, Thailand:

The flight was being prepared by 5 cabin crew members and 3 ground staff members for a flight to Chiang Mai. The Thai Prime Minister was one of the 149 passengers waiting to board the plane. 27 minutes before scheduled departure time, a fire erupted in the cabin, killing a flight attendant and injuring 6 others. The fire was put out in an hour, but by then the aircraft had been gutted. Subsequent investigation discovered that the centre tanks pumps had been left running when tank was dry which caused the explosion. Accident very similar to Philippine Airlines 737-300 accident on 5 Nov 90.

4 Apr 2001; C-GDCC, 737-200F Adv, 20681/319, FF 6 Mar 73, Royal Cargo Airlines; St John's, Canada:

The weather at St John's was as follows: wind 050/35G40kts; visibility 1 statute mile in light snow and blowing snow; ceiling 400 feet overcast; temperature -1°C; dew point -2°C. The crew decided to make an ILS approach onto Runway 16 since it was the only runway with a serviceable ILS.

The aircraft touched down at 164 KIAS (Vref +27Kts), 2300 to 2500 feet beyond the threshold. Radar ground speed at touchdown was 180 knots. The wind at this point was determined to be about 050°M at 30 knots. Shortly after touchdown, the speed brakes and thrust reversers were deployed, and an engine pressure ratio (EPR) of 1.7 was reached 10 seconds after touchdown. Longitudinal deceleration was -0.37g within 1.3 seconds of touchdown, suggesting that a significant degree of effective wheel braking was achieved. With approximately 1100 feet of runway remaining, through a speed of 64kts, reverse thrust increased to about 1.97 EPR on engine 1 and 2.15 EPR on engine 2. As the aircraft approached the end of the runway, the captain attempted to steer the aircraft to the right, toward the Delta taxiway intersection. Twenty-two seconds after touchdown, the aircraft exited the departure end of the runway into deep snow. The aircraft came to rest approximately 75 feet beyond and 53 feet to the right of the runway centreline on a heading of 235°M. One engine was sheared off and one main gear was damaged. The 737 has been written off. No fatalities.

22 May 2001; C-GNWI, 737-200C Adv, 21066/413, FF 2 May 75, First Air; Yellowknife, NWT, Canada:

As the aircraft approached Yellowknife, the spoilers were armed, and the aircraft was configured for a visual approach and landing on Runway 33. The computed Vref was 128 knots, and target speed was 133 knots. While in the landing flare, the aircraft entered a higher-than-normal sink rate (1140fpm reducing to 400fpm at touchdown), and the pilot flying (the First Officer) corrected with engine power and nose-up pitch. The aircraft touched down on the main landing gear and bounced twice. While the aircraft was in the air, the captain took control and lowered the nose to minimize the bounce. The aircraft landed on its nose landing-gear, then on the main gear.

The aircraft initially touched down about 1300 feet from the approach end of Runway 33. Numerous aircraft rubber scrub marks were present in this area and did not allow for an accurate measurement. During the third touchdown on the nose landing-gear, the left nose-tire burst, leaving a shimmy-like mark on the runway. The aircraft was taxied to the ramp and shut down. The aircraft was substantially damaged. There were no reported injuries.

16 Sep 2001; PP-CJN, 737-200 Adv, 21012/392, FF 6 Jan 75, Varig; Santa Genoviva, Brazil:

Heavy landing during rain at Santa Genoveva airport. The aircraft touched the left side of the runway with the right landing gear, at 500 meters (1,640 feet) from Runway 14, and then touched with the left gear, leaving the runway right after that, when it collided with the landmark electricity boxes. The aircraft had its nose gear retracted; suffered a break in the right landing gear and thus a loss of the right engine. With this, it ended up by touching the ground with the tip of the right wing. No fatalities.

14 Jan 2002; PK-LID, 737-200, 2363/218, FF 5 Nov 69, Lion Airlines; Pekanbaru, Indonesia:

The First Officer (PF) started the take-off run. The Captain called "V1" and "ROTATE" and the FO rotated the control column to 15deg nose up. The aircraft's nose was lifted up but the aircraft did not get airborne. The FO felt the stick shaker. The Captain added power which increased the speed to V2+15 (158kts) but the aircraft still did not get airborne. The Captain aborted the take-off but the aircraft nose went down hard and opened the front left door. The aircraft veered to the right of the approach lights and stopped after hitting trees 275 meters from the end of runway.

The crew did not perform the Before Take-off Checklist properly and inadvertently tried to take-off with flap up rather than the scheduled flap 5. There was no take-off config warning because the associated CB was found to be unable to latch in. No fatalities.

16 Jan 2002; PK-GWA, 737-300, 24403/1706, FF 7 Apr 89, Garuda Indonesia; Yogyakata, Java:

The aircraft was descending from FL320 when the aircraft entered a heavy thunderstorm with both engines at flight idle. Both engines lost power while passing FL180 in heavy precipitation and turbulence. Three unsuccessful attempts were made to relight the engines and one unsuccessful attempt was made to relight the APU. The crew then decided to carry out a flapless and gear up emergency ditching in a shallow, 1 metre deep, part of the Benjawang Solo River. One stewardess was killed in the rear of the aircraft which broke off during touchdown. Similar occurrences (Boeing 737-300 double engine flameout while descending in heavy precipitation with engines at flight idle) happened May 24, 1988 and July 26, 1988. Following these incidents OMB 89-1 & AD 89-23-10 were issued to require minimum rpm of 45% and to restrict the use of autothrottle in moderate/heavy precipitation; engine modification was provided for increased capacity of water ingestion. eg spinner redesigned.

7 May 2002; SU-GBI, 737-500, 25307/2135, FF 24 Sep 91, Egyptair; Tunis, Tunisia:

Landing gear failed to extend on approach to Tunis. Crew went around but aircraft crashed into a hillside about 6km from the airport on second non-precision approach. Reports said "The control tower had lost contact with the plane a few seconds before the crash, just after a distress call from the pilot."
Weather was foggy and rainy at the time, with a sandy wind, called the "Khamsin," blowing from the Sahara desert.
METAR 16.00Z DTTA 071600Z 09023G36KT 5000 SCT012 FEW023CB BKN026 18/18 Q1000 RERA RMK CB/NE/E/SE

26 Jan 2003; PP-SPJ, 737-200 Adv, 21236/461, FF 26 May 76, Vasp; Rio Branco, Brazil:

The First Officer (PF) was approaching Runway 06 at Rio Branco Airport when it flew into fog. The crew continued the approach below MDA and collided initially with a tree and then touched down 100 meters short of the threshold. At the time of impact, both engines cut out and the Captain took control. The aircraft skidded some 600 meters and came to rest on the taxiway. The undercarriage was torn off and the aircraft struck several small trees causing damage to both engines, forward & mid fuselage and horizontal stabilizers. No injuries.

6 Mar 2003; 7T-VEZ, 737-200 Adv, 22700/885, FF 9 Jun 82, Air Algerie; Tamanrasset, Algeria:

The aircraft took-off at 150kt and the First Officer (PF) called for the undercarriage to be raised (but this was not done). Immediately after rotation, the first stage of the No1 engine (JT8D-17A) HP turbine suffered a major uncontained failure. Several seconds after the left engine failure, there was a significant unexplained power reduction on the No2 engine, and the Captain took control. He maintained the same rate of climb, but the speed decreased toward the stall and the aircraft descended, generating a GPWS "Don't sink" alert. The aircraft was near MTOW and only reached 400ft, then veered and stalled striking the airport perimeter fence, tail-first, 600m beyond the runway. Tamanrasset was hot (23C) and high (4500ft).
The Captain came under scrutiny in the report for various aspects of his operation. Most importantly as to why he took control from the F/O (who had 5000hrs experience) 8 sec after the engine failure and allowed the speed to decay to the stall with the same rate of climb being maintained. He also did not retract the gear on the F/O's request after take-off nor did he allow her to retract the gear after he had taken control.
The report stated "The accident resulted from the loss of an engine during a critical phase of flight, from the failure to raise the landing gear after the engine failure, and from the taking of control by the Captain before he had completely identified the nature of the failure."
102 of the 103 people on board died. The sole survivor was a young soldier, seated in the last row and with seat belt unattached, who according to his statement, was ejected from the plane by the impact and escaped from the accident.

8 Jul 2003; ST-AFK, 737-200C Adv, 21169/429, FF 29 Aug 75, Sudan Airways; Port Sudan, Sudan:

15 minutes after takeoff, the Captain reported an engine failure, and that it had been shutdown. He elected to return to Port Sudan. The crew made an ILS approach to Runway 35 but went around because they were right of centreline. Power was increased and the gear retracted but the aircraft appeared to have gone out of control during the go-around. The aircraft crashed in flat wasteland about 3 miles from the airport. Night visibility was 4000m in sand. 115 of the 116 people on board died. The sole survivor, a two year old boy lost a leg in the crash.

19 Dec 2003; TR-LFZ, 737-300, 23750/1431, FF 10 Aug 87, Air Gabon; Libreville Airport, Gabon:

The aircraft had been holding for approx 30 mins before making its approach due to heavy rain. It landed on Runway 16 (3000m) but overran by approx 500m, stopping beyond the airport boundary fence. The engines and landing gear separated from the airplane. No serious injuries to the 118 passengers and 6 crewmembers on board were reported. The investigation is being conducted by the government of Gabon. The aircraft was damaged beyond repair.

03 Jan 2004; SU-ZCF, 737-300, 26283/2383, FF 9 Oct 92, Flash Airlines; Sharm el Sheikh, Egypt:

The Aircraft departed from Runway 22L at Sharm el Sheikh. The weather was night visibility 10K+, 17C, light winds. The aircraft took off, climbed normally and began a left hand turn as scheduled. But at 2000ft the turn slowly reversed to the right until at its maximum altitude of 5460ft it was banked 50 degrees. It continued to roll to 111 degrees and 43 degrees nose down, rapidly loosing height and hit the sea at 416kts, 2 minutes after takeoff. No mayday call was made.

The accident report was inconclusive. No technical fault was found with the aircraft but it is believed that the crew were not aware that the autopilot was not engaged and became disorientated.

The Egyptian operated charter flight was bound for Cairo for a crew change and then on to Paris. All 135 pax and 13 crew died.

12 Aug 2004; 3X-GCM, 737-200C Adv, 23469/1266, FF 6 Aug 86, Air Guinee Exp; Lungi, Sierra Leone

The Aircraft had just got airborne from Lungi airport in Sierra Leone's capital Freetown at 14:23 local time bound for Banjul. It crashed into a swamp three miles from the runway, leaving one wing partly submerged in mud. Eyewitness reports say that the left wing caught fire after take-off and that the left engine subsequently exploded shortly before impact. About 50 passengers were treated in hospital for shock and minor injuries but all 126 survived.

28 Nov 2004; PH-BTC, 737-400, 25424/2200, FF 9 Jan 92, KLM; Barcelona Airport, Spain:

The aircraft had a bird-strike in the area of the nose landing gear just before rotation from Amsterdam. The crew raised the gear and had no abnormal indications in the flight deck so they continued to Barcelona (their intended destination). During the landing roll the airplane started deviating to the left. The crew applied right rudder, braking and nose wheel steering tiller but could not keep the aircraft on the runway. It left the runway at around 100 knots, hit some obstacles from building works and suffered major damage. An emergency evacuation was carried out with minor injuries to some passengers. There was no fire but the airplane was subsequently declared a Hull loss.

Examination of the aircraft revealed that the cables and pulleys of the Nose Wheel Steering (NWS) system sustained damage from the bird-strike, resulting in a left steering command after nose wheel touch down. (NWSB cable was found broken, and NWSA cable was found jammed in a pulley on the nose landing gear.) This resulted in loss of directional control after the rudder was no longer aerodynamically effective during landing rollout. Also, the cables were severely worn in the trunnion seal area.

The FCTM now contains the following advice: "Aggressive differential braking and/or use of asymmetrical reverse thrust, in addition to other control inputs, may be required to maintain directional control."

04 Jan 2005; PK-YGM, 737-200C, 20206/249, FF 7 Apr 70, Tri-MG; Banda Aceh, Indonesia:

The aircraft was making a normal landing at Banda Aceh airport when it hit a water buffalo that had strayed onto the runway. The port main gear collapsed and the port engine and landing gear were badly damaged. The Republic of Singapore Air Force was called in to use their Chinooks to "float" the aircraft off the runway using airbags.

There were no injuries and the aircraft was declared a write off and scrapped several months later.

03 Feb 2005; EX-037, 737-200, 22075/630, FF 4 Jan 80, Kam Air; Near Kabul, Afghanistan:

The crew were approaching Kabul which was in a snowstorm. They were told to expect a VOR/DME approach to Runway 29, descend to FL130 and maintain VFR. Three minutes later it disappeared from radar. The aircraft struck a ridgeline 50ft below the crest of the mountain ridge at 9600ft. The FDR did not record and the CVR was never found. All 104 pax and 8 crew died.

14 Aug 2005; 5B-DBY, 737-300, 29099/2982, FF 27 Dec 97, Helios Airways; Grammatiko, Greece:

The aircraft departed Larnaca at 06:07 GMT for Athens. As the aircraft climbed through 16,000 ft, the Captain contacted the company Operations Centre and reported a Take-off Configuration Warning and an Equipment Cooling system problem. At 06:26 the crew said that they had solved the problem and requested a climb to 34,000ft. Radio contact was lost with the aircraft at approximately 06:37, 30 minutes after its departure, although it did squawk 7700. Greek F16s intercepted the aircraft at 07:20 and reported that the Captain was not visible and that the F/O appeared to be slumped over the controls. Two mayday calls were recorded on the CVR at 08:54, some reports say that a cabin crew member with a PPL licence was in the flight deck at impact. The aircraft crashed into mountains at 09:03 GMT after running out of fuel approx 19NM North of LGAV near the village of Grammatiko, the passenger oxygen masks had deployed. All 115 pax and 6 crew died.

The final report issued 10 Oct 2006 states: "The direct causes were:

1) Non-recognition that the cabin pressurization mode selector was in the MAN (manual) position during the performance of the Pre-flight procedure, the Before Start checklist and the After Takeoff checklist.

2) Non-identification of the warnings and the reasons for the activation of the warnings (Cabin Altitude Warning Horn, Passenger Oxygen Masks Deployment indication, Master Caution).

3) Incapacitation of the flight crew due to hypoxia, resulting in the continuation of the flight via the flight management computer and the autopilot, depletion of the fuel and engine flameout, and the impact of the aircraft with the ground.".

It also acknowledges that a contributory cause was the "omission of returning the cabin pressurization mode selector to the AUTO position after non-scheduled maintenance on the aircraft."

24 Aug 2005; OB-1809P, 737-200 Adv, 22580/787, FF 4 Aug 81, TANS; Near Pucallpa, Peru:

The aircraft was approaching Pucallpa's runway 02 in a storm. The F/O was under training and had only 60h on type and the covering F/O was in the cabin rather than on the jump-seat. The aircraft descended to 987ft AGL and entered intense hail bombardment that caused the crew to lose situational awareness. Shortly after the crew disengaged the autopilot, the 737 entered a sharp descent exceeding 1,700ft/min and struck terrain 34s later, 3.8nm (7km) from the Pucallpa VOR, travelling 1,500m through trees. The accident killed 40 of the 98 persons on board. Peru's accident investigation board, the CIAA, attributed the crash to the crew's decision to continue the approach despite a non-stabilised approach. A "lack of airmanship", the absence of the assigned first officer, and a failure to adhere to standard operating procedures were contributing factors.

05 Sep 2005; PK-RIM, 737-200 Adv, 22136/783, FF 20 Jul 81, Mandala; Medan, Indonesia:

The aircraft crashed shortly after take-off, 500m beyond the runway, into a residential area. Eyewitness reports of a bang followed by shaking and the aircraft tilting before impact would be consistent with an engine failure after take-off. Investigators have found a damaged fan blade, The crash killed 102 of the 117 people on board and at least 47 on the ground.

9 Oct 2005; VT-SID, 737-400, 24705/1971, FF 10 Dec 90, Air Sahara; Mumbai, India:

The aircraft sustained substantial damage after it overran runway 27 at Mumbai landing in poor weather. The gear collapsed and the aircraft became stuck in mud from which it took 2 days to move. No fatalities.

22 Oct 2005; 5N-BFN, 737-200 Adv, 22734/818, FF 13 Nov 81, Bellview Airlines; Lagos, Nigeria:

The aircraft took off from Lagos airport at 2045 local time in heavy thunderstorms. Three minutes later the aircraft made a distress call and was lost from radar. The wreckage was found in Lissa about 20 miles northwest of Lagos the following day.

Fidelis Onyeyiri, director general of the Nigerian Civil Aviation Authority said, 24 hours after the crash, "Our preliminary appraisal suggests that the aircraft might have started stalling after passing flight level 130, lost control, then nosedived into the ground and created a huge crater into which it disappeared." Eyewitness reports say the aircraft exploded before impact. By 13 Nov the FDR & CVR had not been recovered. There were several high ranking officials on board and accident investigators are not ruling out sabotage. All 111 passengers and 6 crew were killed.

15 Jun 2006; OO-TND, 737-300F, 23515/1355, FF 27 Feb 87, TNT; Birmingham, UK:

After an uneventful cargo flight from Liege, Belgium, with two flight crew on board, the aircraft entered a holding pattern, as the weather at its planned destination of Stansted precluded making an approach. Approximately 30 minutes later, the commander initiated a diversion to Nottingham East Midlands airport, where the weather conditions required the crew to plan and conduct a Category IIIA approach to Runway 27. In the late stages of this approach, the autopilot momentarily disengaged and re-engaged, and the aircraft deviated from both the glideslope and localiser. It landed heavily on a grass area to the left of the runway threshold, whereupon the right main landing gear detached from the aircraft. After scraping the right engine, outer flap track fairing and right wing tip on the ground, the aircraft became airborne again and made an emergency diversion to Birmingham Airport. The aircraft landed on Runway 33 on its nose and left landing gears, and the right engine. There were no injuries or fire.

30 Sep 2006; PR-GTD, 737-800SFP, 34653/2039, FF 22 Aug 06, GOL; Brazil:

The 737 collided with an Embraer "Legacy" executive jet at 37,000ft whilst en-route from Manaus to Brasilia. The Legacy landed safely but all 149 passengers and 6 crew aboard the 737 were killed. The 18 day old 737-800SFP had only flown 234 hours since new. Initial reports suggest that ATC believed the Legacy to have been at FL360, the correct FL for that direction. Preliminary findings in the ongoing investigation indicate that, for reasons yet to be determined, the TCAS in the Legacy was not functioning at the time of the accident, thereby disabling the system's ability to detect and be detected by conflicting traffic. In addition, CVR data indicates that the flight crew was unaware that it was not functioning until after the accident.

29 Oct 2006; 5N-BFK, 737-200Adv, 22891/988, FF 27 Sep 83, ADC; Abuja, Nigeria:

The aircraft crashed shortly after taking-off from Abuja in a rainstorm. It came to rest about 1 mile from the end of the runway in several pieces. 96 of the 105 passengers & crew were killed.

24 Dec 2006; PK-LIJ, 737-400, 24682/1824, FF 15 Feb 90, Lion Air; Makassar, Indonesia:

On arrival, the aircraft was executing a visual approach via a left downwind pattern to runway 31. The crew manoeuvred in a close pattern due to proximity of terrain on under the base and final approach areas of runway 31. The weather was 10 km, calm, SKC, 29C. According to a written report from the crew, when they selected the flaps from 15 deg. to 30 deg. on final, they observed that the flaps indicator indicated an asymmetrical condition. The crew re-selected the flaps back to 15 and elected to continue approach and landing. The report said that the crew referred to the QRH for the situation and they also checked the actual landing distance for flaps 15 landing configuration. The captain was PF. According to ground witness reports, on landing the aircraft was not on centerline, it bounced twice, and swerved down the runway. The aircraft came to rest beyond the runway in the overrun area. The passengers were evacuated with no reports of injuries. According to a report from the local airport authority, the aircraft sustained substantial damage; the right main landing gear was detached, the left main gear protruded through the left wing structure, and some fuselage skin was wrinkled. There was a significant ground scar on the runway surface. The aircraft was damaged beyond repair.

1 Jan 2007; PK-KKW, 737-400, 24070/1655, FF 11 Jan 89, Adam Air; Indonesia:

The aircraft was in the cruise at FL350 in marginal VMC conditions when it developed an IRS fault. The crew were preoccupied with the IRS malfunction for the last 13mins of the flight and subsequently lost control of the aircraft after one of the IRSs was switched to ATT. The aircraft reached 100deg bank, 60deg nose down and 3.5g; it broke up at 490kts in the dive. The aircraft had a history of "154 recurring defects directly and indirectly relating to the IRS between Oct and Dec 2006". All 96 passengers and 6 crew perished.

13 Jan 2007; PK-RPX, 737-200C, 20256/238, FF 4 Feb 70, RPX Airlines; Kuching, Malaysia:

The aircraft touched down 20-30m short of the runway whilst landing in fog. It skidded along the side of the runway for a further 1,000m during which time the undercarriage and No2 engine detached. No fatalities.

21 Feb 2007; PK-KKV, 737-300, 27284/2606, FF 18 Apr 94, Adam Air; Surabaya, Indonesia:

Aircraft damaged beyond economical repair after a 5g landing following an unstable approach. The aft fuselage bent down several degrees causing the fuselage to crack aft of the wings. The undercarriage remained intact. No fatalities.

7 Mar 2007; PK-GZC, 737-400, 25664/2393, FF 5 Nov 92, Garuda; Yogyakata, Java:

After a high energy, unstabilized approach, the aircraft landed at a speed of 221kts with flap 5 set. It bounced twice, snapping off the nosegear and broke through the perimeter fence where it was destroyed by impact forces and subsequent fire. There were 21 fatalities and 50 serious injuries among the 133 people on board.

5 May 2007; 5Y-KYA, 737-800, 35069/2079, FF 9 Oct 06, Douala, Cameroon:

The aircraft took off from Douala, Cameroon at 00:05L in a rainstorm after waiting an hour for a thunderstorm to pass. At 1500ft the Captain called for the autopilot to be engaged but this did not happen. The aircraft slowly rolled rolled 35 degrees to the right when the crew noticed and rolled further right and pulled back. The aircraft reached 118 degrees AoB right and impacted at 48 degrees nose up and 3.75G. It is believed that the crew were not aware that the autopilot was not engaged and became disorientated. All 114 passengers & crew killed.

28 Jun 2007; D2-TBP, 737-200Adv, 23220/1084, FF 28 Jan 85, M'banza Congo, Angola:

Overran runway on landing and hit a building. Nose section detached from the fuselage and came to rest about 50 meters away from the rest of the aircraft. 5 killed.

20 Aug 2007; B-18616, 737-800, 30175/1182, FF 11 Jul 02, Okinawa, Japan:

The aircraft was destroyed by fire shortly after docking. The cause was a detached bolt from the slat track down-stop which punctured the fuel tank near the #1 engine. According to the JTSB the washer that should have held the nut in place probably detached during maintenance 6 weeks before the incident. The design of the nut has since been changed to limit the likelihood of detachment. All 157 pax & 8 crew evacuated safely with no fatalities.

24 Aug 2008; EX-009, 737-200Adv, 22088/676, FF 16 Jun 80, Itek Air; Bishkek, Kyrgyzstan:

Iran Aseman Airlines flight 6875 from Bishkek Manas, Kyrgyzstan to Tehran Iran was carried out by Itek Air. The aircraft was returning to Manas on a visual approach after developing a pressurisation fault shortly after takeoff. Preliminary FDR information suggests that the first approach was too fast (250kts at 6.5d and 185kts at 2.5d at 400m height). The crew made a left orbit with 30 degree bank and speed continued to decrease to 155kts. After a minute the plane was downwind at very low altitude. The plane contacted the ground 7.5km from Runway 08, with gear down and flaps 15. According to preliminary analysis of FDR and accident site there were no signs of a technical malfunction except for the original pressurization fault. 65 of the 90 persons on board died.

27 Aug 2008; PK-CJG, 737-200Adv, 23320/1120, FF 23 May 85, Sriwijaya Air; Jambi, Indonesia:

The airplane had been landing towards northwest in heavy rain and marginal conditions, when the brakes failed. The airplane went about 250 meters past the runway and 3 meters below runway elevation. The right hand wing received damage, both engines and the main landing gear detached. Initial reports state that the flaps were at 15 and the thrust reversers were stowed. Jambi Sultan Taha has a single 2000 x 30m asphalt runway and no ILS. Passengers reported that the captain made an announcement before landing of a possible problem and "not to worry".

30 Aug 2008; YV102T, 737-200Adv, 21545/525, FF 6 Jul 78, ConViasa; Latacunga, Ecuador:

The 30-year old Boeing 737-200 disappeared while en route from Maiquetia to Latacunga. It was being ferried to a new owner after storage at Caracas with 3 crew on board. The aircraft struck Iliniza volcano, elevation 17,000ft. There is no evidence of poor weather in the area around Latacunga Airport at the time. All 3 occupants were killed.

13 Sep 2008; VP-BKO, 737-500, 25792/2353, FF 22 Aug 92, Aeroflot Nord; Perm, Russia:

The crew were approaching Perm at night and in cloud, with the autopilot and autothrottle disconnected when they became disorientated and lost control of the aircraft. The Russian investigators noted that the throttles had to be staggered to give equal thrust and that the crew were possibly fatigued and not used to the western artificial horizon display. All 82 passengers and 6 crew were killed.

2 Oct 2008; EI-DON, 737-300, 23812/1511, FF 10 Feb 88, KD Avia; Kaliningrad:

The crew made a go-around from its first approach due to a flap problem. The crew made several orbits whilst executing the QRH, but appear to have forgotten to lower the undercarriage for landing. No injuries, but the aircraft was an insurance write-off.

21 Dec 2008; N18611, 737-500, 27324/2621, FF 31 May 94, Continental; Denver, USA:

The aircraft started to swing to the left at approx 90kts on its take-off roll. The crew aborted the take-off and the aircraft reached 119kts before departing the left side of the runway where it caught fire. No fatalities.

25 Feb 2009; TC-JGE, 737-800, 29789/1065, FF 24 Jan 02, Turkish Airlines; Amsterdam, Netherlands:

The aircraft was making a coupled ILS to runway 18R when the Captains radio altimeter erroneously indicated ground level. The autothrottle retarded the thrust levers as though for an autoland which was not noticed by the crew. 1m40s later at 100kts the stickshaker activated and the F/O started to recover; the Captain took control but did not notice the autothrottle again retard the thrust levers. The aircraft impacted tail first and broke into 3 pieces. 9 of the 134 on board died.

29 Apr 2009; TL-ADM, 737-200Adv, 22264/753, FF 3 Apr 81, Trans Air Congo; Near Kinshasa, Congo:

The Boeing 737 was flying from Bangui, capital of Central African Republic, to Zimbabwe for a maintenance check and was carrying 2 crew and 5 mechanics when it crashed at 0600h local time around 210km (125 miles) east of Kinshasa. All 7 on board died.

22 Dec 2009; N977AN, 737-800W, 29550/1019, FF 30 Nov 01, American Airlines; Kingston, Jamaica:

The crew made an approach to Runway 12 during a rainstorm with a 14kt tailwind because the opposite runway did not have an ILS. The aircraft landed 4100ft into the 8900ft runway and bounced before settling onto the runway. It overran the runway end at 62kts and came to rest on a beach. The plane's fuselage was cracked, its right engine broke off from the impact and the left main landing gear collapsed. No fatalities.

25 Jan 2010; ET-ANB, 737-800W, 29935/1061, FF 18 Jan 02, Ethiopian Airlines; Beirut, Lebanon:

The aircraft was in a climbing turn after take-off at night. It reached 9,000ft before descending into the Mediterranean Sea. All 8 crew and 82 passengers died.

13 Apr 2010, PK-MDE, 737-300, 24660/1838, FF 16 Mar 90, Merpati; Manokwari, Indonesia.

Overran Runway 35 (2000m) while landing in rain and mist and came to a stop in a river about 170 meters past the end of the runway. The airplane received substantial damage, the fuselage broke up in at least two parts. Wx: 360/5 1000m, rain, 24C. All 97 passengers and 6 crew survived.

22 May 2010, VT-AXV, 737-800W, 36333/2481, FF 20 Dec 2007, Air India Express; Mangalore, India.

The aircraft overran the runway on landing and slid down a ravine, where it caught fire. 158 of the 166 occupants died.

02 Nov 2010, PK-LIQ, 737-400, 24911/2033, FF 08 Apr 91, Lion Air; Pontianak, Indonesia.

Overran Runway 15 (2250m) while landing in rain and high winds. It came to a stop in a field about 25 meters past the end of the runway all landing gear detached. No fatalities.

30 Jul 2011, 9Y-PBM, 737-800, 29635/2326, FF 06 Jul 07, Caribbean Airlines; Georgetown, Guyana.

Overran Runway 06 (2270m) while landing in rain. Broke through the perimeter fence, fell onto a perimeter road and broke up, the front section separating just ahead of the wing root. 162 PoB, 2 serious injuries, no fatalities.

20 Aug 2011, C-GNWN, 737-200 Combi, 21067/414, FF 08 May 75, First Air; Resolute Bay, Canada.

Crashed during approach to Resolute Bay in poor visibility and low cloud, 12 of the 15 occupants died.

20 Apr 2012, AP-BKC, 737-200Adv, 23167/1074, FF 13 Dec 84, Bhoja Air; Islamabad, Pakistan.

Crashed 2 miles short of the runway in thunderstorms, All 127 of the occupants died.

13 Apr 2013, PK-LKS, 737-800, 38728/4350, FF 5 Feb 13, Lion Air; Bali, Indonesia.

Crashed just short of the runway after descending below minimums in rain on a non-precision approach, No fatalities.

17 Nov 2013, VQ-BBN, 737-500, 24785/1882, FF 18 Jun 90, Tatarstan Airlines; Kazan, Russia.

Crashed just short of the runway in poor weather, All 50 on board are believed to have died.

8 May 2014, YA-PIB, 737-400, 26077/2425, FF 20 Jan 93, Ariana Airlines; Kabul, Afghanistan.

The aircraft overran the end of Runway 29 for about 300 meters striking ILS aerials and detaching the landing gear. There were heavy rain showers at the time of landing.

FMC Updates

The following chapter lists the features of every 737 FMC software update.

PDCS - Performance Data Computer system

Feb 1980 Onwards

Fitted to 737-1/200s from Feb 1980 onwards, this was not technically an FMC since it had no navigation functions.

The PDCS has the following AFM limitations:

- Do not use the PDCS information unless the engine configuration displayed on the PDCS is the same as the engine configuration of the airplane.

Fuel management and range calculations presented by the PDCS have not been evaluated by the FAA.

- Verify that the representative takeoff EPR limits displayed on the CDU and EPR indicators agree with the predetermined limits obtained from the flight manual.

Flight Management Computer, (FMC)

1984 Onwards

CERT - Prototype aircraft

1984+ This, logically, had no update number and was simply known as the "CERT" model, the model originally used for certification.

Update 1 – Non EFIS Aircraft

Update 1.0. Fitted to aircraft manufactured after 1 July 1985.

1) Added 22,000lb thrust engine data and related software.
2) Added pre-planned en-route holding pattern.
3) Improved message logic to eliminate nuisance messages
4) Improved waypoint bypass logic.
5) Expanded navigation database facility.
6) Improved software timing.
7) Software filter incorporated to minimise the effect and conditions that caused fuel indication or processing anomalies.

Update 1.1 - 29 Aug 1986

1) LEGS page course jump.
2) Wrong direction turn after engaging LNAV eliminated.
3) Erroneous position of created waypoints corrected by eliminating MAG VAR errors.
4) Software restarts have been further reduced.
5) FMC navigation database memory failures have been virtually eliminated by an error checking self-correcting routine.
6) Forecast en-route wind entry problem, when CRZ WIND was deleted on PERF INIT page, has been corrected.
7) Bank angle and rate in LNAV have been made much less aggressive by limiting roll rate to 2 degrees per second on initial LNAV engagement and correcting errors in the DIRECT TO mechanism. The course cut limit has also been reduced to 8 degrees to minimise S-turn manouvres.
8) Cruise to VNAV path descent transition manouvre has been smoothed out.
9) The VNAV climb to cruise transition has been improved to avoid movement of the throttle.
10) Use of manually tuned navaids and autotune has been modified to limit the use of VOR stations to 40nm to reduce VOR induced navigation errors.
11) VERIFY POSITION message occurrences should be reduced because of the 40nm restriction on use of VOR data.
12) DAA fail on sensor status page no longer latches after DAA CB is pulled out.
13) A worldwide MAGVAR model has been mechanised (same as IRS) to eliminate all course and bearing errors.
14) Wrong target speed highlighting on DES page has been corrected.
15) Hold pattern dimensions are limited to airspace restrictions and they account for navigation error.

Update 1.2 - 24 Oct 1986

1) Freeze of automatic climb derate selection at takeoff.
2) Co-located last waypoint and destination airport in a flight plan.
3) FMC performance data change due to electrical power loss or software restarts.

Update 1.3 - 14 Jun 1988

1) VERIFY POSITION message display criteria changed.
2) Revised the IRS/radio reasonable criteria to improve performance in regions with few navaids.
3) Three new messages: ENTER IRS POSITION, IRS MOTION, CYCLE IRS OFF-NAV, related to IRS alignment can be displayed when the appropriate IRUs are installed.
4) RED-TO N1 annunciation revised so that it is displayed only when the selected temperature N1 value is less than the full rated N1 value.
5) BUFFET ALERT message logic improved to minimise nuisance messages.
6) Navigation logic improved so that while in DME/DME update mode, the FMC will search for a better pair of DME stations to tune to every two minutes.
7) Bearing calculations on the FIX INFO page changed to reflect a great circle path.
8) Added suppression of wind calculation and display whenever true airspeed is less than 100kts.
9) Software restart logic revised.
10) Added suppression of radio updating whenever groundspeed is less than 100kts.
11) Improved flight plan predictions whenever there are no waypoints in cruise.
12) SPD REST field on the DES page revised to display blanks if the destination airport does not have a speed restriction.
13) Revised logic to allow the FMC to tolerate very small values of estimated wind during cruise.
14) Added the capability to define a place/bearing/distance temporary waypoint by referencing a multi-defined waypoint name.
15) Revised the display of multi-defined waypoints on the SELECT DESIRED WPT page.
16) Added common single-track database tape cartridge capability when a mixed fleet of U1.3 and U3 series of FMCs exists.

Update 1.4 - 18 Nov 1991

1) DME INHIBIT and VOR INHIBIT lines added on the REF NAV DATA page to provide for entry of up to two DMEs and two VORs that will not be used by the FMC for navigation updating.
2) IDENT page to include display of the software update version, in addition to the existing OP PROGRAM number.
3) ENGINES field on the IDENT page to display thrust rating instead of engine model number.
4) Added a POS SHIFT page to allow display of position differences between the computed FMC position and each available sensor position. In addition, crew updating of the FMC to the radio or IRS positions displayed on this page is allowed.
5) TAKEOFF REF page 2 added to allow selection of fixed takeoff derates in addition to the assumed temperature method of takeoff thrust derating.
6) Revised the data provided to the autothrottle to include the engine thrust rating instead of the engine model number.
7) Incorporated the performance option for full descent path predictions using high (flight) idle engine thrust.
8) Added capability to select alternate fixed climb derates.
9) Navigation program reduced maximum range for VOR updating to 25 nautical miles.
10) Autotune logic increased maximum allowable range to a DME station to 200 nautical miles, independent of altitude.
11) Navigation program reduced probability of radio update contamination from misplaced or excessively biased DMEs.
12) Navaid pair selection logic changed to include range to visual horizon in addition to crossing angle.
13) Navigation program changed to revert to flying a constant offset from the reference IRS position during periods of IRS only navigation.
14) Navigation program revised to prevent navigation to an excessive drifting IRS.
15) IRS NAV ONLY message logic revised to cause message display after two minutes of operation without radio navigation updating after an IRS position shift or an FMC navigation program reset has occurred.
16) FMC is now compatible with a universal format navigation data base.
17) Navigation filter constants revised to reduce the effect of VOR/DME radio updating on the filter position computation.
18) Navigation receiver selection improved performance when only one receiver is autotuned.
19) Navigation program eliminated one dimensional DME updating.
20) Navigation program eliminated the Service Volume limits for autotuned navaids.
21) Added a velocity divergence test to determine if automatic re–initialization of FMC navigation is required.

Update 1.5 - 19 May 1992

1) A wind modelling anomaly is corrected.
2) An anomaly in the nav data base unpacking routine is corrected.

Update 1.6 - 30 Mar 1995

1) FMC restarts, due to the FMC navaid search algorithm in a high density of navigation aids, eliminated.
2) APPROACH REF page now displays ILS frequencies to two decimal places.

Update 2 - Never existed

Update 3 – EFIS Aircraft

Update 3 was developed to provide an FMC interface for the new EFIS equipped 737-300 and/or an Alternate Navigation System. Subsequently they were certified for use on aircraft with electro-mechanical flight instruments.

Update 3.0 - 14 Jun 1988

1) New EFIS interface which allows EFIS equipment to provide mode and range logic to the FMCS which provides background and dynamic data to the EFIS to generate a map display.
2) ANCDU interface feature.
3) 192k database memory (optional)
4) EFIS BITE.
5) Along track offsets
6) Crew entered course for intercept.
7) Activate prompt on LEGS page.
8) Bank angle and roll rate limits.
9) Distance to go in 1/10 nm.
10) Course path terminators.
11) Descent forecast winds
12) Thrust cutback (optional)
13) Flight path angle on approach.
14) Vertical deviation HI/LO display.
15) Deletable step altitude.
16) Flight path angle independent of vertical bearing for display.
17) Predicted speed with altitude constraints.
18) Disallow step altitude equal to cruise altitude.
19) Thrust anticipation at CLB-CRZ transition.
20) Pitch/thrust coordination.
21) Calculation of QRH speeds (optional).
22) Supplemental nav data base.
23) VOR navaid inhibit.
24) VOR update mode inhibit (optional).
25) Gate positions.
26) Localiser updating.
27) Scanning DME (optional).
28) VOR only navaids in nav database as navaids.
29) VOR-DME minimum range criteria.
30) Duplicate frequencies on autotune candidate station list.
31) Cruise wind propagation.
32) VOR-DME updating range limit reduction.
33) Extended service volume for DME updating.
34) FMC position update on runway.
35) TOGA switch activated position update at runway intersection (optional).
36) Interface for clock date input.
37) New "USING RSV FUEL" message.
38) C/F degree program (optional).
39) Flight number entry (optional).
40) ISA deviation entry on DES FORECAST page.
41) Numeric entry of runway.
42) Revised runway entry procedure.
43) Eliminate ERASE on FIX page.
44) FIX page header.
45) Runway length in feet and meters.
46) Slash key rule change for Mach/CAS.
47) Step descent display.
48) Selected Reference Point (SRP) radial/distance.
49) Fuel filter improvement.
50) Multiple holds.
51) Non-ICAO airports.
52) Numbered procedures/transitions.
53) Aspirated TAT probe (optional).
54) Software restarts.
55) Nav database correction table.
56) Secondary fault storage.
57) Analogue discretes - BITE test.
58) Scanning DME BITE test.
59) DAA failures.

Update 3.1 - 14 Jun 1988 (Same date as U1.3)

This was the EFIS version of U1.3 and had all the same improvements plus the following extras:
1) Criteria for switching from one IRU to the other revised to improve performance in regions with few navaids.
2) ALTITUDE CONSTRAINT message logic modified so the message is cleared when the set logic is no longer true.
3) NO ACTIVE ROUTE message deleted. The DIR/INTC CDU key now active whenever an origin airport has been entered even though the route may not be active.
4) EFIS FIXED OUTPUTS information in BITE now provided on two pages instead of three.
5) Display of the INTC CRS for an INTC LEG entry in line 6R of the RTE LEGS page has been revised.
6) EHSI display of MCP selected course radials can be inhibited for manually tuned VOR stations.
7) The maximum enterable gross weight limit changed from 135,000 pounds to 150,000 pounds.
8) Direct-to logic modified to allow a direct-to a hold exit waypoint.

9) SEL TEMP logic modified so that the temperature format that is displayed on the optional TAKEOFF page is in the format specified for entries.
10) CDU and EADI display of V1, VR and VREF are reset when a new gross weight or ZFW are entered.
11) The ability for the pilot to inhibit two DME stations identified as not usable for navigation has been provided. The inhibit can be cleared manually at any time or automatically on flight completion.
12) A 250kt minimum descent speed has been provided for the CAS portion of the ECON descent profile.
13) REF NAV DATA page revised to suppress the runway header and data field.

Update 4

Update 4 was developed for the 737-400 but could also be used on the 737-300. It incorporated both hardware and software changes from U3. Where the software changes were retrofitted into U3 computers, they became known as U4.0, the hardware changes could not be retrofitted. Where the U4 was fitted in its entirety, they were know as 4.0E and was referred to as the "Operations Eagle" FMC. The hardware changes included a revised power supply & circuit cards which result in less cards in the box and a weight reduction.

Update 4.0 / 4.0E - 6 Feb 1989

1) New RTA function to give capability to designate the desired time of arrival at a flight plan waypoint.
2) New position shift function to give capability to update the FMC navigation position to either of the IRU positions or to the computed radio position by crew manual selection via the CDU.
3) IDENT page displays engine thrust values such as 23.5K or 22K pounds instead of the full engine identification.
4) Added the capability to select one of two possible fixed takeoff thrust derates available via crew selection on the CDU.
5) True North INBD CRS on the HOLD page (when appropriate), suffixed with a T.
6) Various MAG/TRUE switch bugfixes.
7) Navigation filter constraints revised to reduce the effect of VOR/DME radio updating on the filter position computation.
8) IDENT page shows the version of the FMC installed for ease of identification.
9) The range of allowable position bias, which may be applied via the runway position update feature, is limited to 10nm to avoid navigation filter problems with very large (possibly inadvertent) instantaneous position shifts.
10) FMC navigation receiver selection improved performance when only one receiver is autotuned.
11) Decreased assumed position uncertainty for RWY POS UPDATE. This should preclude erroneous radio data, shortly after takeoff, from contaminating the FMC navigation position.

Update 5

Update 5 was developed for the 737-500 but could also be used on the -300 and -400. It used 5.0 software and Ops Eagle hardware. They are available with either 96K, 192K or 288K word databases.

Update 5.0 - 1 May 1990

1) 737–500 aerodynamic data added to the performance data base.
2) 18.5K engine data added to the performance data base.
3) Added the capability to allow VNAV to remain engaged without LNAV being engaged when the airplane is in HDG SEL.
4) Added the ability to intercept a selected course with no DIR/INTC key push.
5) High speed disk loader interface capability added for loading of the nav data base to allow airlines to standardize on a single loader type and decrease the loading time for larger navigation data bases.
6) Revised the step climb algorithm to provide information relative to the first possible step climb point rather than attempting to find the optimum step point.
7) Calculated MAX ALT now cruise mode dependent and displayed on the CRUISE page. Previously, maximum cruise altitude was only displayed as a part of the "MAX ALT XXX" scratchpad message if maximum altitude exceeded.
8) Improved the calculation of engine–out maximum altitude to properly account for engine bleed configuration. Added the capability for the flight crew to select via the CDU which engine is out when selecting the engine–out mode.
9) Three new messages to support the changes to the step climb and maximum altitude calculations: "MAX MACH .XXX"/"MIN MACH .XXX" or "MAX CAS XXX"/"MIN CAS XXX" and "UNABLE MACH .XXX".
10) Added provisions for a position comparison with the GPS. An IRS with GPS is required to activate the function. When GPS is installed, the position shift function will allow moving the FMC position to the GPS position if desired by the crew.
11) PROGRESS page now display fuel at the history waypoint.
12) Modified the block speeds used for flap extension targets to account for the HGW conditions for the 737–3/400.

13) Suppressed the background data display of waypoints which appear in terminal procedures as DME distances but are otherwise unnamed (unless they are part of the procedure being flown). Waypoints can still be displayed by selection on the EFIS control panel.

14) Allow control of certain options without the need for program pins via PERF FACTORS page.

15) High engine idle descents in the FMC descent path predictions (PERF FACTORS option).

16) Various navigation improvements. These improvements will increase the percentage of time the FMC is able to achieve radio updating, speed up the return to full radio updating after obtaining a navigation update from a marginal navaid, and increase the ability to detect erroneous updating and in some cases re–initialize automatically. Additionally, the FMC will gradually washout IRS compensation after long periods without radio updating.

17) localizer activity monitor added to avoid using false zero ILS deviation input signals caused by broken wires. Also, a localizer rate monitor was added to prevent erroneous position calculation and display of position when the localizer is switched by the airport controller from front course to back course.

18) Variable cruise CG added in calculating maximum altitude (buffet margin). This has an effect on the display of max altitude messages and on the range of cruise altitude entries which are allowed when operating under CAA rules.

19) FMC time and date output to ARINC 429 buses 1, 2, 8 and 9 for use by other avionics systems.

20) Deleted the automatic creation of an 8nm final approach waypoint. This is replaced with an optional crew selection of a runway extension fix waypoint.

21) FMC to recognize glideslope intercept (GSI) waypoints which occur prior to the final approach course fix (FACF).

22) Runway position update function to include an option for entry of runway distance remaining rather than offset distance (PERF FACTORS option).

23) Runway position update function will now accept entries of airplane displacement from the runway landing threshold or distance remaining in feet as well as meters (PERF FACTORS option).

24) Incorporated the EFIS display of reference runways feature for procedures which do not include the runway as a flight plan waypoint, but nonetheless are associated with a specific runway.

25) Bugfix to eliminate CDU "lockups" associated with flight plan editing during IRS transition from ALIGN to NAV mode.

26) Bugfix to eliminate mislocated altitude waypoints resulting from entering cruise wind on the PERF page.

Update 6

Update 6 required the following modifications to get the full use from its functions:
- -4 FCC.
- An MCP with speed & alt intervention buttons.
- Additional interface wiring between the FMC, VHF comms and ACARS.

It was available with either 288K or 1 million word databases.

Update 6.0 - 10 Jun 1992

1) Speed Intervention. This allows crew to override the FMC target speed without disengaging VNAV through the use of the MCP SPD INTV on/off pushbutton, the MCP IAS/MACH select pushbutton and the MCP SPEED select knob.

2) Altitude Intervention. Using the MCP ALT select knob and the MCP ALT INTV pushbutton, the crew has the capability to:
   a. Delete the next altitude constraint while in climb or descent.
   b. Increase the cruise altitude while in climb or cruise.
   c. Initiate either a step climb or a transition to descent phase while in cruise phase.

3) Selection of the "ABEAM PTS" prompt on LEGS page 1/x prior to execution of a "Direct–To" creates a series of waypoints in the new active flight plan that correspond to the positions abeam the waypoints of the previous flight plan.

4) An eight character alphanumeric flight number can now be entered on the CDU RTE page 1/2.

5) Missed approaches are allowed in the navigation data base and are flyable by the FMC with LNAV engaged. Upon transition onto the missed approach, VNAV will disconnect (if engaged), the FMC will transition into the climb phase and CRZ ALT will be reset to the appropriate value. VNAV may then be re–engaged.

6) The FMC provides calculated distance, time, and fuel remaining for a diversion to an alternate destination. It also provides a nearest airport search function that automatically finds the five nearest airports from the aircraft and calculates the distance, time, and fuel remaining for a diversion to one of these airports.

7) A MESSAGE RECALL page on the CDU is added which allows for display of CDU messages which had previously been cleared but whose set logic is still valid.

8) The FMC provides the raw position data from each navigation sensor to the EFIS for display. The positions are displayed on the EFIS when the POS SHIFT page on the CDU and the EFIS PLAN mode on the EFIS control panel are simultaneously selected.

9) Procedure holds and procedure turns are allowed in the navigation data base and are flyable by the FMC with LNAV and VNAV engaged.

10) Two new CDU pages added; TEMP NAV SUMMARY and SUPP NAV SUMMARY to allow review of the current contents of the temporary and supplemental navigation databases.

11) Computation and display of FAA (or UKCAA) certified QRH takeoff speeds as a result of taking into account runway slope, takeoff winds, and runway condition (wet or dry)

12) Provision for one million word navigation database.

13) New PLAN FUEL entry is provided on the PERF INIT page which overrides the total fuel value in the gross weight calculation. Plan fuel is cleared after engine start.

14) Display of thrust rating in the takeoff derate prompts on TAKEOFF REF page 2/2 and in the header for the FMC calculated takeoff N1 on TAKEOFF REF page 1/2 and 2/2 is provided for more meaningful takeoff derate N1 data.

15) Runway centerline (RC) intercept waypoints are provided as part of the FMC missed approach function.

16) Adaptable FMC/ACARS Williamsburg interface which provides for receipt (uplink) and transmission (downlink) of information by the FMC from/to an airline's ground computer via the ACARS management unit (MU), VHF transceiver and the service providers network. Downlinks can be sent by the FMC manually (CDU prompts) or automatically (triggers). Downlink messages can contain any number of downlink elements chosen from a master hardcoded list. The FMC/ACARS interface is adaptable to specific customer's needs through the use of custom tables which can be obtained from Smiths Industries. The custom tables are loadable through the ARINC 615 data loader.

Update 6.1 - 10 Jun 1992

1) Bugfix reduced thrust command anomaly.

2) Bugfix ACARS uplink/downlink anomalies

3) Bugfix altitude intervention anomaly

4) Lower entry limit for maneuver margin changed from 1.15 to 1.20 for commonalty with FMCs on other Boeing types.

Update 6.2 - 10 Jun 1992

1) Bugfix FMC aircraft invalid ACARS configuration

2) Bugfix Supplemental Navigation Data Base anomaly

3) PERF INIT page now blanks the displayed fuel value with the loss of total fuel signal from the fuel summation unit.

4) Bugfix FMC automatic downlink anomaly.

5) QRH Data Error Correction.

6) Bugfix destination ETA anomaly.

7) Bugfix FMC fail anomaly.

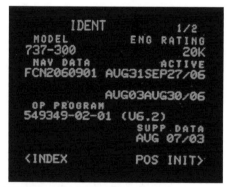

8) Revised the total fuel quantity processing as follows:
a. Added an advisory CDU scratchpad message "CHECK FMC FUEL QUANTITY" that is displayed when a sudden drop in fuel quantity greater than 1000 pounds is detected by the FMC.
b. Modified the performance downmode logic associated with detection of an invalid total fuel quantity signal to provide faster recovery from the performance downmode.
c. Incorporated logic in the FMC to ignore the total fuel quantity signal from the DAA for 20secs after a power transient.
d. Added a rate limit to the total fuel quantity signal, upstream of the total fuel quantity lag filter.

9) Logic correction for triggering runway change and destination change automatic FMC/ACARS downlinks.

10) Correction to prevent disabling of downlinks when the FMC receives a power transient after having just transmitted a no acknowledge (NAK) or a synchronize (SYNC) to the ACARS MU.

Update 7

Update 7 was a big hardware improvement on previous computers because it was a "4 modular concept unit (4MCU)" which required only half of the equipment rack space of the previous 8MCU. Additional benefits expected from the 4 MCU FMC include: improved reliability, on–board operational software programmability, reduced weight/size, and increased growth capacity. One noted feature is that the 4 MCU FMC supports on–board loadable operational flight program (OFP) software, which can be loaded via an ARINC 615 data loader using 3.5 inch 1.4 megabyte compatible data disks. It was available with either 256K or 1 million word databases.

Update 7.0 - 4 Nov 1993

1) The set logic for issuing the "VERIFY POSITION" message has been revised as follows:
The FMC radio position data differs from both the left and right inertial reference unit positions by more than 4nm for 80

secs (was 150 secs) or if the FMC radio position data differs from the FMC system position by more than 2.8nm enroute, and 1.7nm in terminal areas (was 2nm for both) for 80 seconds (was 150 secs).

2) The message "CHECK FMC FUEL QUANTITY" has been added to indicate that the FMC has detected a decrease in fuel quantity of 1500 pounds or greater for longer than 120 secs.

3) When large font takeoff speeds are displayed on the "TAKEOFF REF" page and a GW, or ZFW change has been made, the new message "VERIFY TAKEOFF SPEEDS" will be displayed.

4) The "NAV STATUS" page has been added to provide information regarding the status of all navaids being tuned. This change consolidates navaid tuning information on one page.

5) "NAV STATUS 2/2" page is renamed "NAV OPTIONS". This page provides the capability to inhibit the use of up to two each DME and VOR stations. This page also provides the capability to inhibit FMC position updates based on VOR, DME or GPS position data. Selections are automatically cleared at flight completion. This capability was previously provided on the "REF NAV DATA" page. The "REF NAV DATA" page now provides access to the "NAV OPTIONS" page.

6) The option is available to enter the flight number on the "RTE" page. The flight number will also be displayed on the "PROGRESS" page. In addition, at the bottom of the "PROGRESS" page, the radio updating modes and stations have been removed and are now located on the "NAV STATUS" page. Display of the current wind has been added to the bottom of the "PROGRESS" page.

7) "NAV SUMMARY" pages have been added to display what is contained in the "REF NAV DATA" and "SUPP NAV DATA" databases. The number of "NAV SUMMARY" pages depends on data contained in each database section.

8) The "POS REF" and "POS SHIFT" pages have been modified to allow for future display of dual FMC positions and GPS positions. Display of navigation sensor "raw data" has been added to the EHSI when the EFIS is in the "PLAN" mode and the CDU is on the "POS SHIFT" page. Capability of displaying position data from the right channel GPS, when installed has been added. Header for current radio position relative to FMC position is now displayed as "RADIO". Previous FMC updates displayed this data as either "DME/DME" or "VOR/DME". Positions of the sensor data (L/R IRS, L/R GPS) has been re–arranged to display left channel sensor data on the left side of the page and right channel sensor data on the right side of the page. Navaid station identifications and frequencies used to calculate relative position have been moved to the "NAV STATUS" page, where this information is displayed in an expanded format to show the tuning activity and FMC usage of all VOR and DME channels. GPS information is displayed only if GPS equipment is installed and GPS operation is enabled by the FMC software option code.

9) The 4 MCU hardware allows update of the FMC operational flight program via ARINC 615 portable and airborne data loaders while the FMC is installed in the airplane.

10) The Update 7.0 version has loadable performance defaults via the ARINC 615 data loader. This allows operators the capability to customize their performance factors, performance initialization and performance limit values, and then load these factors into the performance options data base using an ARINC 615 data loader. In addition, performance option codes are no longer enterable through the CDU. Option codes are now loadable through the ARINC 615 data loader. Option code disks and performance defaults disks must be obtained through Boeing.

11) Changes to FMC BITE Pages:
FMC SENSOR STATUS page added symbol generator BITE status and an elapsed time indicator "ETI HOURS" display.
FMC ANALOG DISCRETE page added the airplane model and engine thrust rating ("MODEL/ENG") to page 3/4.
FMC FLIGHT N page added date on which the first fault displayed on page 1/X was detected.

Update 7.1 - 3 Mar 1994

1) Flight crew capability for selecting parallel lateral path offsets of up to 99.9 nm, either to the left or right of the original path.

2) For dual FMC aircraft, the two identical FMCs installed each provide an independent navigation solution based on their associated onside selected navigation inputs which are then combined to yield a composite navigation solution. Each FMC; via its intersystem bus, monitors differences between itself and the other FMC with respect to flight plan data, performance data and current aircraft state.

3) For dual FMC aircraft, an FMC source select switch has been added which is located on the aft overhead (P5) panel. The source select switch has three positions, "Both on L", "Normal", and "Both on R". When in the "Normal" position, the FMCS is operating in a dual configuration. The "Both on L" and "Both on R" positions selects either the left or right FMC respectively.

4) Allows the FMC to provide path descent guidance when flying a STAR, approach, or approach transition that contain legs with vertical angles. VNAV is now operational for approach with enterable approach Vref speeds for flaps greater than 15 degrees. The associated message, "APPR VREF NOT SELECTED", has been added.

5) "END OF OFFSET" will be displayed 2 minutes prior to passing the offset leg termination.
"INVALID OFFSET" will be displayed for any flight plan mod that causes the start of offset to become non–offsetable (i.e. delete start waypoint from flight plan and current leg is not offsetable). This entry can result from an ACARS uplink.
"NO OFFSET AT LEG XXXXX" will be displayed if entry of a lateral offset start or end waypoint is for a leg that is non–offsetable.
"OFFSET ENDS ABEAM XXXXXX" will be displayed if an offset path terminates prior to returning to the original path.

6) New Dual FMC Related Messages:
"DUAL FMC OP RESTORED" will be displayed when movement of the FMC source select switch position has caused a resynchronization and dual operation is restored.
"SINGLE FMC OPERATION" will be displayed when the primary FMC has determined that the secondary FMC is not available.

7) The "OFP CROSSLOAD" page has been added to provide for the initiation and status display of a transfer of the OFP from one FMC to the other over the intersystem bus on airplanes equipped with a dual FMC installation.

8) The "DATABASE CROSSLOAD" page has been added to provide for the initiation and status display of a transfer of the various data bases from one FMC to the other over the intersystem bus on airplanes equipped with a dual FMC installation.

9) Provisions have been incorporated to accept GPS position data. Given valid GPS position data, the FMC can use this data to update its navigation solution.

10) The FMC software options (i.e. high idle descent, runway remaining) are now set by loading of the appropriate software options data base disk as required. The applicable software options data base disk can be obtained from Boeing.

11) Navigation additions:
– LOC–DME navigation position update.
– Capability to detect erroneous navigation aids used for FMC radio position determination when three or more navaids are tuned on airplanes equipped with dual scanning DMEs.
– Required Navigation Performance (RNP) and actual navigation performance (ACTUAL) displays to the "POS SHIFT" and "LEGS" pages. The displays allow the crew to make navigation accuracy appraisals based on the FMC position update history.
– Optional capability to allow manual entry of RNP.
– Capability in dual FMC installations for each FMC to develop its own navigation solution using different navaid sets.
– Capability to support VOR–DME radio updates on airplanes equipped with scanning DME.

Update 7.2 - 26 Jan 1995

1) Bugfix unexpected track and wind changes due to navaid integrity check errors
2) Bugfix downmodes to single FMC operation due to nuisance discrete miscompares
3) Addition of green circles to indicate intermediate altitude T/D
4) Addition of Missed Approach Procedures to allow LNAV/VNAV missed approaches and procedure turns/holds.
5) Single and dual FMC installations now have the capability to use full time GPS updating for position determination.
6) Improvement to FMC prediction logic to establish down route time, fuels, phase transition points, and appropriate messages.
7) Compatibility with the FANS MCDU incorporated.

Update 7.3 - 11 Oct 1995

1) Display logic for the message "UNABLE REQD NAV PERF–RNP" has been changed to allow it to remain displayed on the EHSI when the message is cleared from the CDU scratch pad.
2) The default RNP values (in nautical miles, nm) have been changed to agree with the values specified in the draft advisory circular AC120–CNS

FMC Operating Environment	Old RNP Value	New RNP Value
Oceanic	12.6 nm	12.0 nm
Enroute	2.8nm	2.0nm
Terminal	1.7nm	1.0nm
Approach	0.5nm	0.5nm

3) Display logic for "VERIFY RNP" changed to issue the message when a GPS approach is flown using a default RNP.
4) Display logic for "VERIFY RNP VALUE" changed to issue the message immediately if the manual RNP entry is smaller than the current ANP.
5) The logic to inhibit display of FMC messages associated with navigation performance has been revised to inhibit their display during approach for the following two cases:
A. Digital flight control system VOR/LOC mode active.
B. EHSI mode is VOR/ILS with procedure navaid tuned on both VHF NAV radios.
6) Allowed two holds at a waypoint given that one of the holds is in the missed approach and the other is prior to the runway.
7) Bugfix FMC software restarts due to ARINC microcontroller restarts.
8) Allow easy deletion of the active procedure turn by deleting the (INTC) point.
9) Allow a hold and a procedure turn to be entered on the same waypoint, separated by a discontinuity.
10) Add provisions to the FMC to extract NDB approaches from the NAV database.

11) Allow up to four numeric characters to be transmitted on ARINC label 261 when an alpha–numeric flight number is entered in the FMC CDU.

12) Bugfix message "TAI ON ABOVE 10 DEG C" message correlates with the outside temperature only.

13) Change to allow the more restrictive altitude constraint to be used when linking approaches and transitions.

14) Change to attempt to restart dual operation after a period of 5 minutes after a downmode to single when the FMC source select switch is set to either the "Both on L" or "Both on R" position.

15) Applies to HOLD patterns in descent mode, when an approach speed has been entered on the APPROACH REF page. The displayed active speed on the HOLD page will be changed to match the guidance target speed when flaps are set to 15 degrees or more.

16) Improved lateral path tracking when the airplane is slightly out of trim.

Update 7.4 - 9 May 1996

1) The FMC will store information related to the occurrence of an FMC software restart or downmode to single FMC operation. This information includes information regarding the airplane present state and a history of the FMC computations completed prior and at the time of the downmode or restart. The FMC will also have the capability to download its fault log information to a 3.5 inch data disk using the airborne or portable ARINC 615 data loaders.

2) Support for both 512 and 256 byte sector size memory loading.

3) The ARINC 429 microcontroller restart logic software has been improved to reduce ARINC 429 latched failed conditions.

4) Reduce the occurrence of downmodes to single FMC operation and FMC software restarts.

5) Calculate DME slant range using WGS–84 earth model coordinates in lieu of a spherical earth model. This allows consistent calculation of ANP with respect to RNP which is based on the WGS–84 earth coordinate system.

6) Allow tuning procedure navaids in areas where another navaid with the same frequency exists in the local navaid data base.

7) QNH value is now used in the cabin rate computations on the "DESCENT FORECAST" page.

Update 7.5 - 23 Feb 1998

1) Several software fixes to resolve in–service reports of downmodes from dual to single FMC operation and reports of FMC software restarts have been incorporated. These fixes are based on data provided by operators using the Update 7.4 BITE download function.

2) Additional information has been added to the data previously collected in the Update 7.4 software. The additional data is being recorded to assist in determining root causes of rare in–service FMC downmodels/restarts which could not be duplicated and resolved during lab testing based upon download data provided by Update 7.4 FMC BITE data.

3) In the event the fuel input to the FMC is not available, total fuel weight can now be manually entered by the pilot to allow restoration of the VNAV guidance function.

4) The Select Desired Waypoint page will now display up to twelve waypoints that have duplicate identifiers, in order, based upon their distance from the airplane's position.

5) The Engine–Out mode has always been advisory only with driftdown altitude and minimum drag speed calculated. However, the option to activate Engine–Out VNAV had also been displayed on earlier software versions, which when activated, resulted in loss of VNAV guidance. The Update 7.5 software has been revised to remove the option to activate the Engine–Out VNAV mode.

6) The Update 7.5 software has been revised so that it is not required to overfly the runway waypoint to provide missed approach lateral navigation guidance. The Update 7.4 FMC software required the runway waypoint to be overflown before a missed approach could be flown by the FMC.

7) The 150 knots speed restriction is now assigned to all final approach fix points for all approach types. Previously, it was only assigned for ILS approaches.

8) The default software option has been changed to allow manual entry of the RNP value. The Update 7.4 FMC default software option did not allow manual entry of the RNP value.

9) The "Navigation Database Out of Date" message would display for the first navigation data base cycle in the year 2000. The update 7.5 FMC software corrects this problem.

10) Bugfix erroneous DME updates to the FMC radio position were reported in service when one radio was manually tuned to the destination airport ILS well before arrival at the destination airport.

Update 8

Update 8 had the 4MCU hardware improvements and all the software features of U6 and U7. It was available with either 256K or 1 million word databases.

Update 8.0 - 8 Feb 1994

Released at the same time as, and has identical features to Update 7.1.

Update 8.1 - 30 Jan 1995

Released at the same time as, and has identical features to Update 7.2.

Update 8.2 - Not used

Probably omitted to put Update 8 in step with Update 7.

Update 8.3 - 16 Oct 1995

Released at the same time as, and has identical features to Update 7.3.

Update 8.4 - 9 May 1996

Released at the same time as, and has identical features to Update 7.4.

Update 9

Never existed.

Update 10

Update 10.0 was developed for the 737-NG series. However, the system design also incorporated backward compatibility to 737–3/4/5 series (both EFIS and non–EFIS). Its navigation database is from 256K to 4 million words depending upon update version.

Update 10.1 - 13 Nov 1998

1) New takeoff thrust bump feature added to allow the flight crew to select (bump to) the next higher engine thrust rating when more takeoff thrust is desired for operational reasons. The thrust bump selection is made on the N1 LIMITS page.
2) The TAKEOFF REF and N1 LIMITS pages are modified to accommodate the takeoff thrust bump feature and to provide commonality among other Boeing aircraft. Several display items have been relocated between the TAKEOFF REF pages and the N1 LIMITS page and new features have been added. The N1 LIMITS page now has an On–Ground format and an In–Air format. The following provides the detailed changes:
a. A WET SKID RESISTANT (SK–R) runway condition feature has been added to the RW COND field on page 2 of the TAKEOFF REF page. The WET SK–R runway condition is applicable to the 737–6/7/800 models, but not the 737–3/4/500 models. When the RW COND data field is displayed (ACARS AOC Datalink option enabled and/or Takeoff Speeds option enabled), the selection prompt is displayed as DRY/WET/SK–R.
b. TAKEOFF SPEEDS DELETED message is displayed when the active departure runway, runway wind, runway slope, runway heading, takeoff flaps, runway condition, takeoff outside air temperature, selected temperature, or engine thrust rating are changed.
c. A default FLAPS value is no longer required. If a valid FLAPS default value does not exist, the TAKEOFF REF page will display BOX prompts to alert the flight crew that FLAPS data is a required entry.
3) The default size of the NDB has been increased from 256K words to 1 million words of memory.
4) An overfly indicator is contained in the NDB to adjust MISSED APPROACH sequencing. If the OVERFLY indicator is set, then the runway must be flown over prior to sequencing the missed approach, If the overfly indicator in the NDB is not set, lateral guidance will be provided direct to the missed approach once active.
5) A new label has been added to the Model/Engine Data Base (MEDB), to allow different N1 Limit values to be used during a go–around.
6) The FMC MEDB has been updated to include aero and propulsion data for the 737–800 airplane.
7) The FMC VMO and MMO has been lowered to 335 and .805 to reduce the potential of entering into an overspeed condition during cruise and descent.

Update 10.2 - 13 Nov 1998

1) Navigation Data Base increased to 2 million words.
2) FMC now recognises LDA (localizer directional aid), SDF (simplified directional facility) and IGS (instrumental guidance system) approaches that are stored in the NDB.
3) FMC descent path algorithm changed to change the pitch mode from "VNAV PATH" to "LVL CHG" approximately 7 knots below Vmo/Mmo speed limits, instead of at the limit, to reduce the potential of entering into an overspeed condition.

4) Two additional FMC program pins have been reserved for potential growth in the number of model/engine configurations.

5) Colour display option on the LCD (liquid crystal display) FMC display unit.

6) Improved capability to hold target speed during large bank angle turns at high altitudes. The following related FMC warning messages have been added:
"UNABLE HOLD AIRSPACE" is an alerting message issued when the lateral predicted hold path using the bank angle limit causes protected airspace to be exceeded.
"LNAV BANK ANGLE LIMITED" is set when LNAV is engaged and the aircraft is within 5 minutes of the turn waypoint, the leg inbound to the turn is active and the geometry indicates that the airspace is violated.

7) The VSPEED UNAVAILABLE message will now be displayed if any of the independent variables used in the calculation of VSPEEDs fall outside of the tabular data set boundaries.

8) ARINC label 360 is added which contains the active flight number, the active Origin and the active Destination.

9) ARINC label 277 is added which contains the sensors being used in the FMC navigation solution. These include VOR, LOC, MLS, GPS, IRS and DME.

10) Feature added to automatically correct an altitude mismatch which may occur at the approach boundary. The situation arises when the first approach waypoint is linked with an enroute waypoint and the two have different altitude constraints.

11) TO SHIFT or RWY REMAINING features on the Takeoff Ref page are now inhibited when GPS is primary.

12) QUADRANT, RADIAL, and TGT ALT data fields added to the RTE HOLD page.

13) MEDB updated to include aero and propulsion data for the 737–600 airplane.

Update 10.2A - 18 Dec 1998

1) Bugfix to correct FMC lockup when the "DES" key is pressed quickly after executing a flight mod and prior to the FMC completing its descent calculations.

2) Bugfix to correct improper initialization after a long term power interrupt or FMC restart

Update 10.3 - 14 Aug 2000

1) Polar Navigation capability added. Also added GPS track data to the PROGRESS page to provide aircraft heading data at the poles if both IRUs fail. GPS position is used as the center point on the EFIS POS SHIFT display if both IRUs fail.

2) Geometric Path Descent (GPD) option to allow flight along a computed constant gradient path from one altitude restriction to the next during descent.

3) Engine-Out SIDs option. EOSIDs can be added to the navigation data base and can be selected on the departure page. The EOSIDs can be automatically inserted into the flight plan during climb with flaps extended when the ENG OUT prompt on the CLB page is selected.

4) Quiet Climb Option – This supports noise abatement procedures by providing a quiet climb cutback N1 target between a pre-programmed thrust reduction altitude and the restoration altitude. The default thrust reduction and restoration altitudes are provided in the MEDB and may be changed through the Loadable Performance Defaults Data Base (LDDB), manually changed at the CDU, or uplinked via ACARS.

5) VNAV may now remain engaged within the allowable 2 x RNP Crosstrack Deviation window, or when LNAV is engaged to reduce VNAV disconnects during RNP operation.

6) Planned Fuel values are now retained during test of the engine packs.

7) The runway symbol on the map remains displayed during a missed approach operation, even after it has been overflown.

8) FMC automatically transitions to the CLB page when on the TAKEOFF REF page and weight goes off wheels.

9) Revised block operating speeds for the 737-3/4/500 increased as follows:

Flap Position	Gross Weight		
	Less Than 117K Lbs	117K to 138.5K Lbs	Greater Than 138.5K Lbs
5 Degrees	180 Knots	190 Knots	200 Knots
10 Degrees	170 Knots	180 Knots	190 Knots

10) IRS NAV ONLY message is eliminated and the UNABLE REQD NAV PERF-RNP message is displayed when ANP exceeds RNP for the specified delay time.

11) VNAV will no longer be disconnected at glideslope intercept point unless the glideslope is armed.

12) CLB N1 Limit values are now displayed as well as Derate CLB N1 values.

13) Pilot defined company routes option which adds an additional data base consisting only of company routes and waypoints of up to 20 flight plans (maximum of 150 waypoints each up to a total of 1000 waypoints).

14) Various bugfixes including the following:
A. Improved dual FMC operation to reduce the number of vertical and fuel quantity miscompares between two FMCs.
C. Improved FMC restart logic to reduce the number of restarts experienced in service.

D. Improved FMC software restart logic to retain all flight plans through the first software restart when on.

E. Corrected a condition where the FMC used a derated N1 for Climb/Go-Around when an assumed temperature derated takeoff was performed.

F. Corrected a condition where the FMC would provide the wrong set-landing altitude under certain conditions.

G. Corrected a condition where a non-selected IRU could impact the FMC navigation position in a dual FMC installation.

H. Corrected a condition where the FMC would erroneously display "Database Crossload Fail" when crossloading a navigation database file greater than 1 meg between FMCs.

Update 10.4 - 14 Sep 2001

1) Increased navigation data base capacity to 3.5 Mega-Words and the MEDB capacity to 1.5 Mega-Words

2) Enhanced VNAV as follows:
   - VNAV will now remain in PATH mode (rather than revert to VNAV SPEED) when the FMC cross track error exceeds RNP during idle descent legs and computed gradient legs.
   - When VNAV Speed Intervention is exited during VNAV descent, VNAV will switch to PATH mode.
   - "VNAV DISCONNECT" message displayed if a disconnect during approach due to reversion to LVL CHG.

3) Improved GPS and IRS reasonableness tests have been added to protect the FMC navigation solution from unreasonable data. Dual FMC logic has been improved to better identify and reject a badly drifting inertial reference sensor.

4) Resize Hold Pattern for Descent Altitudes - The default leg time for a hold pattern is now based on the predicted altitude of the hold fix waypoint. When the HOLD point becomes the GO-TO waypoint, the FMC does a final check of the altitude and adjusts the hold leg time accordingly (1- minute leg time below 14,200 feet or 1.5 minutes above 14,200 feet).

5) RTE HOLD page improved. Default values on the RTE HOLD page are now displayed in small font. Large font will be used to display navigation data base or crew entered items. The RTE HOLD page will now display an "EXIT HOLD" prompt after an in-hold edit has been executed. Subsequent selection of the "EXIT HOLD" prompt will delete the new hold.

6) Addition of Data Outputs to Support Future Enhanced Systems and Displays - Several new parameters are now output from the FMC to support future CDS VSD symbology, HGS and EGPWS.

7) APPROACH REF page improved. A new FLAP/SPD field has been added to the APPROACH REF page at LSK 4R. Selection of a VREF speed will result in the selection appearing in the LSK 4R field.

8) New "RW/APP TUNE DISAGREE" or "RW/APP CRS ERROR" messages appear when a tuned frequency or selected course does not agree with the FMC flight plan.

9) New "H/W CONFIG" maintenance page. The displayed data includes hardware part number, serial number, elapsed time indicator, inter-system bus status and processor speed.

10) Bugfix to correct intermittent blanking of route data to the CDS Map display in conditions of high FMC processing.

11) Optional automatic Engine Out SID Selection - The FMC will automatically select an available Engine-Out SID and create a MOD page for the departure runway, if Engine-Out indication is received by the FMC when the airplane is in climb with flaps extended.

12) Optional display of missed approach on the CDS MAP display in cyan.

13) Optional inhibit gross weight entry on FMC CDU "PERF INIT" page, ie only entry of ZFW is allowed.

14) Optional disable takeoff fixed derates ie only assumed temperature takeoff derates are allowed.

Update 10.4A - 17 May 2002

U10.4A was simply a bugfix for a condition where the software could "lock up" if multiple ARINC BITE tests were initiated.

Update 10.5 - 10 Feb 2003

1) The Approach Ref page will now accept landing flap entries of 0, 1, 2, 5, 10 or 25 for the landing flaps, with a corresponding manual Vref entry, in addition to the standard landing flap settings (15, 30, 40) for non-normal procedures such as landing with flap asymmetries.

2) The next altitude constraint is sent to the EFIS/CDS for display on the MAP, independent of the RTE DATA button selection on the EFIS/CDS control panel making this function common with other Boeing model FMCs.

3) Updated MAGVAR (magnetic variation) model with data extrapolated to 2005 values.

4) New ANP calculated value lower limit of 0.02nm.

5) Allow entry of any runway from the destination airport as the reference fix on the FIX and DESCENT pages.

6) TACAN approaches now supported in the FMC Navigation Data Base.

7) Fuel Quantity Lag Filters disabled during on-ground operations.

8) A tropopause breakpoint of 36,000 feet is now incorporated into the planned descent path. This should reduce the tendency to over-speed during path descents that start from cruise altitudes above FL360.

9) The FMC can now switch directly from climb to descent on short flight plans without going through the cruise phase.

10) Thrust Rating Model and Engine Performance Model now supports 737-600 airplanes equipped with 7B22/B2 bump thrust rating and new 737-700 configurations with 7B26/B2 bump thrust rating.

11) Temperature and Airport Altitude are now included in the calculation of Minimum Takeoff Weight. This will primarily benefit 737-BBJ aircraft which have relatively higher Minimum Gross Weight requirements before calculated takeoff speeds are available. With the new method, 737-BBJ aircraft will have calculated takeoff speeds available at lower minimum gross weights when operating at high altitude airports and particularly on hot days.

12) Missed approach path is now predicted full time so that crews can determine turn directions and other data before actually beginning the missed approach.

13) New FPA limits based on calculated limits of Mmo/Vmo minus 5 knots when above the airport speed restriction altitude and limits of the airport speed restriction plus 10 knots for below this altitude. This gives slower acceleration rates toward limit speeds when above the descent path or when encountering unforecast tailwinds during path descents. Occurrences of OVERSPEED DISCONNECT should be reduced.

14) Optional Features:
   • Vertical RNP can be entered/reviewed in the FMC, defining the full scale deviation for the Navigation Performance Scales.
   • Upper and lower limits of a speed band in VNAV path descent. Theses limits are the speeds at which mode changes will occur, to arrest acceleration or deceleration, while flying the idle thrust VNAV path.
   • FMC can transmit Thrust Mode information in support of CDS Double Derate N1 Limits annunciations.
   • Integrated Approach Navigation (IAN) allows FMS RNAV approaches and other non-precision approaches to be flown in an ILS-look-a-like manner, with common flight crew procedures.
   • GPS Landing System (GLS). The FMC selects Local Area Augmentation (LAA) only GLS approach types if they are defined in the Navigation Database, and decodes the GLS channel/deviation input.
   • ATS Datalink. This gives the FMS the following functions:
      • ATC Datalink (ATC DL). Enables direct Controller-Pilot datalink communication via the FMC MCDU.
      • Automatic Dependent Surveillance (ADS). Provides automatic position data reporting.
      • ATS Facilities Notification (AFN). Is the "logon" function used to notify ATC of an aircraft's "address" for the purposes of establishing ATC DL and ADS communication.

15) Various bugfixes including the following:
   • CDU blanking during upload of a large number of flight plan wind data via ACARS.
   • CDS/EFIS Map data blanking when the airplane is flying slightly below path during descent.
   • FMC software restarts after flying direct-to with multiple ABEAM waypoints.
   • Marginal ARINC 429 transmitter operation in the P/N 10-62225-004 hardware.
   • Addition of 429 data labels to support ACARS operation for some operators.
   • Reduce occurrences of down modes from dual to single FMC operation by improving comparison logic.

Update 10.5A - 1 Oct 2004

Big bugfix package (SP1!) including the following:

1. A correction to address an anomaly unique to operating in the Dual FMC configuration where the FMC navigation position became inconsistent with the displayed FMC calculated ANP value resulting in subsequent display of the VERIFY POSITION message.

2. A fix to address a condition that results in improper leg sequencing from an RF leg (Fixed Radius Turn) to the next flight plan leg and may not properly sequence from an AF (DME Arc) leg to another AF Leg for AF to AF leg combinations.

3. Inhibit DME Update Default Option.

4. Bugfix a condition that would erroneously display a Multimode Receiver (MMR) GLS fault as an ILS fault.

5. ANCDU fixed outputs BITE page access.

6. Bugfix to display the MOD Hold symbol when the Hold waypoint has an active altitude constraint attached.

7. Improvements to the FMC Message logic to reduce nuisance display of the following FMC Messages: RESET MCP ALT, UNABLE CRZ ALT, THRUST REQUIRED, USING RESERVE FUEL, CHECK FMC FUEL QUANTITY,

8. Usual bugfixes to reduce occurrences of FMC software restarts, exceptions and data mis-compares.

Update 10.6 - 11 Aug 2005

1. VNAV logic changed to prevent nuisance VNAV DISCONNECT messages during landing.

2. LNAV bank angle reduced to 1/2 track change with a minimum of 8 degrees and a maximum of 23 degrees.

3. SELECT DESIRED WPT page now titled "SEL DESIRED XXXXX". The page has space for names that are 15 characters in length for VHF Navaids, Airports, and Non-Directional Beacons. The Navigation Data Base will include the applicable ARINC 424 name characters.

4. ANP calculation is adjusted to sense the position changes in the primary navigation sensors.

5. A Flight Path Angle field is added to the ARRIVALS page and will let pilots change the runway angle. It will occur when the crew makes a selection of one of the identified runways of the Destination airport.

6. Dual FMC Enhancements will let the two FMCs continue to operate when there are disagreements between them.

7. Geometric Path Descent changed from Optional to Basic. This makes the FMCs function the same as other Boeing FMC equipped airplanes.

8. VERIFY POSITION message is changed to prevent an accidental display during low RNP procedures. A 2-second along-track threshold is added to the message logic.

9. Errors in ETA and Fuel Remaining predictions are corrected.

10. Air Traffic Services Datalink (ATSDL) uplink delay timer change tells a pilot if a datalink clearance is older than a voice message that was sent to replace it.

11. Can now stay in Quiet Climb when data related to V-speeds is entered again.

12. Engine-Out SID is revised to find if the active waypoint is one of the waypoints in the engine-out SID.

13. Altitude Intervention is revised to prevent the deletion of altitude restrictions on approach waypoints by Altitude Intervention. The altitude constraints for base approaches can now only be deleted on the CDU.

14. A QRH Takeoff Speed Logic Check is added to make sure that the QRH takeoff speeds entered by the pilot agree with the formula: V1 <= VR and VR < V2.

15. Speed Propagation Option lets the FMC continue to use the IAS or Mach that was manually entered on the cruise page when the flight continues to the descent page.

16. Optional number of FIX pages increased from 2 to 6.

17. Common VNAV Option, when activated, VNAV SPD is used as the speed on elevator mode for over-speed reversions removing the necessity to disengage VNAV.

18. Performance data added for winglet equipped 737-300 and 737-400 series.

Update 10.7 – 27 Mar 2007

1. UNABLE REQD PERF-RNP message now also displays on the map in addition to the scratchpad whenever an unable RNP condition exists.

2. The UNABLE REQD PERF – RNP alerting times can be modified via the OPC. The new default values are as follows:

Environment	U10.7 Default	
OCEANIC	12.0 nm	60 sec
ENROUTE	2.0 nm	30 sec
TERMINAL	1.0 nm	10 sec
APPROACH	0.5 nm	10 sec

3. The default RNP, Vertical RNP and RNP time to alarm values for oceanic, enroute, terminal and approach are shown in the SW OPTIONS 1/2 Page.

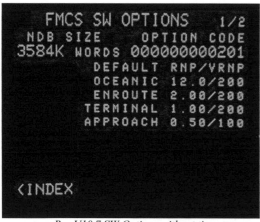

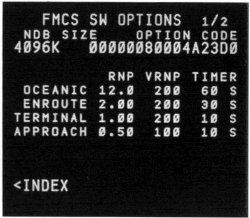

Pre U10.7 SW Options without timer   *U10.7+ SW Options with timer*

4. Lateral ANP Calculation revised to utilize an elliptical threshold to mitigate GPS data latency on ANP calculation.

5. Vertical ANP Calculation is revised to be based on Mach in order to better reflect actual conditions.

6. Hold entries and exits revised to use non-overfly entries and exits in place of fly-over entries and exits.

7. LNAV control laws adjusted to provide improved low speed (<200kts g/s) stability.

8. New Model Engine Database, to support updated EEC software version 7.B.R3 to utilise a higher idle descent profile.

9. VDEV always displayed while in a descent mode.

10. Can now set "AT", "AT or BELOW" and "AT or ABOVE" speed constraints on the LEGS page. Two new alerting level messages will be added:

    1) UNABLE XXX KTS AT YYYYY, is displayed when the next waypoint speed restriction cannot be met.

2) DRAG REQ AFTER XXXXX, is displayed when waypoint speed constraint greater than 10 knots above the predicted speed.

11. New position discrepancy messages as follows:

On-Ground: IRS POS/ORIGIN DISAGREE or FMC POS/RW DISAGREE.

In-Air: VERIFY POS: IRS-IRS or IRS-FMC or IRS-RADIO or FMC-RADIO or FMC-GPS or FMC-FMC.

12. A GPS loss alert is now provided if GPS data is missing or invalid for 30 seconds to the FMC.

13. VNAV/LNAV Armed Before Takeoff (NG only) – Allows the flight crew to arm VNAV and LNAV independently prior to takeoff. If armed prior to F/D Takeoff selection, the VNAV Arm annunciation will appear on the FMA and VNAV will engage at 400 feet with the speed target set to the existing F/D Takeoff speed target, until reaching acceleration height. The acceleration heights are specified on the TAKEOFF REF page 2. If LNAV is armed prior to F/D Takeoff activation, the LNAV Arm annunciation will appear on the FMA and LNAV will engage at 50 feet.

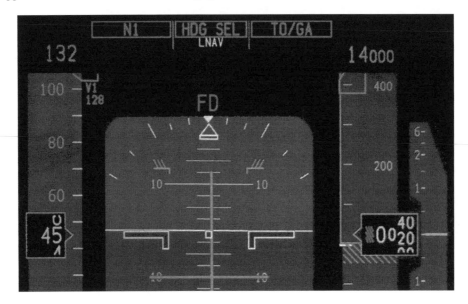

14. FMC Target Speed limited to 5kts below the trailing edge flap placard speed. The flap placard speeds are hard coded for the 737-3/4/500 and are contained in the MEDB for the 737NG. Also, for the 737NG, the FMC target speed is limited to 230kts until all leading edge devices are retracted.

15. LOC updating automatically disabled when the FMC is GPS updating.

16. Over a dozen significant bugfixes.

17. Optional Takeoff Thrust Auto-Select – Allows the FMC to auto-select an appropriate Takeoff Thrust based upon airline policy and the airport/runway entered by the flight crew.

18. Optional Geometric Path Descent enabled in approach phase only instead of active for the entire descent path. If enabled, VNAV will use the traditional air mass type descent vertical path construction to the Initial Approach Fix, after which a geometric vertical path is built to the runway or missed approach point.

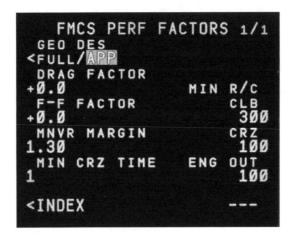

19. Optional FMC-10 Intent data bus.

Update 10.8 – 18 Feb 2009

1. Bugfix for FCOM Bulletin TBC-70 to allow VNAV to be armed for takeoff on NGs, equipped with a newer FCC Collins P4 or later, Honeywell model 710 or later, and CDS BP06 or later software.
2. Improved Engine-Out Logic and bugfixes the unintentional initiation of the FMC engine-out Mode (from U10.7, TBC-68):
   - FMC engine-out mode will not be set when groundspeed is less than 60 knots. If engine-out mode is set while on-ground, the FMC will exit the mode when speed drops below 60 knots.
   - Engine-out speeds will be available if an engine fails on takeoff after 60 knots of ground speed.
   - Predictions display will be blanked on the MCDU pages when engine failure is detected and airspeed is over 60 knots of groundspeed. RTE data, ETA data and TOC data displayed on the ND will be blanked during an engine-out condition.
   - Engine-out will be cleared and the target speed and predictions will return to normal two engine values when the crew selects EXEC and the ALL ENGINE prompt on the CLB page, or CRZ, or DES phase is entered.
   - New "ENTER EO CRZ SPD AND ALT" message will be displayed when the engine-out operation is terminated due to reaching cruise altitude or the pilot depresses the ALL ENGINE prompt button on the climb page.
3. Various improvements to software exception handling logic.
4. Performance data added for winglet equipped 737-500 series. Also added optimum altitude calculation for all winglet equipped 737 Classic aircraft.
5. Navigation database size increased to 16 megabytes.
6. Updated MEDB and logic to support an enhanced 7B26B2 engine rating.
7. Bugfixes:
   - Bus interrupt overflow condition for the dual FMC failures.
   - Updated FMC DISAGREE message logic to now remain displayed until the condition causing the message has been resolved. Also no longer checking for Mach and IAS miscompares on the ground.
   - The following software exceptions were resolved:
     o When reselecting approach after having sequenced procedure hold exit
     o After selecting approach when a GSI is before the FACF
     o When selecting a SID during a missed approach and when the active waypoint is the start of an ARC Leg.
     o Creating the DES page when on the ACT plan, in Descent, and the mod plan is in CRZ
     o When sequencing a double bypass on arc legs
     o When creating the CRZ DES page for long subsequent CRZ segment
     o Upon accessing last LEGS page with PREV key depressed.

Update 10.8A – 7 May 2010

Mainly bugfixes for the withdrawn U10.8 including:
1. Bugfixes VNAV commanding the airplane to descend below the expected path on approach.
2. Bugfixes VNAV descent paths into airports above 10,000ft elevation.
3. Bugfixes left and right FMCs commanding different descent speeds
4. Added new "FMC DISAGREE-VERTICAL" alert message during VNAV path descents.
5. Allows full climb thrust after a derated takeoff thrust is uploaded by ACARS on aircraft that have the FMC PERF option code zero enabled.

Update 11.0 – Expected 2015

U11.0 is expected to have the following features and improvements:

1) Route 2
2) Improvements to LNAV control laws
3) Enable cruise descent to sequence into a PATH descent
4) Use transition altitude and transition level from the airport record
5) Altitude Intervention revision
6) Enhanced takeoff error checking
7) Allow the FMC to change flight phases during holding
8) Connect the EO SID trigger to the branch point
9) Down track RNP alerting
10) Use airport MagVar for non-CF approach legs
11) Enhance FP aux data
12) Optimize the waypoint linked list
13) Improve LG/PP interface
14) Improve bypass (flight plan revision) generation
15) WGS-84 model (all legs)
16) Optional Climb N1 limit default
17) Maximum thrust limit labelling.
18) Runaway IRS enhancement
19) Revised descent block speeds for 737C
20) Modified VNAV armed TOGA logic
21) Vertical path monitor
22) Takeoff threshold length reduction
23) Message queue increase
24) Revised RESET MCP APP MODE logic

TECHNICAL SPECIFICATIONS

Dimensions - Metric:

Series:	100	200	300	400	500	600	700	800	900
Fuselage:									
Aircraft length (m)	28.65	30.53	33.40	36.45	31.01	31.20	33.60	39.50	42.10
Fuselage length (m)	27.66	29.54	32.30	35.23	29.80	29.88	32.18	38.08	40.68
Fuselage height (m)	4.01		4.01			4.01			
Fuselage width (m)	3.76		3.76			3.76			
Accommodation:									
Cabin width (m)	3.53		3.53			3.53			
Typical max seating capacity	96	130	134	159	122	130	148	184	189/215
Max certified seating capacity	124	136	149	188	149	149	149	189	189/215
Hold cargo volume (m³)	18.40	24.78	30.20	38.90	23.30	21.40	28.40	45.10	52.00
Wing:									
Span (m)	28.35		28.88			34.32			
Span with winglets (m)	N/A		31.11			35.79			
Gross Area (m²)	102.00		105.40			124.58			
Aspect Ratio	8.83		9.16			9.45			
Taper Ratio	0.27		0.24			0.16			
Root Chord (m)	7.32		7.32			7.88			
Tip Chord (m)	1.60		1.62			1.25			
M.A.C.(m)	3.80		3.41			3.96			
Dihedral (°)	6.00		6.00			6.00			
¼ Chord Sweep (°)	25.00		25.00			25.02			
High Lift Devices:									
Flap Span/Wing Span	0.74		0.72			0.60			
Flap Area/Wing Area	0.29		0.29			0.30			
Fin:									
Aircraft Height (m)	11.23		11.13			12.60			
Fin Height (m)	6.15		6.15			7.16			
Fin Area (m²)	20.81		23.13			26.44			
Rudder Area (m²)	5.22		5.22			5.22			
Aspect Ratio	1.64		1.81			1.91			
Taper Ratio	0.29		0.31			0.27			
¼ Chord Sweep (°)	35.00		35.00			35.00			
Horiz Stabilizer:									
Span (m)	10.97		12.70			14.35			
Tailplane Area (m²)	28.99		31.40			32.78			
Elevators Area (m²)	6.55		6.55			6.55			
Aspect Ratio	4.15		4.04			6.16			
Taper Ratio	0.26		0.26			0.20			
Dihedral (°)	7.00		7.00			7.00			
¼ Chord Sweep (°)	30.00		30.00			30.00			
Undercarriage:									
Track (m)	5.23		5.25			5.76			
Wheelbase (m)	10.46	11.38	12.40	14.27	11.07	11.23	12.60	15.60	17.17
Nose Wheel Diameter (In)	24.00		27.00			27.00			
Nose Wheel Tread Width (In)	7.75		7.75			7.75			
Max Nose Tyre Pressure(psi)	145	145	165	184	194	205	205	205	205
Max Main Wheel Diameter (In)	40.00		42.00			44.50			
Main Wheel Tread Width (In)	14.50		14.50			14.50			
Max Main Tyre Pressure (psi)	157	183	203	217	202	205	205	205	205

Dimensions - Imperial:

Series:	100	200	300	400	500	600	700	800	900
Fuselage:									
Aircraft length (ft)	94'0	100'2	109'7	119'7	101'9	102'6	110'4	129'6	138'2
Fuselage length (ft)	90'7	96'11	105'7	115'7	97'9	97'9	105'7	124'9	133'5
Fuselage height (ft)	13'2		13'2			13'2			
Fuselage width (ft)	12'4		12'4			12'4			
Accommodation:									
Cabin width (ft)	11'7		11'7			11'7			
Typical max seating capacity	96	130	134	159	122	130	148	184	189/215
Max certified seating capacity	124	136	149	188	149	149	149	189	189/215
Hold cargo volume (ft³)	650	875	1065	1375	825	755	1000	1595	1835
Wing:									
Span (ft)	93'0		94'9			112'7			
Span with winglets (ft)	N/A		102'1			117'5			
Gross Area (ft²)	1098		1135			1341			
Aspect Ratio	8.83		9.16			9.45			
Taper Ratio	0.27		0.24			0.16			
Root Chord (ft)	24'0		24'0			25'10			
Tip Chord (ft)	5'3		5'4			4'1			
M.A.C.(ft)	12'6		11'2			13'0			
Dihedral (°)	6.00		6.00			6.00			
¼ Chord Sweep (°)	25.00		25.00			25.02			
High Lift Devices:									
Flap Span/Wing Span	0.74		0.72			0.60			
Flap Area/Wing Area	0.29		0.29			0.30			
Fin:									
Aircraft Height (ft)	36'10		36'6			41'3			
Fin Height (ft)	20'2		20'2			23'6			
Fin Area (ft²)	224		249			285			
Rudder Area (ft²)	56		56			56			
Aspect Ratio	1.64		1.81			1.91			
Taper Ratio	0.29		0.31			0.27			
¼ Chord Sweep (°)	35.00		35.00			35.00			
Horiz Stabilizer:									
Span (ft)	36'0		41'8			47'0			
Tailplane Area (ft²)	312		338			353			
Elevators Area (ft²)	70		70			70			
Aspect Ratio	4.15		4.04			6.16			
Taper Ratio	0.26		0.26			0.20			
Dihedral (°)	7.00		7.00			7.00			
¼ Chord Sweep (°)	30.00		30.00			30.00			
Undercarriage:									
Track (ft)	17'2		17'2			18'9			
Wheelbase (ft)	34'4	37'4	40'10	46'10	36'4	36'10	41'4	51'2	56'4
Nose Wheel Diameter (In)	24.00		27.00			27.00			
Nose Wheel Tread Width (In)	7.75		7.75			7.75			
Max Nose Tyre Pressure(psi)	145	145	165	184	194	205	205	205	205
Max Main Wheel Diameter (In)	40.00		42.00			44.50			
Main Wheel Tread Width (In)	14.50		14.50			14.50			
Max Main Tyre Pressure (psi)	157	183	203	217	202	205	205	205	205

737-1/200

Weights

Series:	737-100		737-200		737-200Adv	
Version:	Basic	HGW	Basic	HGW	Basic	HGW
Std Weights (Kg):						
Max. ramp	44,362	50,349	45,723	52,617	52,617	58,333
Max. take-off	43,999	49,896	45,360	52,390	52,390	58,106
Max. landing	40,688	44,906	43,092	46,720	46,720	48,535
Max. zero-fuel	37,059	40,824	38,556	43,092	43,092	43,092
Typical DOW	26,581	28,123	27,170	27,125	28,395	29,620
Max payload	10,478	12,701	11,385	15,967	14,697	13,472
Std Weights (Lbs):						
Max. ramp	97,800	111,000	100,800	116,000	116,000	128,600
Max. take-off	97,000	110,000	100,000	115,500	115,500	128,100
Max. landing	89,700	99,000	95,000	103,000	103,000	107,000
Max. zero-fuel	81,700	90,000	85,000	95,000	95,000	95,000
Typical DOW	58,600	62,000	59,900	59,800	62,600	65,300
Max payload	23,100	28,000	25,100	35,200	32,400	29,700
Fuel Capacity:						
Liters	13,399	17,865	13,096	18,092	19,531	22,596*
Kilograms	10,758	14,345	10,515	14,527	15,682	18,143*
US Gallons	3,540	4,720	3,460	4,780	5,163	5,973*
Pounds	23,718	31,624	23,182	32,026	34,590	40,019*

*Aircraft with 810gal (3,065ltr) auxiliary fuel tank in aft cargo compartment.

Performance

Series:	737-100	737-200
Loadings:		
Thrust Loading (kg/kN)	343.17	367.91
Wing Loading (kg/m²)	433.33	513.63
Thrust/Weight Ratio	0.297	0.2771
Take-off Distance:		
ISA +20°C, s.l.,MTOW, Flap 5	1859m / 6100ft	2030m / 6660ft
Landing Distance:		
ISA, s.l, MLW, Flap 40	1470m / 4820ft	1400m / 4600ft
Speeds (kt/Mach):		
V2 F5 at basic MTOW	143	147
Vref F40 at basic MLW	127	131
Vmo/Mmo	350/0.84	350/0.84
Long range cruise:		
TAS / Mach	420 / 0.73	420 / 0.73
Ceiling (ft)	35,000 / 37,000 with mod	35,000 / 37,000 with mod
L.R. Fuel flow (kg/lbs per hr)	2400kg / 5300lbs	2827kg / 6200lbs
Range with max payload (nm)	1720	2645

The following drawings depict an example of each series of the 737 together with a representative sample of options available for the series.

Boeing 737-100
Length 94ft 0in, Short engine nacelles, Clamshell reversers, Aux inlet doors, Short Krueger flaps, 3 horn rudder.

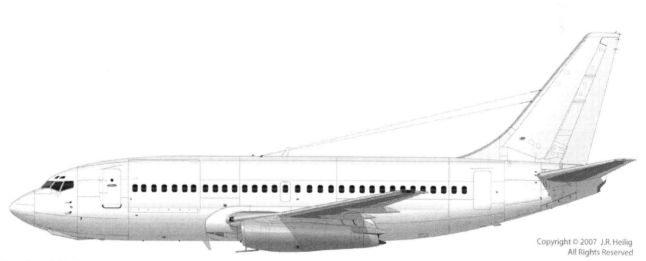

Boeing 737-200Adv
Length 100ft 2in, Gravel kit, Target reversers, Side cargo door, HF antenna, Krueger flaps extended to wing root, 2 horn rudder, integral rear door airstairs.

737-3/4/500

Weights

Series:	737-300		737-400		737-500	
Version:	Basic	HGW	Basic	HGW	Basic	HGW
Std Weights (Kg):						
Max. ramp	56,700	63,504	63,050	68,266	52,617	60,781
Max. take-off	56,473	63,277	62,823	68,040	52,390	60,554
Max. landing	51,710	52,527	54,885	56,246	49,896	49,896
Max. zero-fuel	47,628	48,308	51,256	53,071	46,494	46,494
Typical DOW	31,480	32,904	33,190	33,643	31,312	31,312
Max payload	16,148	15,404	18,067	19,428	15,182	15,182
Std Weights (Lbs):						
Max. ramp	125,000	140,000	139,000	150,500	116,000	134,000
Max. take-off	124,500	139,500	138,500	150,000	115,500	133,500
Max. landing	114,000	115,800	121,000	124,000	110,000	110,000
Max. zero-fuel	105,000	106,500	113,000	117,000	102,500	102,500
Typical DOW	69,400	72,540	73,170	74,170	69,030	69,030
Max payload	35,600	33,960	39,830	42,830	33,470	33,470
Fuel Capacity:						
Liters	20,102	23,827*	20,102	23,827*	20,102	23,827*
Kilograms	16,141	19,131*	16,141	19,131*	16,141	19,131*
US Gallons	5,311	6,295*	5,311	6,295*	5,311	6,295*
Pounds	35,583	42,177*	35,583	42,177*	35,583	42,177*

*Aircraft with 1,000gal (3,785ltr) auxiliary fuel tank in aft cargo compartment.

Performance

Series:	737-300	737-400	737-500
Loadings:			
Thrust Loading (kg/kN)	353.48	325.35	318.28
Wing Loading (kg/m²)	596.29	645.15	497.05
Thrust/Weight Ratio	0.2884	0.3133	0.3203
Take-off Distance:			
ISA +20°C, s.l.,MTOW, Flap 5	2109m / 6920ft	2665m / 8745ft	2003m / 6572ft
Landing Distance:			
ISA, s.l, MLW, Flap 40	1396m / 4580ft	1540m / 5050ft	1357m / 4450ft
Speeds (kt/Mach):			
V2 F5 at basic MTOW	148	159	143
Vref F40 at basic MLW	133	137	128
Vmo/Mmo	340/0.82	340/0.82	340/0.82
Long range cruise:			
TAS / Mach	430 / 0.745	430 / 0.745	430 / 0.745
Ceiling (ft)	37,000	37,000	37,000
L.R. Fuel flow (kg/lbs per hr)	2250kg / 4960lbs	2377kg / 5240lbs	2100kg / 4630lbs
Range with max payload (nm)	2950	2800	2950

Boeing 737-300QC
Length 109ft 7in, Side cargo door.

Boeing 737-400
Length 119ft 7in, 4 overwing exits, 2 sidewall risers, tailskid.

Boeing 737-500
Length 101ft 9in.

737-NG

Weights

Series:	737-600		737-700		737-800		737-900		
Version:	Basic	HGW	Basic	IGW	Basic	HGW	Basic	HGW	ER
Std Weights (kg):									
Max. ramp	56,472	66,224	70,307	77,791	70,760	79,333	74,616	79,243	85,366
Max. take-off	56,246	65,997	70,080	77,564	70,534	79,016	74,389	79,016	85,139
Max. landing	54,648	55,657	58,604	60,781	65,317	66,361	66,361	66,814	71,350
Max. zero-fuel	51,483	51,709	55,202	57,152	61,689	62,732	62,732	63,639	67,721
Typical DOW	36,378	36,378	37,648	37,648	41,413	41,413	42,901	42,901	44,676
Max payload	15,105	15,558	17,554	17,554	20,276	21,319	19,831	20,738	23,045
Std Weights (Lbs):									
Max. ramp	124,500	146,000	155,000	171,500	156,000	174,900	164,500	174,700	188,200
Max. take-off	124,000	145,500	154,500	171,000	155,500	174,200	164,000	174,200	187,700
Max. landing	120,500	120,500	129,200	134,000	144,000	146,300	146,300	147,300	157,300
Max. zero-fuel	113,500	114,000	121,700	126,000	136,000	138,300	138,300	140,300	149,300
Typical DOW	80,200	80,200	83,000	83,000	91,300	91,300	94,580	94,580	98,495
Max payload	33,300	34,300	37,500	38,700	44,700	47,000	43,720	45,720	50,805
Brake Category:	D	E	A / B	F	C	C	G	G	G
Fuel Capacity:									
Liters	26,022	26,022	26,022	26,022	26,022	26,022	26,022	26,022	29,663*
Kilograms	20,894	20,894	20,894	20,894	20,894	20,894	20,894	20,894	23,817*
US Gallons	6,875	6,875	6,875	6,875	6,875	6,875	6,875	6,875	7,836*
Pounds	46,063	46,063	46,063	46,063	46,063	46,063	46,063	46,063	52,508*

* Aircraft with two auxiliary fuel tanks.

Performance

Series:	737-600	737-700	737-800	737-900
Loadings:				
Thrust Loading (kg/kN)	396.89	389.89	365.51	336.79
Wing Loading (kg/m²)	522.39	556.98	627.77	634.11
Thrust/Weight Ratio	0.2568	0.2615	0.2789	0.3027
Take-off Distance:				
ISA +15ºC, s.l.,MTOW, Flap 5	1616m / 5300ft	1744m / 5720ft	2100m / 6890ft	2591m / 8500ft
Landing Distance:				
ISA, s.l, MLW, Flap 40	1342m / 4400ft	1418m / 4650ft	1646m / 5400ft	1662m / 5450ft
Speeds (kt/Mach):				
V2 F5 at basic MTOW	129	136	153	162
Vref F40 at basic MLW	126	131	141	141
Vmo/Mmo	340/0.82	340/0.82	340/0.82	340/0.82
Long range cruise:				
TAS / Mach	450 / 0.785	450 / 0.785	450 / 0.785	450 / 0.785
Ceiling (ft)	41,000	41,000	41,000	41,000
L.R. Fuel flow (kg/lbs per hr)	1932kg / 4260lb	2070kg / 4565lb	2187kg / 4820lb	<2300kg / 5070lb
Range with max payload (nm)	4500	4400	4000	3200

Boeing 737-600
Length 102ft 6in.

Boeing 737-700/BBJ
Length 110ft 4in, winglets, non-standard window configuration, extra comms antennae.

Boeing 737-800
Length 129ft 6in, winglets, 4 overwing exits, 2 sidewall risers, tailskid.

Boeing 737-900ER
Length 138ft 2in, winglets, 4 overwing exits, 2 sidewall risers, type II emergency exit, two position tailskid.

ABOUT THE AUTHOR

Chris Brady is the author and producer of "The Boeing 737 Technical Site" at www.b737.org.uk. It began in 1999 after a small set of pilots' technical notes he had written for his own use began to be circulated amongst his colleagues. Responding to increasing demand, he set up a small website to distribute them, which quickly became known throughout the pilot community. In the following years Chris devoted many hours building up this extensive and entirely free pilots' information service in the interest of improving the depth of knowledge and understanding of the aircraft amongst 737 pilots worldwide. Throughout this time, pilots, trainers and engineers from many countries have not only emailed their appreciation of the site, but have also often submitted notes and articles which, once checked, have been added to it. The website, which attracts over 15,000 visitors per week, is now the most authoritative open source of information freely available about the 737, and is a valuable contribution to air safety.

Chris began his career in general aviation as an instructor and examiner from PPL to CPL level, C of A test pilot on Group A and B aircraft and Chief Pilot. During seven years at British Midland, he became a Training Captain and groundschool instructor, re-writing technical and performance training manuals and has conducted over 200 post maintenance check flights on every series of 737 Classic and NG. He is currently easyJet's senior safety investigator and Chairman of the UK Flight Safety Committee.

He has now produced this book as an invaluable guide for all who study, work and train on this hugely successful airliner.

INDEX